Advancements in Textile Finishing

Mohammad Shahid · Santosh Biranje ·
Mohd Yusuf · Ravindra Adivarekar
Editors

Advancements in Textile Finishing

Techniques, Technologies, and Trends

Springer

Editors
Mohammad Shahid
Department of Chemistry
Koneru Lakshmaiah Education Foundation (KLEF)
Guntur, Andhra Pradesh, India

Santosh Biranje
Department of Fibres and Textile Processing Technology
Institute of Chemical Technology
Mumbai, Maharashtra, India

Mohd Yusuf
School of Science and Technology
Glocal University
Saharanpur, Uttar Pradesh, India

Ravindra Adivarekar
Department of Fibres and Textile Processing Technology
Institute of Chemical Technology
Mumbai, Maharashtra, India

ISBN 978-981-96-6384-2 ISBN 978-981-96-6385-9 (eBook)
https://doi.org/10.1007/978-981-96-6385-9

This Springer imprint is published by the registered company Springer Nature Singapore Pte Ltd.
The registered company address is: 152 Beach Road, #21-01/04 Gateway East, Singapore 189721, Singapore

Preface

The textile industry is undergoing significant changes as it faces increasing demands for sustainability, improved functionality, and innovative solutions. This book, *Advancements in Textile Finishing: Techniques, Technologies, and Trends*, explores recent developments in textile finishing, an essential part of textile production that determines the final properties and performance of fabrics. Textile finishing plays a crucial role in enhancing the appearance, feel, and durability of fabrics. However, as the industry shifts towards more sustainable and environmentally responsible practices, new approaches and advanced technologies are being adopted to reduce the reliance on harmful chemicals and minimize environmental impact. This book covers both traditional and emerging textile finishing techniques, focusing on their contribution to sustainability and improved textile functionality. In Chap. 1, the book introduces "Mechanical Finishing Techniques in Textiles for Sustainable Development", highlighting how mechanical processes are reducing the need for chemicals and promoting eco-friendly alternatives. Throughout the book, the theme of sustainability continues, with each chapter exploring how different finishing techniques and technologies are being used to improve both the performance and environmental impact of textiles. Chapter 2 provides a detailed overview of the "Mechanical and Chemical Processes of Textile Finishing Industries", discussing the processes used in transforming raw materials into finished textiles and the environmental challenges associated with these techniques. This chapter lays the foundation for understanding the complexities of textile manufacturing. Other chapters address important topics such as sustainable yarn sizing (Chap. 3), wellness and cosmeto textiles (Chap. 4), functional finishes for outdoor textiles (Chap. 5), and advances in polymeric materials for textile finishing (Chap. 6). These chapters explore how technological advancements are enhancing textile performance, improving comfort, and addressing environmental concerns. Chapters 9 and 10 focus on "Microencapsulation in Textile Finishing" and "Enzyme-Based Textile Finishing", both of which are newer approaches that offer sustainable alternatives to traditional methods. These techniques allow for more efficient and eco-friendly textile treatments while improving functionality. Bio-based and eco-friendly materials are discussed in Chap. 8, "Bio-Derived Polymers for Eco-Friendly Textile Finishing", which examines how renewable resources can replace petroleum-based chemicals in textile finishing processes. The use of biotechnology for surface modification of polyester fibers is explored in Chap. 11, while Chap. 12 discusses advancements in plasma technology for textile surface modification. The final chapter, Chap. 13, looks at the future of textiles, focusing on the integration of "Conductive and Smart Textiles", which have the potential to revolutionize applications in fields such as healthcare, military, and consumer electronics. While these innovations offer exciting possibilities, the chapter also addresses the environmental concerns associated with their production and use. This book provides an academic overview of the advancements in textile finishing, offering valuable insights for researchers, industry professionals, and students. It presents both established and emerging techniques, with a focus on their practical applications and potential to address the evolving demands of the textile industry.

Guntur, India — Dr. Mohammad Shahid
Mumbai, India — Dr. Santosh Biranje
Saharanpur, India — Dr. Mohd Yusuf
Mumbai, India — Prof. Ravindra Adivarekar

Contents

About the Editors

Dr. Mohammad Shahid Ph.D., is an Assistant Professor in the Department of Chemistry, Koneru Lakshmaiah Education Foundation, India. Before joining KLEF, he worked at various esteemed institutions, including Jamia Millia Islamia, New Delhi; Soochow University, China; University of Glasgow, UK; and the Institute of Chemical Technology, Mumbai. He earned his Master's degree in Chemistry from Shibli National College, Azamgarh, in 2006 and his Ph.D. in Chemistry from Jamia Millia Islamia, New Delhi, in 2014. Dr. Shahid has received several prestigious fellowships to support his research pursuits, including grants from UGC, India, China Postdoctoral Fellowship, EU Marie Curie Fellowship, and Dr. D. S. Kothari Fellowship. He has contributed to over 50 scientific publications with ~4000 citations and h-index of 24. In addition, has also edited and authored several books with renowned international publishing houses.

Dr. Santosh Biranje Ph.D., is an Assistant Professor in the Department of Fibres and Textile Processing Technology at the Institute of Chemical Technology (ICT), Mumbai, India. He completed his Ph.D. at ICT Mumbai, specializing in Fibres and Textile Processing Technology, with a focus on biopolymer extraction and modification. Following his doctoral studies, Dr. Biranje pursued postdoctoral research at Jiangsu University, China, under the mentorship of Profs. Sun and Liu Jun. During this period, he pioneered the development of 3D-printed wound healing material scaffolds made from nanocellulose, contributing to medical materials and regenerative medicine. In April 2022, he joined an Assistant Professor position at ICT Mumbai, where his research focuses on extracting nanocellulose for sustainable textile applications. Dr. Biranje's academic and professional trajectory underscores his unwavering commitment to advancing research in fibrous materials and biopolymers.

Dr. Mohd Yusuf Ph.D., is an Associate Professor in the School of Science and Technology at Glocal University, Saharanpur, India. He earned his Ph.D. in 2012 from Jamia Millia Islamia, New Delhi, India. Dr. Yusuf's research interests include green and sustainable materials, functional biomaterials, composite and applied materials, nanomaterials, wastewater treatment, and energy storage technologies. He has published over 70 research articles, reviews, books, and chapters in renowned international publishing platforms, including Elsevier, Springer, Wiley, and Taylor & Francis. With over 3500 citations and an h-index of 22, Dr. Yusuf is recognized for his significant contributions to the field.

Dr. Ravindra Adivarekar Ph.D., is a distinguished professor in Fibre Chemistry at the Institute of Chemical Technology (ICT), Mumbai, where he has dedicated over 25 years of service. He holds multiple degrees, including an honorary Fellowship of the Textile Association (FTA). Dr. Adivarekar is renowned for his expertise in textile processing and served as the Head of the Department of Fibres and Textile Processing Technology from 2008 to 2021. He currently holds the position of Dean of Human Resource Development (HRD) at ICT. His extensive research interests span eco-friendly coloration, natural fibre composites, medical textiles, nanotechnology in textile processing, natural dyes, microbial dyes, and the application of enzymes. A committed mentor, Dr. Adivarekar has guided numerous students and managed research projects with total funding exceeding 18 crores. He has contributed significantly to the academic community with over 200 research papers published in national and international journals, more than 70 conference proceedings, and three patents. Beyond his academic achievements, Dr. Adivarekar is an esteemed member of several textile-related organizations and has received significant recognition for his contributions to the industry. He actively engages in advisory roles and government committees, addressing environmental issues in textile manufacturing.

Chapter 1
Understanding Mechanical Finishing Techniques in Textile Processing

Mohammad Neaz Morshed

Abstract Mechanical finishing techniques play an important role in current industrial textile processes due to their extensive use in improving textile performance, appearance, texture and functionality. The technique involves physical (mechanical) processes to modify the textile structure and surface, enhancing the fabric's functionality and aesthetic appeal without the need for chemical treatments. Driven by the increasing demand for specialized treatments, enhanced functionality as well as sustainable finishing of textiles, the field of mechanical finishing has seen significant advancements in recent years. The push for sustainable textile processes further boosted the advancement of mechanical finishing techniques as eco-friendly alternative to chemical finishing. Mechanical finishing techniques reduces the reliance on synthetic and toxic chemicals, minimize water and energy consumption, thereby minimizing environmental impact and promoting responsible manufacturing practices in textiles. As the industry continues to innovate, mechanical finishing remains a critical component in producing textiles that meet the evolving needs of both consumers and industries. This introductory chapter will present a general overview of various mechanical finishing techniques in textile, their principles as well as recent progress and prospects. Historical development of mechanical finishing techniques in textiles, mechanical forces for different mechanical finishing techniques, and emerging applications and potential challenges in mechanical finishing has also been discussed in this chapter.

Keywords Finishing · Textiles · Mechanical · Textile process · Sustainability

M. N. Morshed (✉)
Department of Textile Technology, Faculty of Textiles, Engineering and Business (Including the Swedish School of Textiles), University of Borås, Borås, Sweden
e-mail: mohammad_neaz.morshed@hb.se

M. Shahid et al. (eds.), *Advancements in Textile Finishing*,
https://doi.org/10.1007/978-981-96-6385-9_1

1.1 Introduction

Textile finishing encompasses a wide range of treatments applied to fabrics before they reach the final customer, aiming to enhance their appearance, functionality, and overall value. These processes can be temporary, like pressing bed sheets before packaging, or permanent, such as applying UV-blocking treatments to tent fabrics [18]. A simple definition of finishing is "the sequence of operations, other than scouring, bleaching and coloration, to which the fabrics are subjected after leaving the loom or knitting machine". Most finishes are applied to woven, nonwoven, and knit fabrics. Finishing is mostly done in fabric before the development of the product; however, in many instances, finishes are also applied to yarn (e.g., silicone finishing on sewing yarn) or garments for various applications [5, 19]. The primary objectives of textile finishing are to improve the attractiveness and serviceability of textiles, thereby increasing customer satisfaction [12]. When finishing takes place in a wet state, it is called wet finishing, and while finishing in a dry state, it is called dry finishing. Dry finishing has several advantages, including its environmental sustainability, as it uses machinery and heat rather than chemicals. Dry finishing utilizes relatively little water and produces little waste, as opposed to wet finishing, which can consume considerable amounts of water and energy; thus, it produces a significant amount of effluent and waste [16].

On the basis of process, textile finishing treatments are divided into two main types: (a) chemical finishing and (b) mechanical finishing. Under these two main types, textile finishing can be further divided into multiple sub types based on the durability and purpose of the treatments (see Fig. 1.1). Chemical finishing involves adding chemicals to textiles to achieve specific effects, typically without altering the textiles' appearance. Here, water often serves as the medium for applying chemicals to textiles. Heat is then utilized to evaporate the water and activate the chemicals. Examples of chemical finishing techniques include applying softening agents to improve the fabric's hand feel [22], antimicrobial treatments to inhibit microbial growth, water repellent finishes for outdoor textiles, flame retardant treatments for safety [1], and stain-resistant coatings for easier cleaning [23]. Additionally, UV protection finishes, wrinkle resistance finishes, and color fixatives are commonly applied chemical finishes on textiles that enhance functionality and durability [24].

Mechanical finishing encompasses the treatments that are performed on yarn or fabric to enhance its appearance, performance, or 'hand feel' [6] using physical, particularly mechanical processes. This can include altering properties such as fabric luster [3], smoothness, softness, residual shrinkage, and overall feel [15]. At large, current textile processing utterly relies on key mechanical finishes such as calendaring, exercising, compressive shrinkage, raising, brushing, shearing, cropping, and so on for enhancing the appearance, texture, strength, and functionality of textile products [11]. As a result, almost all the textile products which made it to the market underwent some form of mechanical treatment. Nevertheless, it is often that textile finishing incorporates both chemical and mechanical treatments. Certain mechanical finishes require the use of chemicals; shrink-proofing wool fabrics require

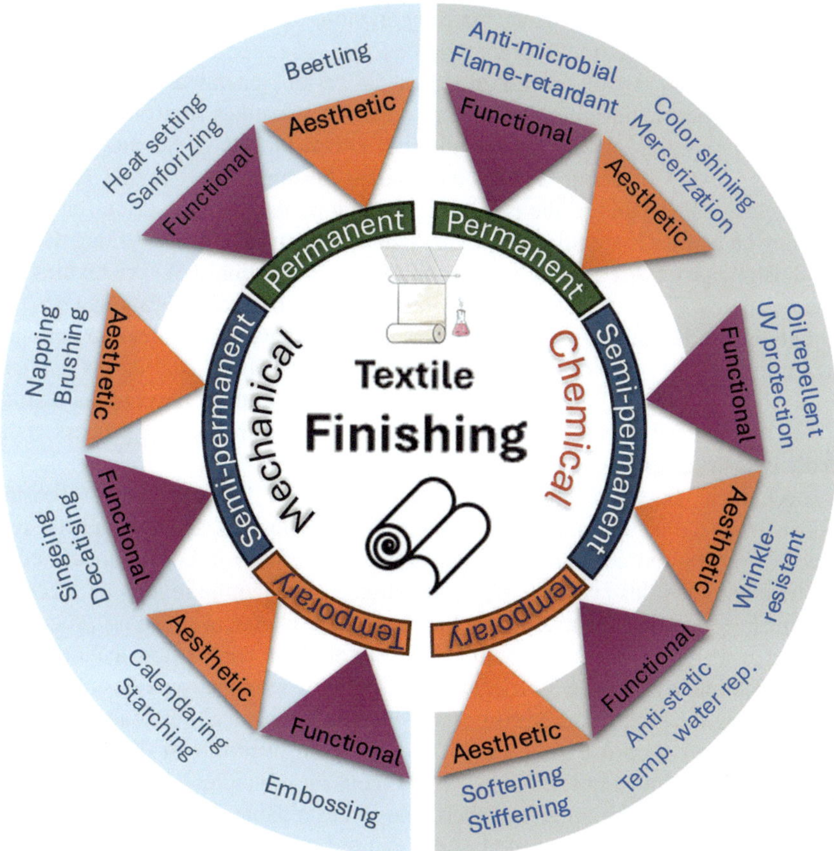

Fig. 1.1 Overview of textile finishing on the basis of the process (Mechanical, Chemical), durability (Temporary, Semi-permanent, Permanent), and purpose (Aesthetic, Functional)

the use of reductive and fixing agents, while fulling requires milling agents. On the other hand, chemical treatments depend on mechanical equipment for preparation and fabric handling equipments [3], making the process classification more complex than a simple division. Consequently, categorizing textile finishes based solely on processing methods proves challenging. Consequently, the classification of textile finishing as either chemical or mechanical now depends on the dominant process involved in enhancing the fabric. While both methods utilizes mechanical equipments; the key difference between them lies in whether the machine or the chemical that caused the intended modification on the fabric. If the desired fabric alteration is mainly driven by chemical treatments (such as shrink-proofing or softening agents), the process is classified as chemical finishing. Conversely, if the modification results predominantly from physical processes (such as calendering or brushing), it falls under mechanical finishing. Thus, the categorization is determined by the principal means of achieving the final fabric properties.

Over the years, significant development of various chemical and mechanical finishes has been achieved to develop conventional as well as smart and functional textiles. Accounting to recent publications recorded in Web of Science under the keywords "textile finishing" AND "mechanical finishing" and "textile finishing" AND "chemical finishing," the extend of research publication on various chemical finishing on textiles is significantly higher than that the mechanical finishing. Chemical finishing processes often involve reactions and interactions between textiles and various chemical agents, leading to various applications and innovations [10, 15]. Technological developments in polymers and textiles have resulted in a remarkable stream in the utilization of chemical-coated (finished) fabrics for various applications. Chemical-coated fabrics have an important place among conventional textiles as well as technical textiles and constitute an important technological process in the textile industry [2].

Mechanical finishing techniques like brushing, calendaring, and shearing contribute to the tactile and visual characteristics of textiles and play a vital role in achieving desired aesthetics and performance attributes in conventional, smart and functional textiles [17, 21]. While the number of scientific reports on mechanical finishing may be comparatively lower than the chemical finishing processes, ongoing research in this area continues to explore new techniques, technologies, and applications to enhance the mechanical properties and functionalities of textiles. By analyzing the progress and prospects of various mechanical finishing techniques employed in textile processing, the chapter gives an overview of different industrial mechanical finishing techniques with a focus on principles, advantages, disadvantages of these techniques, and their role in sustainable development in textile processes. The chapter also presents a compelling synopsis of the historical development of mechanical finishing of textiles, starting from ancient and medieval times to modern eras, including emerging trends, principles, advantages, disadvantages and application potentials. The chapter intends to benefit students, researchers, and industry stakeholders to gain understanding into various mechanical finishing techniques for sustainable development, ultimately contributing to the continued advancement of textile technology.

1.2 Historical Development of Mechanical Finishing Techniques in Textiles

Mechanical finishing techniques play a vital role in enhancing textiles' quality, functionality, and performance. From creating soft and cozy fabrics to adding texture and visual interest, these techniques offer various possibilities for textile manufacturers. The history of mechanical finishing of textiles is characterized by a continuous evolution driven by technological advancements, industrialization, and changing consumer

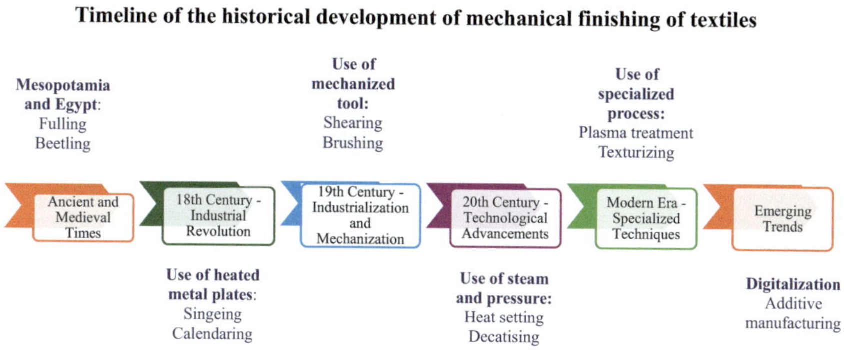

Fig. 1.2 Synopsis of historical development of mechanical finishing of textiles from ancient and medieval times to modern eras, including emerging trends

preferences. From ancient fulling mills to modern heat-setting equipment, mechanical finishing processes have played a crucial role in enhancing textiles' quality, appearance, and functionality throughout history. A timeline of the historical development of mechanical finishing of textiles can be found in Fig. 1.2. Starting from ancient and medieval times, techniques like fulling and beetling were employed to cleanse and thicken wool and linen fabrics, laying the groundwork for subsequent innovations. Fulling, practiced in civilizations such as Mesopotamia and Egypt, involved soaking wool fabric in water and then pounding it with a wooden tool containing fuller's earth, a clay that helped remove impurities and compact fibers. Beetling, developed during the Middle Ages, saw fabric subjected to heavy wooden hammers to flatten and smooth its surface, resulting in a lustrous finish primarily for linen textiles.

The Industrial Revolution marked a pivotal moment in the history of textile finishing, with the advent of mechanized processes enabling greater efficiency and consistency. Singeing, which gained popularity in the eighteenth century, utilized flames or heated metal plates to burn off protruding fibers from cotton and linen fabrics, creating a smoother surface. Calendaring, introduced in the seventeenth century but widely adopted during this period, involved passing fabric through heated rollers to compress and polish it, resulting in a glossy finish essential for cotton, linen, and silk textiles. The nineteenth century witnessed further mechanization and the development of new finishing techniques to meet the demands of an industrializing world. Brushing and teaseling became prevalent, with fabric passed over brushes or teasel plants to raise fibers, creating a soft, fuzzy texture desired for woolen fabrics. Embossing also emerged during this period, using engraved rollers or plates to imprint raised patterns onto fabric surfaces, adding decorative elements and enhancing aesthetic appeal.

In the 20th century, technological advancements continued to drive innovation in mechanical finishing. Processes like decatising, which utilized steam and pressure to set fabric and enhance luster, and heat setting, essential for stabilizing synthetic fibers like polyester and nylon, became standard practices in textile manufacturing.

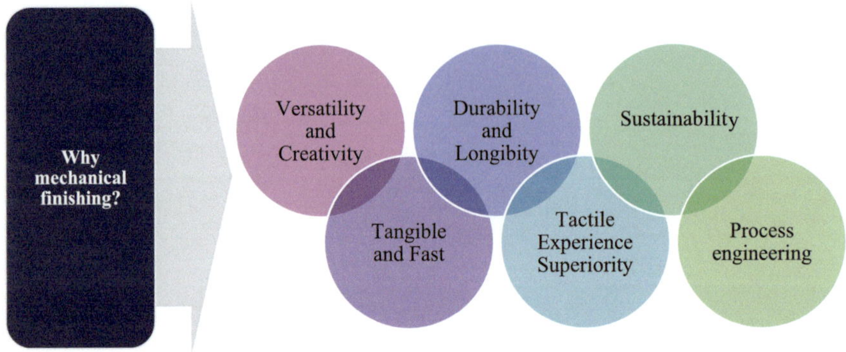

Fig. 1.3 Common advantages of mechanical finishing of textiles

The latter half of the century saw the rise of specialized techniques such as creping, which involved compressing fabric under tension to create a crinkled texture, and Sanforizing, a pre-shrinking process to prevent further shrinkage during washing. In the modern era, the textile industry has embraced advanced technologies and sustainable practices to meet evolving consumer preferences and environmental concerns. Techniques like air jet texturizing, which uses high-pressure air jets to create textured yarns, and plasma treatment, employing ionized gas to modify fabric surfaces, offer new avenues for customization and eco-friendly finishing. Emerging trends like three-dimensional (3D) printing and nanotechnology hold promise for further innovation, shaping the future of mechanical finishing and the textile industry as a whole with creativity, efficiency, and a commitment to sustainability.

Mechanical finishing in textiles is intriguing for its versatility, immediacy, and hands-on nature, offering tangible results and fostering a deep connection the sustainability due to no use of chemicals (see Fig. 1.3). Durability is another crucial aspect, with mechanical finishes typically integrating into the fabric structure to provide long-lasting benefits. Moreover, these techniques are often favored for their compatibility with sustainable practices, requiring minimal water, energy, and chemical usage compared to some chemical finishing methods. As a cornerstone of textile production, mechanical finishing represents a dynamic intersection of engineering and innovation, shaping the future of textiles with creativity, and technological advancement.

1.3 Fundamentals of Mechanical Finishing Techniques in Textiles

The fundamentals of mechanical finishing techniques in textiles encompass a range of processes aimed at enhancing the properties, appearance, and performance of fabrics through physical manipulation. There are variety of chemical finishing process

Table 1.1 Summary of use of mechanical forces for different mechanical finishing techniques

Mechanical finishing	Mechanical forces
Calendaring	Pressure exerted by heated rollers
Heat setting	Tension and heat applied to the fabric to set its shape
Compacting	Compression and pressure applied to the fabric
Brushing	Abrasion and friction caused by rotating brushes
Embossing and debossing	Pressure applied by engraved rollers or plates
Sanding, napping	Abrasion caused by sandpaper or rotating brushes
Shearing, sueding and sanforizing	Cutting or shearing of raised fibers on the fabric surface
Napping	Brushing or teaseling to raise fibers
Beetling	Beating of the fabric with heavy wooden hammers
Creping	Compression and tension applied to the fabric

namely, calendaring, heat setting, compacting, brushing, embossing, debossing, sanding, napping, shearing, sueding and sanforizing, napping, beetling, creping and more. These techniques rely on mechanical forces such as pressure, abrasion, or manipulation to achieve desired effects [15] as summarized in Table 1.1. Key fundamentals include precise control over variables like pressure, temperature, and speed to ensure optimal results.

The historical development of mechanical finishing techniques in textiles has led to the establishment of several key principles that underpin their diverse range. These principles encompass various aspects such as physical manipulation, surface treatment, and customization, each contributing unique effects and characteristics to the finished fabric. Through centuries of innovation and refinement, these principles have evolved to meet changing consumer preferences, technological advancements, and sustainability considerations. They serve as foundational elements guiding the application and execution of mechanical finishing techniques, ensuring consistent quality, functionality, and aesthetic appeal in textile products [20].

1.4 Large Scale Industrial Application of Mechanical Finishing Techniques in Textiles: Overview

The global textile finishing market, including mechanical and chemical finishing processes, was valued at approximately USD 8.17 billion in 2020 and is projected to grow at a compound annual growth rate (CAGR) of 3.6% from 2021 to 2028. This growth underscores the increasing demand for finished textiles in various applications, from fashion to technical fabrics. Mechanical finishing techniques are extensively applied on a large scale in the textile industry, significantly enhancing the efficiency and quality of textile production. The global textile market was valued at approximately $1.8 trillion in 2023, with billions of meters of fabric produced annually [8]. Mechanical finishing is applied to a significant portion of these textiles to enhance their properties. Mechanical finishing is used across various segments, including apparel, home textiles, technical textiles, and industrial fabrics. Each segment has specific requirements and employs different mechanical finishing techniques accordingly. With advancements in technology, the efficiency and capacity of mechanical finishing processes have increased, allowing for greater volumes of textiles to be processed more quickly and with improved consistency. The importance of mechanical finishing is further highlighted by its ability to improve specific fabric properties without the extensive use of chemicals, aligning with the industry's shift towards more sustainable and eco-friendly practices. Commonly used large scale industrial mechanical finishing techniques includes calendaring, heat setting, compacting, brushing, embossing and debossing, sanding, napping, shearing, sueding and sanforizing, napping, beetling, creping, etc. Bellow presents an overview of various mechanical finishing techniques with a focus on principle, purpose, advantages, and disadvantages.

1.4.1 Calendaring

Calendaring is a process whereby a fabric is compressed by passing it between two or more rollers under controlled conditions of time, temperature, and pressure to alter its handle, surface texture, and appearance. The calendaring effect on the fabric is usually temporary and disappears after the first washing. Semi-permanent luster is sometimes achieved by padding fabric in a sparingly soluble polyvinyl acetate emulsion before calendaring, where the solution acts as a binding agent [3]. The major changes to the fabric induced by the calendaring process are reduced fabric thickness, increased fabric luster, increased fabric cover, smooth, silky surface feel, reduced air porosity, and reduced yarn slippage.

1.4.1.1 Principle of Calendaring Finishing

The basic principle of calendaring is to expose the cloth to the combined effect of moisture, heat and pressure until the fabric acquires a very smooth and light-reflecting surface and gets a good luster as presented in Fig. 1.4. The calendaring machine consists of multiple rollers, which can be made of different materials such as steel, cotton, or synthetic fibers. The type of rollers used, and their arrangement (nip pressure, temperature, and speed) depend on the desired finish. Common configurations include: (a) *Simple Calender:* Uses smooth, hard rollers for a basic smooth finish; (b) *Friction Calender:* Includes one roller that moves faster than the others to create a high-gloss finish. (c) *Embossing Calender:* Has engraved rollers that produce patterns on the fabric.

The rollers are heated to the appropriate temperature. The temperature must be carefully controlled as it impacts the fabric's finish. Higher temperatures can produce

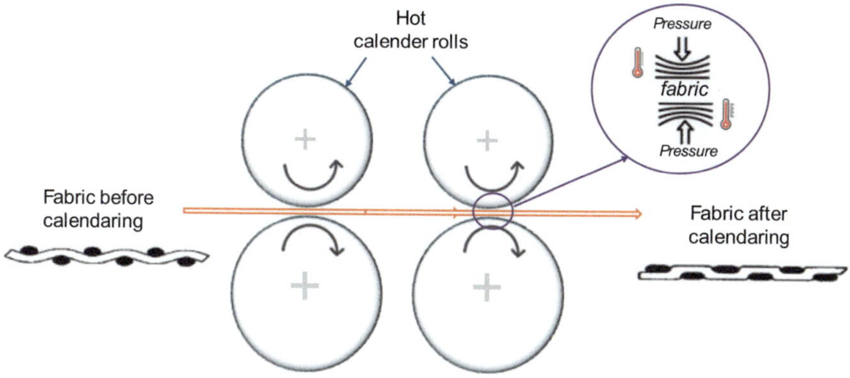

Fig. 1.4 Schematic presentation of calendaring finishing

a higher gloss but may risk damaging the fabric if not controlled properly. The pressure between the rollers is adjusted. The amount of pressure applied is crucial as it influences the degree of smoothness and glossiness of the fabric. High pressure results in a more pronounced effect. The fabric passes through the rollers. As it moves through, it is compressed and heated, which flattens the fibers and imparts the desired finish. Depending on the machine setup, the fabric may pass through several sets of rollers for more pronounced effects. After passing through the rollers, the fabric may be cooled using air or water jets to set the finish. This cooling process helps stabilize the fibers and maintain the achieved smoothness and glossiness. Finally, the finished fabric is wound onto rolls for storage or further processing. This step ensures the fabric remains flat and smooth without wrinkles or creases [3].

Various parameters influence the outcome of the calendaring process. Fabric construction, such as the type of weave or knit, plays a significant role, as does the fiber content, which determines the fabric's inherent properties. The finish applied to the fabric before calendaring can affect the final result, along with the moisture content, which impacts the fabric's flexibility and smoothness. The temperature of the calender bowls is critical for achieving the desired surface effects, while the running speed of the fabric through the calender and the differential bowl speed (the degree or ratio of friction between the rollers) influence the process's efficiency and the fabric's final texture. The pressure applied by the calender bowls determines the extent of flattening and luster, and the bowl composition (whether metal, paper, or synthetic materials) affects the interaction with the fabric. Each of these parameters must be carefully controlled to produce fabrics with specific characteristics and high-quality finishes [15].

1.4.1.2 Purposes of Calendaring Finishing

Calendaring finishing serves several key purposes in the textile industry. Calendaring finishing in textiles primarily aims to improve fabric smoothness, appearance, and feel by flattening surface fibers to achieve a glossy finish. It increases fabric density and strength, reduces thickness and pilling, and allows for special effects like embossed patterns. The process also enhances fabric performance by improving reflectivity, moisture management, and thermal properties, making fabrics more durable and suitable for various applications.

1.4.1.3 Advantages and Disadvantages of Calendaring Finishing

There are both advantages and disadvantages in calendaring finishing of textiles.

Advantage of the calendaring finishing process are as follows;

- Enhanced Smoothness and Luster: Calendaring imparts a smooth and glossy finish to the fabric, enhancing its appearance and feel.
- Improved Fabric Density: The process compresses the fabric, making it denser and reducing porosity. This can improve the fabric's durability and strength.
- Increased Fabric Strength: By aligning fibers and flattening the surface, calendaring can improve the tensile strength of the fabric.
- Better Print Clarity: The smooth surface resulting from calendaring allows for clearer and more precise printing on the fabric.
- Soft Hand Feel: The process can make fabrics feel softer and more luxurious to the touch.
- Controlled Shrinkage: Calendaring helps in stabilizing the fabric, controlling shrinkage during subsequent washing and use.
- Enhanced Resistance to Soiling: The smooth and compact surface is less prone to attracting and retaining dirt and stains.
- Aesthetic Appeal: Calendaring can produce various finishes like embossed effects, adding aesthetic value to the textiles.

Disadvantage of calendaring finishing process is as follows;

- Temporary Effects: The smooth and glossy finish achieved through calendaring is often temporary and can diminish after washing and use.
- Risk of Fabric Damage: The high pressure and heat applied during calendaring can weaken the fibers, leading to potential damage or reduced fabric lifespan.
- Limited to Specific Fabrics: Not all fabrics respond well to calendaring. Delicate fabrics or those with low melting points can be damaged by the process.
- Environmental Concerns: The process may require significant energy consumption and can involve chemicals that are harmful to the environment if not managed properly.
- Maintenance Requirements: Fabrics that have undergone calendaring may require special care and maintenance to preserve the finish.

1.4.2 *Embossing*

Embossing is a permanent relief surface made by a heavy metal press that translates a pattern to a textile. Embossing in textiles involves creating a raised or three-dimensional pattern on a material's surface through the application of heat and pressure. This process uses engraved rollers or molds that impress the design into the fabric, leaving a tactile pattern that contrasts with the surrounding surface. The technique is versatile and can be applied to various textile products to enhance their aesthetic appeal and functionality. Beyond aesthetics, embossing can also serve functional purposes in textiles. It can improve the fabric's durability by compacting fibers and enhancing its resistance to abrasion and wear. Additionally, embossed patterns can alter the way light interacts with the fabric, creating subtle shading effects or highlighting specific areas of the design.

One of the primary advantages of embossing is its ability to add texture and visual interest to fabrics. By altering the surface structure, embossing can simulate textures such as leather, wood grain, or intricate patterns like floral designs and geometric motifs. This decorative effect makes embossed textiles desirable for applications in fashion, upholstery, and interior decor where unique textures and designs are valued.

1.4.2.1 Principle of Embossing Finishing

Embossing in textile fabrication involves passing the fabric through a calender system where a heated metal roller, engraved with a specific pattern, presses against a softer bowl made of compressed paper or cotton on a metal core. This calender nip applies significant pressure to imprint the pattern onto the fabric's surface. Using a soft bowl beneath the engraved roller ensures that the embossed effect mirrors the pattern accurately, enhancing its definition on the textile. Design possibilities for embossed patterns are extensive and depend largely on the creativity of the designer. One common embossing method, known as Schreiner embossing, utilizes a trihelically engraved steel roller resembling a finely threaded screw. The angle of engraving is critical for achieving optimal luster in Schreiner fabrics, which are often used as linings in garments. The pressure applied during embossing is carefully controlled hydraulically and can reach up to 40 tones per width of the textile substrate, ensuring precise and consistent embossed patterns across the fabric [4].

Debossing is a finishing technique used in textiles to create a depressed or sunken design on the surface of a fabric. Unlike embossing, which raises the pattern, debossing presses the design into the material, resulting in a recessed effect. This process is typically achieved by using a metal die or mold that is heated and pressed with significant pressure onto the fabric. The heat softens the fibers of the fabric, allowing them to conform to the shape of the die or mold, while the pressure ensures that the design is deeply imprinted into the fabric.

1.4.3 Heat Setting

Polymeric fibers in their manufactured condition are oriented semi-crystalline structures where macromolecules are rarely found in their equilibrium state. When the fibers are turned into yarns and the yarns into fabrics, more instabilities are introduced [7]. Heat setting is a crucial finishing process in textiles that involves subjecting synthetic fibers or fabrics to controlled heat and pressure to stabilize their dimensions and shape. This process is particularly important for synthetic fibers like polyester, nylon, and acrylic, which can deform under heat or tension during manufacturing or use. Heat-setting can be carried out at three different stages in a processing sequence, i.e., in grey condition, after scouring, and after dyeing. The stage of heat-setting depends on extent of contaminations and types of fibers or yams present in the fabric. Stenters are widely used for heat-setting finishing of textiles.

1.4.3.1 Principle of Heat Setting Finishing

Hearle made the first attempt to explain the theory of heat setting synthetic fibers. He described several methods for arranging fabrics in a particular way. These methods, which do not include chemical stabilization techniques, concentrate on affecting hydrogen bonds, crystallization, strong dipole connections, and chain stiffness. Stress, moisture content, and temperature all have an impact on these variables [10].

In general, heat setting is carried out to bring the material to its thermal equilibrium the material wouldn't changes its morphology it's kept at that temperature for a long. The degree of set, a term often used to describe the extent of heat setting is the measure of how close the material has moved to a thermal equilibrium. A 100% set material is considered to be at its thermal equilibrium at a given temperature. The setting temperature used is above *Tg*. In heat setting, inter-chain bonds, such as hydrogen and dipole bonds, break. This allows the molecular chains to move and adopt new, stress-free positions. New intermolecular bonds then form with the fabric in a relaxed condition at the setting temperature. After cooling, the polymer molecules in the filaments become frozen in place and new bonds forms on cooling as presented in Fig. 1.5. The new bonds are stable up to the heat setting temperature. The reorganized internal polymer structure, and the material's dimensions, will be stable.

1.4.3.2 Purposes of Heat Setting Finishing

The purposes of heat setting are to increase dimensional stability, minimize shrinkage, eliminate internal tensions in the fiber, and lessen the tendency for woven and knitted materials to pucker at the edges [15]. Thermal relaxation would be a more accurate description of the process in this sense. The features of coloring and

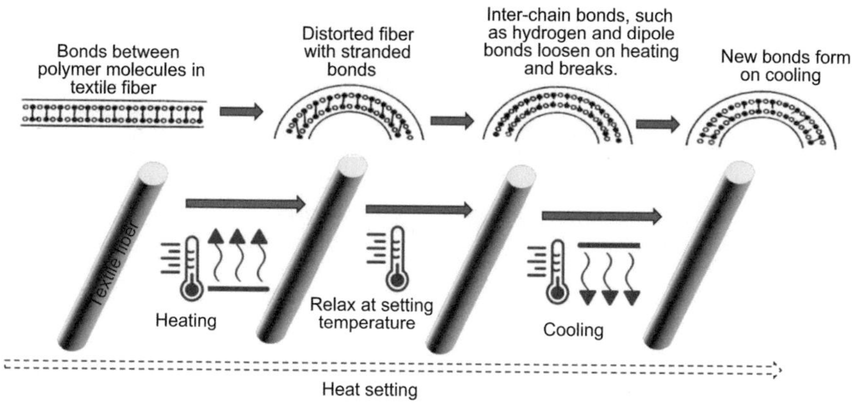

Fig. 1.5 Schematic illustration of heat setting process in textile finishing

mechanical characteristics of synthetic fibers are altered by heat setting. The methodology involves heating the fiber within a temperature range appropriate to it, which is restricted at the upper end by the melting point (softening range) and at the lower end by the corresponding glass transition temperature (which is required to break the secondary bonds).

1.4.3.3 Advantages and Disadvantages of Heat Setting Finishing

There are both advantages and disadvantages in heat setting finishing of textiles. Advantage of calendaring finishing process are as follows;

- Dimensional Stability: By stabilizing the fabric's dimensions, heat setting helps to avoid shrinking or distortion with repeated washings and uses.
- Increased Resistance to Wrinkles: Fabrics become less prone to wrinkles, which enhances their look and lessens the need for ironing.
- Better Texture and Handle: The procedure can make synthetic textiles softer and more pleasant by improving their texture and handle.
- Better Dye Affinity: Applying heat can enhance synthetic fibers' capacity to absorb dye and maintain colorfastness.
- Crease Resistance: This feature keeps the fabric from crinkling and preserves its smooth appearance.
- Enhanced Strength and Durability: By stabilizing the polymer chains, it improves the overall strength and durability of synthetic fabrics.

Disadvantages of heat setting finishing are as follows;

- High Energy Consumption: The process requires high temperatures, leading to significant energy consumption.

- Limited to Synthetic Fibers: Primarily used for synthetic fibers like polyester, nylon, and blends; not effective on natural fibers.
- Potential Yellowing: Some fabrics may yellow or degrade if exposed to excessive heat.
- Environmental Impact: The process can contribute to environmental pollution if not managed properly, especially regarding energy use and emissions.
- Initial Cost: The machinery and setup for heat setting can be costly, leading to higher initial investment.

1.4.4 Compacting

The structure of a fabric influences its tendency to shrink, with more open constructions being more prone to shrinking. Compaction, a mechanical process, reduces fabric shrinkage by mechanically compressing the fabric's structure. Yarns are flexible and bent into irregular shapes during fabrication. In knits, tensions during fabrication and processing often elongate loops. When these loops are shaped more circularly, the fabric becomes more compact and its length decreases, resulting in less shrinkage during washing. In woven fabrics, increased crimp in the warp yarn reduces its length and lowers overall shrinkage. Compacted knits or wovens become denser and thicker, reducing yardage. Softeners and steam during the process facilitate yarns slipping past each other and adjusting. This compression not only improves dimensional stability by limiting both lengthwise and widthwise shrinkage but also enhances the fabric's surface characteristics. Depending on the desired outcome and fabric type, compacting can impart a smoother surface texture or introduce crepe-like effects, thereby enhancing drapability, hand feel, and overall aesthetic appeal. Precise control over parameters such as pressure, temperature, and roller speed ensure consistent and reliable results, making compacting suitable for a wide range of textile applications where durability and aesthetic quality are paramount.

1.4.4.1 Principle of Compacting Finishing

There are two primary methods for fabric compaction. The first method utilizes heated rolls based on the shoe principle (see Fig. 1.6). Here, steam is applied as the fabric wraps around a heated feed roll made of steel, with just enough moisture to ensure smooth movement. The fabric passes between two heated shoes with specialized surfaces that grip it in a controlled manner, and the gap between these shoes is set to achieve compaction. A rubber delivery roll turns at a slower speed than the feed roll and the shoes, effectively slowing down the fabric to facilitate compaction. As the fabric contacts the heated shoe surfaces, it slows down, shortens, and compresses due to surface friction in the compaction zone. Excessive compaction can result in a corrugated appearance, and the fabric often emerges slightly thicker than before

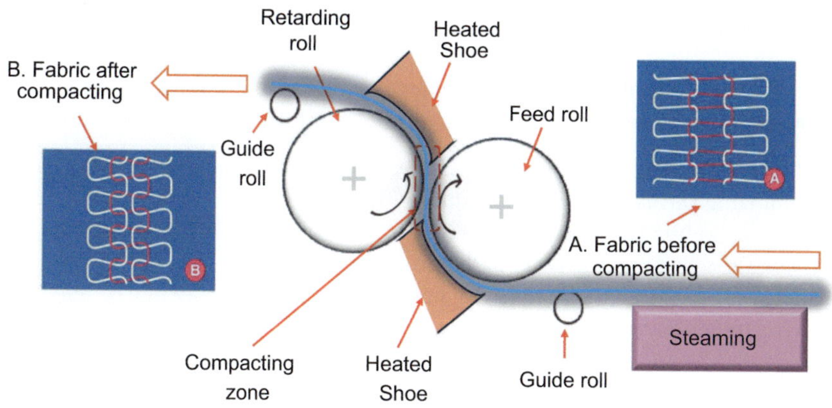

Fig. 1.6 Schematic representation of principle of compacting finishing

entering the machine. An alternative system incorporates a compaction blade positioned beneath one heated shoe between the feed and delivery rolls. This setup achieves compaction by precisely slowing fabric movement through surface contact in the compaction zone. The second compaction method involves belt compaction, using either a rubber or felt belt stretched over a pressure roll. The fabric lays on this belt while it is stretched, and steam is applied as it passes between the belt and a heated Palmer unit. The belt's recoil around the Palmer unit generates compressive shrinkage in the fabric. After compaction, it is recommended to flat-fold the fabric to avoid undesirable tension and stretching that can occur if it is rolled. Following compaction, the fabric is ready for direct cutting and sewing processes [20].

1.4.4.2 Purposes of Compacting Finishing

The primary purpose of compaction in textile manufacturing is to minimize fabric shrinkage through controlled mechanical compression. By subjecting the fabric to heat and pressure, compaction reduces the space between yarns or fibers, effectively stabilizing their positions. This process is crucial for improving dimensional stability, ensuring that the fabric maintains its shape and size over time and through repeated use and laundering. Compaction also enhances fabric density and thickness, which can improve its strength and durability. In knitted fabrics, compaction helps to round out and stabilize loop formations, reducing the potential for elongation and shrinkage during subsequent processing. For woven fabrics, compaction increases crimp in the yarns, thereby decreasing their length and minimizing overall shrinkage.

1.4.4.3 Advantages and Disadvantages of Compacting Finishing

There are both advantages and disadvantages in compacting finishing of textiles. Advantage of calendaring finishing process are as follows;

- Reducing Shrinkage: Compacting improves dimensional stability by significantly minimizing shrinkage of the cloth during washing or use.
- Increased Fabric Density: The procedure results in a greater fabric's density and thickness, which boosts its strength and durability.
- Improved Appearance: Compacting can change the surface texture of the cloth, giving it the desired smoothness or texture.
- Controlled fabric Properties: It enables exact control over the weight, drape, and tactile feel of the cloth.
- Production Efficiency: Compacted fabrics frequently require less post-processing work, which saves money and time during production.

Disadvantages of compacting finishing are as follows;

- Possible Fabric Damage: Excessive pressure or improper compaction parameters might cause harm to sensitive materials or change their intended characteristics.
- Compaction is a Complex Process: Maintaining consistent temperature, pressure, and moisture levels is difficult when compacting.
- Limited Application: Compaction's use in textile processing is limited because not all fabric kinds or architectures are appropriate for it.
- Initial Investment: Specialized machinery and substantial upfront investments are necessary for setting up compaction equipment.
- Environmental Impact: Higher energy use and a larger environmental footprint may result from the compaction process's use of heat and steam.

1.4.5 Sanforizing

Sanforizing is a specialist finishing procedure that shrinks fabrics before they are manufactured into garments or other textile goods. The primary function is to control and reduce fabric shrinkage during future washing and use by consumers. The process is carefully monitored to ensure that the fabric achieves the desired shrinkage level, typically within a range of 1–5%. Sanforizing is particularly important for fabrics made from natural fibers such as cotton, which tend to shrink when exposed to water and heat. By pre-shrinking the fabric through sanforizing, manufacturers can ensure that the final garment or product retains its dimensions and shape after washing, thereby improving the durability and fit for consumers. In addition to reducing shrinkage, sanforizing also enhances the appearance and feel of the fabric by smoothing out wrinkles and improving the fabric's hand feel. This makes sanforized fabrics more comfortable to wear and easier to handle during garment production.

The principle of sanforizing finishing revolves around controlled mechanical and thermal processes to pre-shrink textiles and minimize further shrinkage during subsequent use and laundering [4]. During sanforizing process, fabric is fed under controlled tension onto a rubber belt or conveyor system within the sanforizing machine. This tension ensures the fabric remains flat and stable throughout the process, preventing distortion or uneven shrinkage. As the fabric progresses through the machine, it passes through moisture applicators or steam chambers. Controlled amounts of moisture are applied to the fabric surface. This moisture serves to relax the fibers and make them more pliable for the subsequent shrinking process. After moisture application, the fabric moves between heated cylinders or drums. These cylinders are steam-heated to a precise temperature. The combination of heat and moisture causes the fabric's fibers to swell and relax. The rubber belt or conveyor system gently stretches the fabric widthwise before it encounters the heated cylinders. This pre-stretching accommodates the inevitable shrinkage that will occur when the fabric is exposed to heat and moisture. Once the fabric passes through the heated cylinders, it enters a cooling zone where it gradually cools down. During this cooling phase, the fibers set in their shrunken state, stabilizing the fabric's dimensions. Throughout the process, parameters such as tension, moisture content, and temperature are closely monitored and adjusted as needed to achieve consistent shrinkage control.

Sanforizing and *compacting* are both processes used in textile finishing, but they serve different purposes and achieve different outcomes, especially concerning shrinkage control. Both sanforizing and compacting processes aim to enhance fabric properties, their main difference lies in shrinkage control. Sanforizing is dedicated to pre-shrinking fabrics to specific dimensions, whereas compacting focuses on improving fabric characteristics while minimizing but not eliminating shrinkage. Each process caters to different textile requirements based on the desired end-use and fabric performance expectations.

1.4.6 Brushing

Brushing is an essential finishing process in the textile industry aimed at enhancing the surface characteristics of fabrics. Using specialized machinery equipped with rotating brushes made from materials like horsehair or nylon, the process involves gently raising the fibers on the fabric's surface [4]. As the fabric passes through the machine, these brushes move in the opposite direction to the fabric's travel, effectively lifting individual fibers and creating a soft, fuzzy texture known as a "nap." Unlike more aggressive raising processes, brushing is relatively gentle and does not typically involve a shearing step (discussed in Sect. 4.7) to trim fibers uniformly. This results in a fabric that feels softer and has reduced luster, making it ideal for applications where comfort and warmth are prioritized. Brushing can be applied to various fabrics, including flannels, fleece, and certain knits, enhancing their appeal by improving their tactile qualities and overall aesthetic.

1.4.7 Sanding, Napping, and Shearing

In contrast to calendaring, which typically imparts a smooth, firm, and shiny surface to fabrics, the mechanical finishing processes of sanding, napping, and shearing serve to enhance the tactile and visual characteristics of knit or woven fabrics. These processes are designed to alter the fabric's surface in ways that promote a softer feel and a more textured appearance. Sanding involves subjecting the fabric to abrasive rolls moving at high speeds relative to the fabric's surface. Typically, fabric laid out to its full width comes into contact with one or more abrasive rolls rotating at significantly higher surface speeds than the fabric itself. There are two primary types of sanding mechanisms: the first type involves a multi-role sander, as depicted in Fig. 1.7, which relies solely on surface or tension contact. In this mechanism, the fabric follows a path where it undulates beneath an idler roll, passes over an abrasive roll, and then proceeds to the subsequent sanding units, thus undergoing successive treatments by multiple abrasive rolls. The positioning of idler rolls influences the outcome; lower idler rolls relative to the sanding rolls increase the fabric's angle of wrap, intensifying abrasion. Multi-role sanders offer flexibility by accommodating various levels of aggressiveness and can be adjusted to employ uniform or differing grits across each roll.

The second mechanism, known as a pressure or gap sander, features a single sandpaper roll where additional pressure is applied as the fabric passes through a meticulously adjusted gap between the sanding and pressure rolls. This gap is narrower than the fabric's thickness, ensuring the fabric is firmly pressed onto the sandpaper. The faster-moving abrasive roll then precisely cuts the fabric to a thickness accurate to ten thousandths of an inch. In general, the sanding action not only smooths out irregularities but also creates a subtly textured surface that can resemble sueded leather, offering a softer touch compared to calendaring.

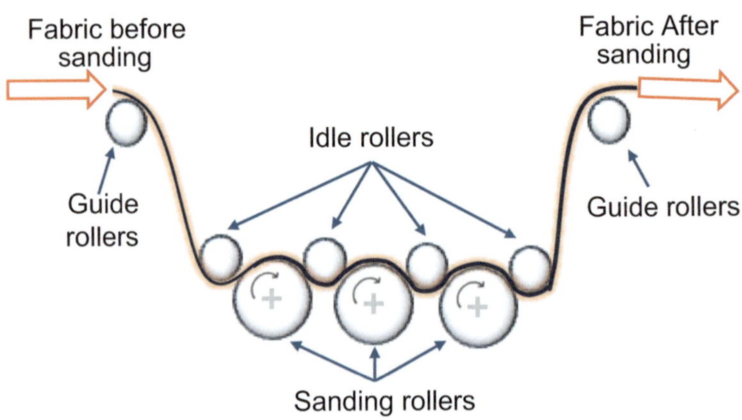

Fig. 1.7 Schematic representation of principle of sanding finishing

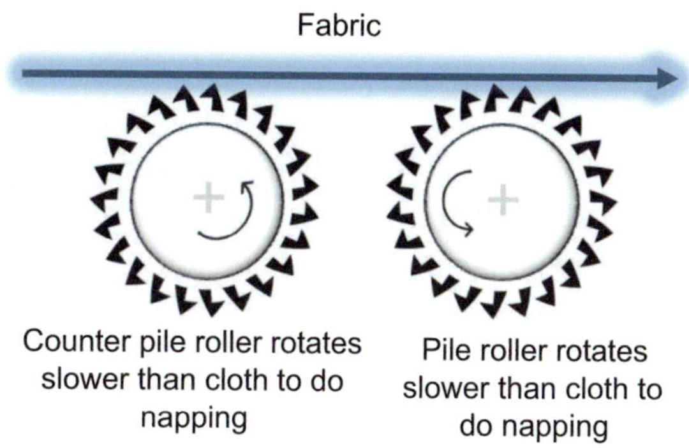

Fig. 1.8 Schematic representation of principle of napping finishing

Napping is a method that introduces pile or fibers to protrude densely from the fabric's surface. Utilizing multiple rolls covered in wire points, napping pulls fibers in an unorganized manner (via the counter pile roll) and arranges them parallel (via the pile roller), resulting in a plush feel that adds depth and softness as illustrated in Fig. 1.8. A fabric effect known as pile is created when a cloth is knapped, causing a large or dense population of fibers to protrude on the surface. This gives it a nice, velvety feel. The napping process utilizes specialized equipment with rolls covered in wire points. These rolls serve distinct roles in altering the texture of fabrics. The "Counter Pile Roll" features wire points that penetrate the fabric yarns directly, extracting fibers in an irregular, random manner from the fabric surface. In contrast, the "Pile Roller" uses its wire points' backs to comb through the fabric, aligning and organizing the lifted fibers to create a smoother appearance. The balance between these actions—counter pile for texture and pile roller for smoothness—is finely tuned by adjusting the speeds of the rolls and selecting appropriate wire types. This control determines whether the fabric retains a textured, uneven surface or achieves a more uniform finish. The combination of these rolls can vary to accommodate different fabric types, ensuring consistent and desired pile heights across knapped textiles.

Shearing complements napping by precisely cutting protruding fibers or loops from the fabric's surface using rotating cylinders and sharp blades as illustrated in Fig. 1.9. This process ensures uniformity in fiber height and spacing, producing a lustrous finish that enhances the fabric's appeal. The cutting mechanism involves multiple spiraling blades that intersect with a rigid, straight shearing blade, creating a scissor-like cutting action. This setup ensures precise cutting of fiber or yarn tips across the fabric's width. Proper alignment of the blades is crucial to prevent damage to both the fabric and the blades themselves. This precision in alignment not only enhances the efficiency of the shearing process but also contributes to achieving consistent and high-quality results in fabric cutting operations.

Brushing, sanding, napping, and shearing are distinct processes employed in textile finishing, each serving specific purposes and yielding unique effects on fabrics as summarized in Table 1.2. Brushing involves using rotating brushes to gently raise fibers on the fabric's surface, creating a soft and fuzzy texture known as a nap. It enhances the fabric's tactile feel and reduces its luster without significant fiber

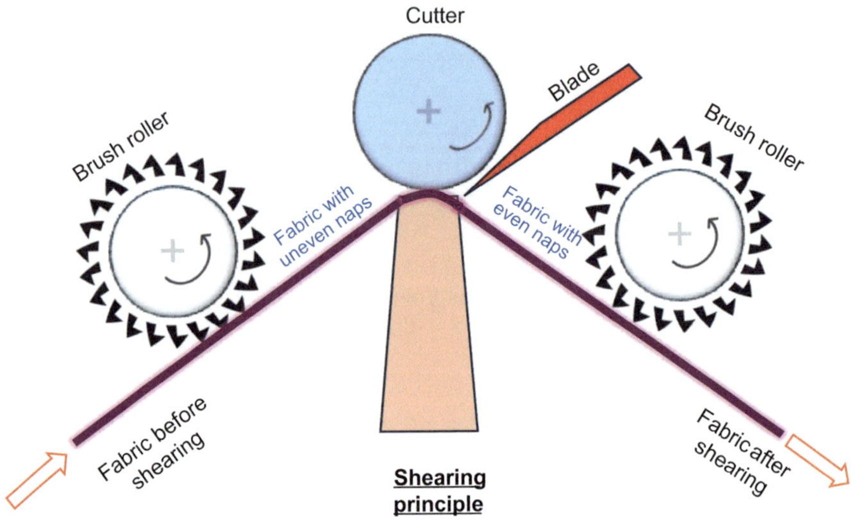

Fig. 1.9 Schematic representation of principle of shearing finishing

removal. In contrast, sanding utilizes abrasive materials to smooth the fabric surface by removing fibers, achieving a flat and uniform texture. Napping utilizes specialized machinery with wire-covered rolls to raise fibers and create a soft nap, enhancing warmth and comfort in fabrics like blankets and flannels. Shearing, on the other hand, trims the raised fibers to a consistent length, resulting in a smooth and even surface. Each of these processes is chosen based on the desired texture, appearance, and functionality required for the fabric, catering to diverse applications in the textile industry.

1.5 Emerging Application of Mechanical Finishing of Textiles

The emerging applications of mechanical finishing in textiles represent a dynamic shift towards integrating advanced technologies and sustainable practices to enhance fabric performance and appeal. These innovations encompass a range of techniques and approaches aimed at improving functional attributes such as moisture management and thermal regulation. For instance, methods like calendaring and napping are being refined to optimize moisture wicking capabilities, thermal resistance and reduce drying times, making fabrics ideal for activewear and outdoor textiles. Literature reports shows, thermal resistance of polypropylene nonwoven fabric manufactured using the mechanical carding process and bonding using the needling process can be improved by the thermal calendaring process [14]. Moreover, techniques such

Table 1.2 Differences between brushing, sanding, napping and shearing finishing

Criteria	Brushing	Sanding	Napping	Shearing
Purpose and effect on fabric	Raise the fibers from the fabric surface, creating a soft and fuzzy texture known as nap	Abrasively smoothing the fabric surface to reduce roughness and irregularities	Abrasively smoothing the fabric surface to reduce roughness and irregularities	Trimming the raised fibers on the fabric surface to achieve a consistent pile height
Equipment and mechanism	Utilizes rotating brushes or napper rolls to lift fibers from the fabric surface	Uses abrasive materials such as sandpaper or sanding drums mounted on rotating cylinders or belts	Involves specialized napper rolls or brushes that lift and align fibers to create a nap	Utilizes blades or knives arranged in a shearing mechanism to cut the raised fibers to a uniform height
Applications	Suitable for a wide range of fabrics desiring softness	Typically applied to sturdy fabrics like denim and canvas	Often used in woven fabrics such as wool, flannel, and fleece	Often used in woven fabrics such as wool, flannel, and fleece, towel
Effect on fabric properties	Increases fabric softness and warmth	Smooths the fabric surface, reducing roughness and irregularities	Enhances fabric warmth, softness, and insulating properties	Improves fabric appearance by creating a uniform pile height
Environmental impact	consumes less energy and water compared to sanding and shearing	Requires more energy and careful management of dust and debris	Consumes less energy and water compared to shearing	Higher energy use and waste production compared to napping

as brushing and sanding are enhancing the thermal insulation properties of textiles, ensuring comfort across varying climates.

Furthermore, the application of mechanical finishing in textiles extends beyond functional enhancements to include aesthetic improvements. Techniques like shearing and sueding are utilized to create luxurious textures and soft hand-feel, enhancing the appeal of fabrics in high-end fashion and home textiles. These processes not only improve the tactile experience but also contribute to the overall durability and longevity of the materials. In addition to performance and aesthetic benefits, the integration of mechanical finishing in textiles is increasingly aligned with sustainable practices. Innovations in this field focus on reducing energy consumption, water usage, and chemical inputs during the finishing process. Techniques such as air finishing, heat pressing and laser texturing are gaining prominence for their minimal environmental impact while achieving desired fabric properties [13].

Moreover, advancements in digital technologies and automation are revolutionizing mechanical finishing in textiles, allowing for greater precision, consistency, and customization. Digital control systems enable fine-tuning of process parameters, resulting in improved quality control and operational efficiency across production scales. Overall, the evolving applications of mechanical finishing in textiles underscore a transformative shift towards enhancing fabric functionality, sustainability, and aesthetic appeal. As technology continues to advance, these innovations promise to reshape the textile industry, offering new opportunities for creating high-performance, eco-friendly, and aesthetically pleasing fabrics for diverse consumer needs.

1.6 Sustainability Aspects of Mechanical Finishing Techniques in Textiles

Sustainability considerations in mechanical finishing techniques within the textile industry are increasingly pivotal, driven by global imperatives to mitigate environmental impact and enhance resource efficiency across production cycles. Unlike chemical finishing methods, mechanical processes often require fewer or no chemicals, thereby reducing environmental impact and health risks associated with chemical use and disposal [4, 10, 15].

Mechanical finishing techniques offer several concrete benefits that enhance the sustainability and efficiency of textile production as presented in Fig. 1.10. By reducing reliance on chemical treatments, these methods contribute significantly to environmental sustainability by minimizing pollution and reducing health risks associated with chemical exposure. Moreover, mechanical finishing processes typically consume less water and generate less wastewater, thereby conserving valuable resources. They also produce less waste material, promoting overall waste reduction and improving resource efficiency in textile manufacturing. Fabrics treated with mechanical finishes tend to be more durable and have an extended lifespan, which reduces the frequency of replacements and minimizes resource consumption over time. Additionally, mechanical processes are often faster and more continuous, enhancing production efficiency and reducing energy consumption per unit of output. Lastly, these finishes make textiles easier to recycle at the end of their life cycle, supporting circular economy principles and reducing textile waste.

Nevertheless, mechanical finishing operations traditionally consume substantial energy due to the high-power requirements of machinery such as calendaring rollers and brushing equipment. Alongside energy consumption, these processes can contribute to environmental pollution through various mechanisms. For example, calendaring and brushing may generate airborne particulates and emissions from machinery operation, potentially impacting air quality in manufacturing environments. Moreover, the use of lubricants, cleaning agents, and other chemicals in maintenance and operation can lead to wastewater contamination if not properly managed.

Fig. 1.10 General sustainability aspects related to mechanical finishing of textiles

However, advancements in technology have enabled the development of energy-efficient machinery and processes. For instance, modern calendaring machines incorporate heat recovery systems to reuse thermal energy, reducing overall energy consumption. Similarly, innovations in brushing and sueding equipment focus on optimizing motor efficiency and reducing power requirements during operation. Integration of renewable energy sources such as solar and wind power further enhances sustainability credentials, minimizing carbon emissions associated with energy-intensive textile finishing processes. Water scarcity and pollution concerns necessitate sustainable water management practices in textile finishing. While processes like dyeing historically consume significant water volumes, mechanical finishing techniques such as embossing offer advantages in water conservation.

Additionally, advanced calendaring and embossing systems employ closed-loop water systems and filtration technologies to reduce wastewater generation and improve water reuse efficiency. The use of chemicals in mechanical finishing processes is a critical area for sustainability improvement. Sustainable practices involve minimizing chemical usage, substituting hazardous substances with eco-friendly alternatives, and adopting technologies that reduce chemical discharge into waterways. For instance, sanforizing processes utilize steam and mechanical pressure to achieve fabric shrinkage, reducing reliance on chemical treatments. By balancing operational efficiency with environmental responsibility, mechanical finishing represents a pivotal pathway toward achieving sustainability in textile manufacturing. By leveraging technological innovation, optimizing resource efficiency, and prioritizing environmental stewardship, mechanical finishing can make substantial contributions to global sustainability goals while meeting consumer demand for eco-conscious products.

1.7 Challenges and Future Perspective of Mechanical Finishing in Textiles

Mechanical finishing in textiles faces several technical challenges that influence its operational efficiency and environmental impact. High energy consumption is a significant concern, particularly in processes such as calendaring, heat setting, and shearing, where large machinery and high speeds are employed. This necessitates efficient energy management strategies to minimize costs and environmental footprint. Environmental sustainability is also a critical issue, despite mechanical finishing generally using fewer chemicals than alternative methods. Processes like sanding and shearing generate considerable dust and waste fibers, necessitating effective dust extraction systems and waste management practices to mitigate environmental impact and ensure workplace safety. Quality control poses another technical challenge, as achieving consistent fabric properties across batches requires precise control over parameters like pressure, speed, and blade alignment. Variations in machine settings and material characteristics can lead to inconsistencies in finish quality, necessitating rigorous monitoring and adjustment protocols. Looking ahead, the future of mechanical finishing in textiles hinges on several critical areas. Sustainability stands at the forefront, compelling innovations that minimize environmental impact through reduced energy, water, and chemical usage, alongside advancements in recycling. Integration of advanced technologies such as AI, automation, and digitalization will revolutionize processes, enhancing precision, customization capabilities, and operational efficiency. Meeting consumer demand for personalized products requires flexible systems that can adapt textures, patterns, and functional properties swiftly. Ensuring consistent quality and performance across batches remains pivotal, driving advancements in uniformity and durability. Cost efficiency and productivity gains will be achieved through streamlined processes and optimized resource management. Compliance with stringent regulatory standards for environmental sustainability and product safety will be non-negotiable. Collaboration across industry sectors and knowledge sharing will accelerate innovation, fostering a robust ecosystem for future mechanical finishing solutions.

Furthermore, advancements in hybrid techniques, such as combining mechanical finishing with plasma treatment or laser processing, offer opportunities to develop textiles with enhanced functionalities like water repellency, antimicrobial properties, or improved dye absorption. These techniques not only expand the performance capabilities of textiles but also contribute to sustainability by reducing chemical usage and enhancing material efficiency. In general, while mechanical finishing in textiles faces technical hurdles related to energy consumption, waste management, and quality control, ongoing advancements in digitalization and hybrid technologies offer promising solutions. Embracing these innovations can enhance process efficiency, product quality, and environmental sustainability, positioning mechanical finishing as a vital component of modern textile manufacturing.

Acknowledgements Authors would like to acknowledge the institutional support from the Department of Textile Technology, Faculty of Textiles, Engineering and Business (including the Swedish School of Textiles), University of Borås, Sweden during preparation of this chapter. Author would also like to thank Prof. Vijay Kumar for spontaneous suggestions and discussion during the writing of the chapter.

Competing Interest Author declares no competing interest in this chapter.

References

1. Attia, N. F., et al. (2024). Recent advances in nanobased flame-retardant coatings for textile fabrics. *Nano-Structures & Nano-Objects, 38*, 101180.
2. Bouasria, A., et al. (2020). Advances in polymer coating for functional finishing of textiles, *In: Shabbir, M., Ahmed, S., Sheikh J. N. (eds) Frontiers of Textile Materials: Polymers, Nanomaterials, Enzymes, and Advanced Modification Techniques,* Scrivener Publishing LLC, pp. 61–86.
3. Choudhury, A. K. R. (2017). *Principles of textile finishing*. Woodhead Publishing.
4. Conway, R. (2016). Technical textile finishing, *In: Horrocks, A. C., Anand, S. C. (eds) Handbook of Technical Textiles,* Woodhead Publishing, pp. 189-210.
5. Fergusson, S. M. (2015). Garment-finishing techniques, *In: Nayak R., Padhye R. (eds) Garment Manufacturing Technology,* Woodhead Publishing, pp. 387–403.
6. Gulrajani, M. L. (2013). *Advances in the dyeing and finishing of technical textiles*. Woodhead Publishing.
7. Gupta, V. (2002). Heat setting. *Journal of Applied Polymer Science, 83*(3), 586–609.
8. GVR_Report. (2024). GVR Report coverTextile Market Size, Share & Trends Report.
9. Textile Market Size, Share & Trends Analysis Report By Raw Material (Wool, Chemical, Silk), By Product (Natural Fibers, Polyester), By Application, By Region, And Segment Forecasts, 2024–2030. Grand view research. from https://www.grandviewresearch.com/industry-analysis/textile-market.
10. Hall, M. E. (2000). Finishing of technical textiles. *In: Horrocks, A. R., Anand, S. C. (eds) Handbook of Technical Textiles,* Woodhead Publishing, pp. 152–172.
11. Hassan, M. M., & Carr, C. M. (2019). A review of the sustainable methods in imparting shrink resistance to wool fabrics. *Journal of Advanced Research, 18*, 39–60.
12. Horrocks, A. R., & Anand, S. C. (2000). *Handbook of technical textiles*. Woodhead Publishing.
13. Jevšnik, S., et al. (2016). Applying heat for joining textile materials, In: *Ishak M. (ed) Joining Technologies*. IntechOpen.
14. Kopitar, D., et al. (2014). Study on the inuence of calendaring process on thermal resistance of polypropylene nonwoven fabric structure. *Journal of Fiber Bioengineering and Informatics, 7*(1), 1–11.
15. Kumar, R. S., & Sundaresan, S. (2013). Mechanical finishing techniques for technical textiles. *In: Gulrajani M. L. (ed) Advances in the Dyeing and Finishing of Technical Textiles*, pp. 135–153.
16. Le Marechal, A. M., et al. (2012). Textile finishing industry as an important source of organic pollutants. *In: Puzyn, T., Mostrag-Szlichtyng, A. (eds) Organic Pollutants Ten Years After the Stockholm Convention-Environmental and Analytical Update*. IntechOpen.

17. Madushika, J. W. A., & Lanarolle, W. D. G. (2022). A review on novel approaches to enhance sound absorbing performance using textile fibers. *The Journal of The Textile Institute, 113*(2), 341–348.
18. Mittal, K. L., & Bahners, T. (2017). *Textile finishing: Recent developments and future trends, Scrivener Publishing LLC.*
19. Ozdil N., Anand S. (2014) Recent developments in textile materials and products used for activewear and sportswear. *Electronic Journal of Textile Technology*, 8(3), 68–83.
20. Stukenbrock, K. H. (2002). Mechanical finishing. *In: Albrecht, W., Fuchs, H., Kittelmann, W. (eds) Nonwoven Fabrics: Raw Materials, Manufacture, Applications, Characteristics, Testing Processes*, Wiley-VCH, pp. 411–420.
21. Suvari, F., & Dulek, Y. (2019). Investigating the effect of raising on the sound absorption behavior of polyester woven fabrics. *Textile Research Journal, 89*(23–24), 5119–5129.
22. Teli, M. D. (2015). Softening finishes for textiles and clothing. *In: Paul, R. (ed) Functional finishes for textiles: Improving Comfort, Performance and Protection*, Woodhead Publishing, pp. 123–152
23. Wang, K., et al. (2023). Non-fluorinated durable water repellent and stain resistant coating. *Applied and Computational Engineering*, 7(1), 438–452.
24. Zhang, Y., et al. (2022). Recent advances in the surface modification strategies to improve functional finishing of cotton with natural colourants-A review. *Journal of Cleaner Production, 335*, 130313.

Chapter 2
Mechanical and Chemical Processes of Textile Finishing Industries

Mohamed Berradi, Omar Berradi, Youssef El Rhayam, Lama Rissouli, Naoual El-Aouni, Ahmed El Yacoubi, Mohamed Rafik, and Abderrahim El Bachiri

Abstract In this chapter, we presented and discussed the various mechanical and chemical processes and sub-processes involved in the manufacture of textile materials such as yarns, fabrics, nonwovens, knits, carpets, and so on, from different raw materials of natural (vegetable, animal, and mineral) and chemical (synthetic) origin. Spinning, winding, warping, weaving, knitting, entangling, and making are all mechanical textile manufacturing processes that require electrical and thermal energy. Pre-treatment, dyeing, printing, and finishing are all wet textile finishing processes that necessitate the use of electrical and thermal energy, heat, water, hydrogen peroxide acids, bases, salts, auxiliaries, dyes, pigments, optical brighteners, additives, and so on. As a result, a large volume of effluent containing these various chemical compounds is produced, and it is sometimes discharged, directly or indirectly, into waterways without prior treatment.

Keywords Mechanical and chemical processes · Manufacture of textile materials · Textile finishing processes · Effluents · Chemical compounds

M. Berradi (✉) · O. Berradi · Y. El Rhayam · N. El-Aouni · M. Rafik
Laboratory of Organic Chemistry, Catalysis and Environment, Department of Chemistry, Faculty of Sciences, Ibn Tofaïl University, Kenitra, Morocco
e-mail: mberradi24@gmail.com

L. Rissouli
Laboratory of Physico-Chemistry Materials, Natural Substances, and Environment, Department of Chemistry, Faculty of Sciences and Techniques, Abdel Malek Essaadi University, Tangier, Morocco

A. El Yacoubi
Laboratory of Chemistry, Environment and Chemistry of Solid Minerals, Department of Chemistry, Faculty of Sciences, Mohamed the First University, Oujda, Morocco

A. El Bachiri
Royal Naval School, University Department-Boulevard Sour-Jdid, Casablanca, Morocco

M. Shahid et al. (eds.), *Advancements in Textile Finishing*,
https://doi.org/10.1007/978-981-96-6385-9_2

2.1 Introduction

The textile industry encompasses a variety of trades that work together to produce semi-finished or fully manufactured products (yarns and fabrics) from natural and/or chemical fibrous materials [1, 16, 76]. It primarily consists of two industries: mechanical manufacturing and textile finishing. The latter phase is the most crucial in the textile business since it employs many chemicals and dyes with a large volume of water [12, 14, 23, 56]. The most important textile finishing techniques are the various stages of dyeing and printing textile articles, and the use of various types of dyes, pigments, additives, chemical auxiliaries, and optical brighteners is critical [10, 11, 31, 36, 58, 62, 78]. This chapter outlines and examines the many procedures needed for the mechanical fabrication and finishing of textile items while bringing to light the different chemicals and raw materials needed for each step and type of manufacturing.

2.2 Textile Material Manufacturing Processes

In the textile industry, some industrial units have multiple manufacturing processes, whereas others specialize in a single type [55, 61]. In this section, we will distinguish two types of textile manufacturing: mechanics and finishing.

2.2.1 Mechanical Processes of Textile Manufacture

These are the mechanical processes required to manufacture various textile articles from natural, semi-synthetic, and synthetic textile fiber materials, including spinning, winding, warping, weaving, knitting, entangling, and making up to those finishing. This produces textile materials in the form of finished and/or semi-finished products, such as yarns, knits, nonwovens, fabrics, carpets, etc. [61, 76]. Figure 2.1 depicts all of the mechanical manufacturing processes for textile materials [23, 61, 76].

2.2.1.1 Spinning

It is a method of assembling fibrous textile materials to obtain yarn on the one hand and shaping the yarn using a die made of metals, glasses, polymers, and ceramic pastes [75].

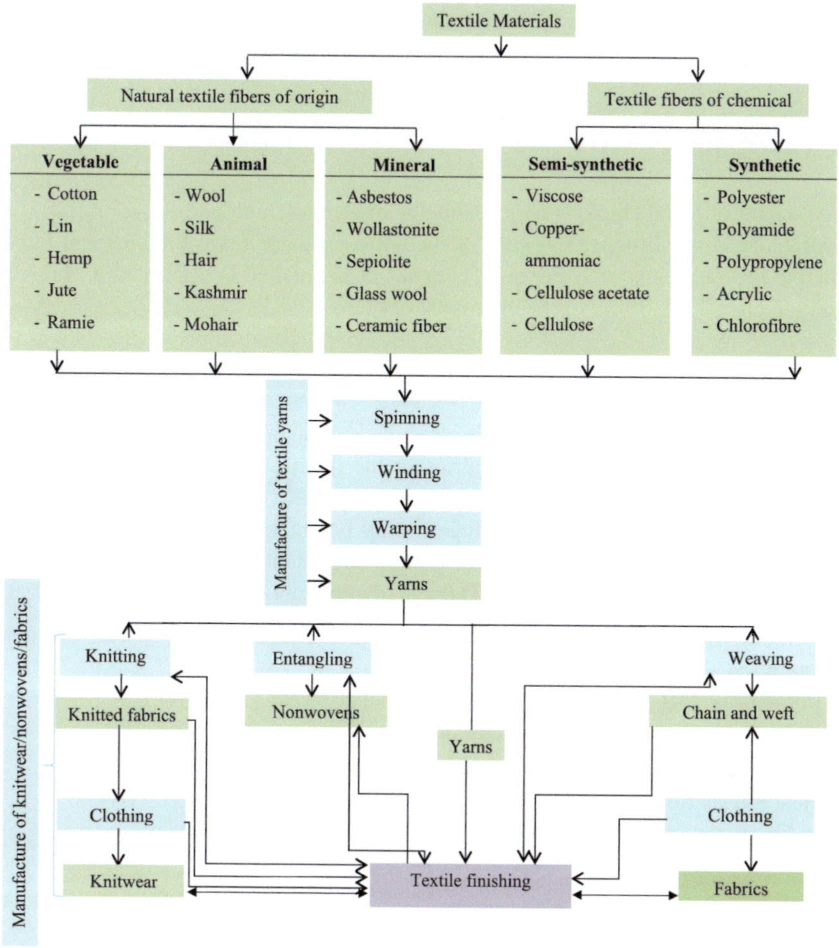

Fig. 2.1 Mechanical processes of textile manufacturing

2.2.1.2 Winding

It is the process of winding textile yarns around the axis of a bobbin in an orderly fashion. It entails transferring yarns from small packages produced during manufacturing to larger cones required for weaving or knitting processes [41].

2.2.1.3 Warping

It is a preliminary and preparatory weaving operation that consists of unwinding the thread reels on the beam (roll) for weaving. To accomplish this, the warp threads are wound in a precise order under the same tension and in parallel. As a result, two types of warping exist: full (direct) warping and fractional (severed) warping [49].

2.2.1.4 Knitting

It is a technique that employs large-diameter single and double-knitting machines to manipulate a one-dimensional yarn or thread to produce a two-dimensional fabric suitable for various sportswear and footwear. It is finished in open width, flat, and ready to cut after knitting in the processes [24].

2.2.1.5 Entangling

It is a tangle-forming process of fabrics in the form of nonwoven (veils) whose fibers or filaments are neither woven nor knitted but are oriented in a definite order or randomly arranged and bonded together by chemical, physical, or mechanical methods [19]. Nonwovens are manufactured in three stages: weaving, consolidation, and finishing. Nonwovens have a wide range of applications and are used in a variety of fields, including:

- Clothing: interlining, clothing reinforcement, and in parkas and anoraks;
- Furnishings: bedspreads and duvets;
- Household sector: product wiping, chamois leather, abrasive pads;
- Building sector: flooring and wall covering;
- Hospital sector and hygiene;
- Industry and public works sector.

2.2.1.6 Weaving

This mechanical process forms textile fabrics by interlacing, usually perpendicularly, all the threads in the form of warp in the longitudinal direction and weft in the transverse direction (pick) [69]. This interlacing is performed using the loom or the weaving machine according to a specific pattern (weave) [49, 69].

2.2.1.7 Confection

It is a textile manufacturing operation that produces clothing textiles, interior textiles, technical textiles, and household linens daily. In most cases, and to give textile materials their distinct forms, mechanical textile manufacturing processes are combined with those of textile finishing, which we will list below.

2.3 Textile Finishing Processes

In the textile finishing industry, there are generally four important and necessary processes, namely pretreatments, dyeing, printing, and primers, which are similar to the diagram in Fig. 2.2 [18, 23, 57, 61].

They are a collection of techniques used in the various stages of textile finishing. They entail providing fabrics with the visual appearance as well as the physical and aesthetic properties desired by the consumer via a number of important processes involving electrical and thermal energy, heat, water, hydrogen peroxide, acids, bases,

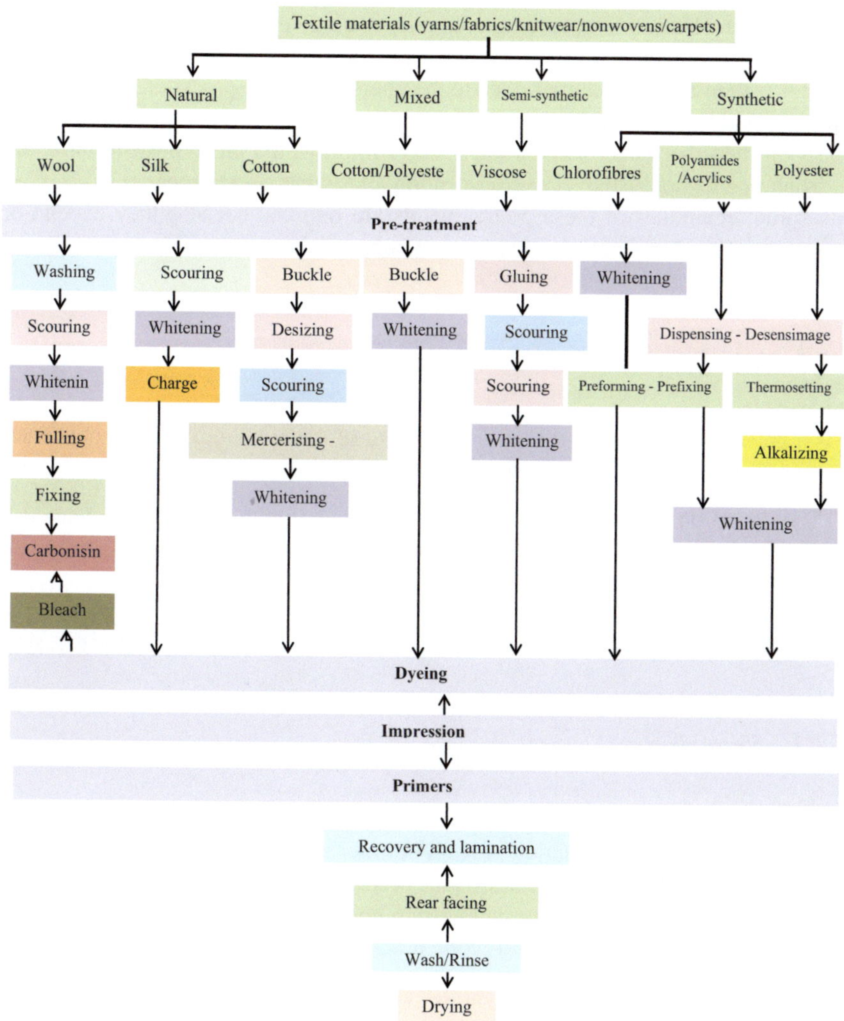

Fig. 2.2 Processes of textile finishing

salts, auxiliaries, dyes, pigments, and optical brighteners [61]. This set of processes produces hot effluents laden with these products' residues and chemical reagents [44, 50]. Pre-treatment, dyeing, printing, and finishing are four important and necessary processes in the textile finishing industry, as illustrated in Fig. 2.2 [18, 23, 61].

2.3.1 Pretreatments

They involve the techniques of washing, scouring, buckling, sizing, desizing, mercerizing-caustification, preforming-prefixing, heat-setting, alkalinization, bleaching, crushing, fixing, carbonizing, chlorination, and loading (Fig. 2.2). These techniques are applied before proceeding to the technique of dyeing textile materials, and this is to remove the debris added to the fibers during the procedures of spinning, warping, and weaving on the one hand, and to release the tensions within synthetic fibers on the other hand. This is intended to improve the uniformity, affinity, and hydrophilic characteristics of dye adsorption and finishing treatments [7, 57]. Note that some techniques of these pretreatments are optional because they depend on the nature of the textile fabric, reagents, and chemicals used.

2.3.2 Washing

It is a process whose purpose is to rid the wool of raw wool or oozing wool. There are two distinct operations, namely:

- The operation of disuniting consists of a treatment with pure water intended to remove the liquid-soluble part of the liquid composed of salts of potash, carbonate, and soap of potash, urea, carbonate of ammonia, and excrement.
- Degreasing consists of a treatment with soapy water to remove the insoluble part of the pure water, suintine. The earthy materials are entrained during degreasing.

2.3.3 Scouring

Its purpose is to remove substances that have adhered to the surface of the textile fibers during the yarn production process in order to facilitate the weaving process. Impurities, oils, sizing agents, pectins, fats, waxes, proteins, and inorganic substances such as alkali metals, calcium or magnesium phosphates, and iron oxides can all be removed. It can be done as a stand-alone procedure or with other treatments [13].

2.3.4 Buckle

This method rapidly passes fabrics over the flames of a gas ramp [30]. There are multi-effect flambers that allow the fabric's two faces, or even each face, to be treated multiple times in a single pass. The fabric must first be passed through a drying drum to ensure consistent hygrometry before it can be buckled.

2.3.5 Gluing

During this process, the warp yarns are coated with a glue film to facilitate the weaving operations. The obtained film must be removed prior to dying to ensure proper unison dyes with the textile yarn. The residual sizing rate must therefore be less than 0.1% because some dyes and primers can react chemically with some sizing materials [12, 14]. Sizing agents are selected based on the type of yarn and textile fabric, their environmental friendliness, their removal ability, cost considerations, and effluent treatment [23]. We divide sizing products into two broad categories: insoluble starch sizing and sizing based on soluble compounds such as polyvinyl alcohol, acrylic resins, and carboxymethyl celluloses. Reagents and chemical auxiliaries used in the sizing of mixed textiles such as cotton and polyester, as well as cellulosic fibers such as viscose, acetate, and triacetate, include anionic surfactants, sodium carbonates, trisodium phosphate, caustic soda, and sodium perborate [8].

2.3.6 Desizing

It is a process applied to the fabric to remove the compounds used to make the yarns more pliable for weaving. It is carried out by soaking the fabric in baths composed of hot water and so-called water-soluble agents such as acids, bases, and enzymes [15, 27, 32, 48]. Regardless of the nature of the desizing agents, the fabric is impregnated with the agents. This allows these agents to degrade or solubilize the sizing material in order to wash away degradation products [13].

It should be noted that enzymatic desizing is the most common due to its ecological nature and characteristics that act at specific locations based on pH, temperature, and concentration that are well defined. Desizing is accomplished primarily through the use of starch [13]. Table 2.1 lists the chemicals used in the desizing of various textile materials [13, 17, 27, 32, 48].

Table 2.1 Reagents and auxiliaries chemically used for the process of desizing

Textile materials	Chemical reagents and auxiliaries
Cotton/polyester	– Cellulases, α-amylases, β-amylases, sodium bromite ($NaBrO_2$), ammonium persulfate ($(NH_4)_2S_2O_8$), detergents (acid medium at pH = 6, sodium carbonate and wetting agents (70–90 °C)
Viscose, acetate and triacetate of cellulose	– Sodium carbonate, trisodium phosphate and caustic soda

Table 2.2 Reagents and chemical auxiliaries used in the scouring process

Textile materials	Chemical reagents and auxiliaries
Cotton, viscose, acetate and triacetate of cellulose	– Wetting agents, NaOH, detergents, sodium hydrosulfite and chelating agents
Wool	– Sodium carbonate, non-ionic synthetic detergents and 1,2-dichlorobenzene

2.3.7 *Scouring*

It is referred to as atmospheric degreasing or pressure degreasing. Its goal is to extract impurities and plant debris from raw fiber (cotton waxes) in order to make it hydrophilic through alkaline washing treatment (Table 2.2) [15, 25, 50].

It requires the use of a more or less concentrated solution of caustic soda, to which the following must be added:

- A reducing agent to prevent cellulose oxidation by air molecular oxygen in an alkaline and closed environment
- Calcium ion complexing agent to avoid insolubilization of organic acids formed from impurities
- A stable dispersant (non-ionic surfactant) in an alkaline medium to remove accompanying substances (pectins, greases, waxes, and inorganic substances) and impurities (pesticides).

2.3.8 *Desizing*

It is a process for removing sizing agents, accumulated oils, and dirt from fibers in order to reduce friction conditions during mechanical processes [15, 28]. In fact, excessive amounts of fat in wool would have a negative impact on dyeing operations and can be eliminated by using chemicals such as surfactants at pH = 10 and anionic detergents in a neutral or alkaline medium [4, 17, 25, 28, 54, 72].

2.3.9 Mercerization–Caustification

Mercerization is a chemical treatment that involves dipping cotton fabric under tension and cold in a highly alkaline medium for brief periods of time (less than four minutes). The resulting tissue is rinsed and neutralized with water or acid [13, 79]. This procedure increases the tissue's affinity for dyes, reagents, and chemical auxiliaries [73]. Mercerization, depending on the method used, also improves the tensile strength, absorption properties, dimensional stability, and overall gloss of the cotton garment. Furthermore, it improves dye growth by reducing dye consumption by 30–50% through improved dye bath exhaustion.

Caustification is a simple and cost-effective pretreatment method that involves desisting, cleaning, bleaching, and caustically treating cotton fabric in a highly alkaline, stress-free, hot environment [73]. The fabric subjected to this process, mercerization-caustification, undergoes significant changes in mechanical, dyeing, and gloss properties, all of which are critical.

2.3.10 Preforming–Prefixing

Preforming, sometimes called hydrofixation, is the process of fixing detergents (anionic and cationic) and sizing agents (cationic softeners) in water under pressure and at around 130 °C for 45 min in an autoclave [33]. This treatment gives the product a full and mellow appearance. In the case of textile materials made of polyamide-6 (or 6.6), the fixing with water vapor entails subjecting the goods to the action of steam at 130 °C for 15–20 min. Prefixing is a polyamide-6 fixing operation that takes about 30 s at 210 °C (nylon 6). The off-white fabric is washed in a washer with a boiling bath packed with trisodium phosphate, carbonate, or caustic soda at a pH of 11 or 12 and approximately 2 g of detergent. The train transports the expressed tissue from the washer to the drying compartment, then to the fixing compartment, then to the cooling compartment, and finally to the winding compartment.

2.3.11 Thermosetting

It is a heat treatment used on polyester-based articles at the knitted or woven stage, often before dyeing [5]. It gives the tissue dimensional stability and wrinkle resistance by changing the dye affinity. Indeed, a heat-fused fabric without tension increases in affinity, whereas a thermofixed fabric under tension decreases in affinity. Thermofixation also improves the behavior of polyester fabrics when dye breakers are present.

Table 2.3 Optical oxidizing agents, reducing agents, and brighteners used in bleaching [17, 52]

Textile materials	Chemical reagents and auxiliaries
Cotton	– Hydrogen peroxide (H_2O_2), sodium chloride ($NaClO_2$), sodium hypochlorite ($NaClO$), peracetic acid, pH buffer, and optical brighteners
Wool	– Oxidizing agents (hydrogen peroxide), stabilizers (pyrophosphate), ammonium (pH between 8 and 9), and optical brighteners
Viscose, acetate, triacetate, and synthetic fibers	– Hydrogen peroxide (alkaline medium), sodium chloride, sodium hypochlorite, and optical brighteners (acid medium)

2.3.12 Alkalizing

It is a stoichiometric saponification process, also known as silkonization, that is used to give polyester articles a silky feel and appearance by immersing them in a medium composed of caustic soda at mass concentrations that vary with temperature. In such a medium, we can proceed to a gradual and superficial dissolution of the textile fiber, i.e., peeling. This operation, also known as "peeling," increases the flexibility of the fiber while decreasing its mechanical properties.

2.3.13 Whitening

It is a process that destroys natural yellow or brown dyes in cotton and discolors the remaining capsule or leaf debris after scouring [13, 32]. It is also known as chemical and optical bleaching. Cotton and other cellulosic fibers are typically bleached using acid and/or alkaline oxidants and optical brighteners [17]. Acidic and basic reducing agents, as well as optical brighteners, are applied to the animal fibers [21, 38, 43, 53]. Synthetic fibers are bleached using oxidizing or reducing agents and optical brighteners based on their chemical composition [17, 52]. Table 2.3 lists all of the oxidizing and reducing agents, as well as optical brighteners that are used in the chemical and optical bleaching of various textile materials.

2.3.14 Fulling

It is a treatment, also known as felting, that is usually performed prior to the dyeing process. It allows you to change the feel of a knit or wool-based fabric by entangling the fibers through mechanical action. Crushing is a characteristic of wool that can be beneficial or detrimental. This property is advantageous for the production of particularly dense and resistant fabrics. When felting occurs during the dyeing or washing process, it causes significant shrinkage in fabrics and knits.

Wool felting in acidic medium yields dry felt, which is used in the industry for headgear, whereas alkaline felting is typically applied to fabrics by impregnating them with a soap solution (100 g/l) and a little sodium carbonate to adjust the pH to 10, at temperatures ranging from 35 to 40 °C [38]. Detergents and emulsified materials are then removed by rinsing and washing at 40 °C.

2.3.15 Fixing

The fixing process, which is required prior to dying in the room, ensures the dimensional stability of the fabric and avoids subsequent withdrawals due to washing. It also keeps the network of threads from moving during subsequent processes (dying and printing) and prevents defects such as breaks and wrinkles [6]. The temperature of the dyeing or printing processes can result in irregular variations in fabric size and form breaks. As a result, exposing them to temperatures higher than those used in these processes will result in a harsh cooling while keeping the fabric off to prevent crease formation. It is possible to dry-fix the finishes. Furthermore, a wet setting may be in the wet phase.

2.3.16 Carbonizing

This procedure aims to remove any vegetable impurities (burrs and seeds) from the fabrics [38]. It is usually done in continuous mode with solutions containing about 50 g/l of 66°Be sulfuric acid and sodium carbonates [38]. Impregnated parts are mechanically expressed, dried, and heat treated (between 110 and 140 °C) before being washed and neutralized. Carbonization alters the dyeing properties; as a result, it must be performed on a regular basis. If it is carbonized on the dyed fabric, the dyes' solidarity must be tested with sulfuric acid.

2.3.17 Bleach

It is a process of increasing dye affinity and achieving dimensional fixation by reducing or canceling shrinkage caused by accidental felting during aqueous treatments. It therefore does not apply to fabrics intended for crushing or to sprained fabrics. With a chlorine, melamine, and hydrochloric acid solution, the wool is chemically attacked, and if the chlorination remains superficial, the fiber can peel off by removing the scales [4, 25, 26, 28].

2.3.18 Charge

It is a silk operation that involves increasing the weight of the silk and binding it to the tissues. It can be done with a high dose of tannin (50–150% by mass of silk) in a weakly acidic bath under boiling for about an hour, or else by organic resins that replace the former tin salt employee. It should be noted that the charge rate is expressed in relation to the weight of raw silk. Thus, calculated, the current load rates vary from 10% (the maximum for parts intended for printing) to 100% for dyed fabrics.

2.3.19 Dyeing

This is a critical process in textile finishing that is used to fix dyestuffs on yarns and/or fabrics in a uniform manner. Various substances and techniques (absorption and impregnation) are used during this stage [7]. Dyeing a textile fiber is typically done in an aqueous solution known as dye liquor or dye bath, so that it does not fade quickly when exposed to light and water [7, 56]. It is generally determined by the nature of the textile materials and the dyes used [7]. Fixing dyes on textile fibers is accomplished through the absorption of reagents and chemical auxiliaries after ionization of the molecule that makes up the fiber. The dye is impregnated into the textile fiber, which is sometimes aided by the addition of sodium salts [7, 56]. In contrast, the dyeing process provides a diverse range of dyes, auxiliaries, and chemical reagents [13, 20, 22, 29, 34, 45, 59, 64, 68], which are grouped in Table 2.4 and tailored to the nature of the textile fibers and the desired color final.

2.3.20 Impression

It is a technique that enables the direct realization of multicolored drawings on a textile material, whether previously bleached or tinted in plain, while using the same pigments, dyes, chemicals, and/or other related materials as those used in the dyeing process [13]. It employs physical and chemical processes that are thus identical to and described for the dyeing process, with the only difference being the specific machines and techniques employed [7, 18]. There are four main methods of textile printing in this technique: block printing, roll printing, screen printing, and thermal transfer printing. Color is typically applied as a thick paste in each of these methods, followed by vaporization, fixing, or heating, and finally washing to remove excess color [13, 56].

Table 2.4 Dyes, auxiliaries, and chemical reagents used in the dyeing process

Textile materials	Dyes	Auxiliaries and chemical reagents
Cotton	Direct	– Neutral electrolytes, humidifiers and equalizers
	Sulfur	– Sodium hydroxide, reducing agents, neutral electrolytes and humidifiers, detergents, and oxidizing agents (sodium or potassium dichromate, perborate, sodium persulfate, oxygenated water)
	Vat	– Sodium hydroxide, sodium hydrosulphite and oxidizing agents (oxygenated water), salts, wetting agents, detergents, and equalizers
	Reactive	– Neutral electrolytes (sodium chloride or sodium hydrosulfite), humidifiers, and alkaline agents (sodium hydroxide, sodium carbonate or sodium bicarbonate)
Wool	Acid	– Equalizers (acetic acid or formic acid (pH 6), retarders (sodium chloride, ammonium sulphate, and sodium sulphate (pH 6–7), and electrolytes
	Mordant	– Sulfuric acid (pH 2), acetic acid (pH 5), leveling agents, sodium acetate, and ammonium sulphate
	Reactive	– Neutral electrolytes (pH 7), humidifiers, detergents, and sodium acetate
Silk	Acid	– Acetic acid (80%), ammonium acetate, crystallized soda ash, and nonionic wetting agent
	Direct	– Very weak neutral or acid medium
Polyester	Dispersed	– Halogenated aromatic hydrocarbons, orthophenylphenol, detergents, caustic soda, and hydrosulfite powder
	Azo	– Caustic soda, alcohol, sodium nitrite, hydrochloric acid, and buffering agents
Cotton-polyester	Vat	– Caustic soda, hydrosulphite powder, detergents, and wetting agents
	Dispersed	– Caustic soda, hydrosulphite of soda, halogenated aromatic hydrocarbons, detergents, and wetting agents

2.3.21 *Primers or Finishing*

They include mechanical finishing and chemical finishing, that is, the operations of coating-lamination, back coating, washing, and drying, with the goal of improving tissue quality and making it more consumer-acceptable [7, 40]. Mechanical finishing is used to alter the fabric's properties, such as its appearance. Thermal processes such as heat adjustment (thermal finishing) are also included, as are typical mechanical processes such as calendering, lining, compression shrinkage, elevation, brushing, shearing, and milling adjustment with restoration and crab [2, 47, 60, 63, 66, 70, 74]. Chemical finishing is used to reduce the moisture content of textile fabrics using phosphorus pentoxide (P_4O_{10}). Fabric impermeability is achieved through impregnation with oil-based solutions and synthetic resins, which prevent water and air from penetrating these tissues [77].

2.3.22 Recovery and Lamination

These are coating operations that modify and improve the physical properties and appearance of the fabric, as well as the potential for creating entirely new products by combining the benefits of fabrics, polymers, foams, and films [37]. The coating can be defined as the application of a continuous or discontinuous layer of a suitable chemical system, the coating formulation of which contains various polymers, qualities, and textiles such as polyvinyl chloride (PVC), polyurethane (PU), acrylic, and polytetrafluoroethylene (PTFE), which are widely used to make versatile textile products such as waterproof protective clothing such as tarpaulins and protective clothing such as electrical insulation, etc. [37, 71].

Laminating consists of two or more layers, one of which is a textile fabric, that are glued tightly together using an added adhesive or the adhesive properties of one or more successive layers. Conventional laminated technical textiles are typically made up of one or more textile substrates joined together with a prefabricated polymer film or membrane [71]. In general, a non-homogeneous composite structure must be created to form a layer of the coating compound on and/or in the substrate [37, 51, 67, 71].

2.3.23 Rear Facing

Textile coatings, also known as "backing" are used to cover textile-surfaced products such as rugs and carpets [7]. Coating materials are typically fluids or fluid precursors, consisting of solid powders deposited on a substrate fabric to form a film and then a bond on the substrate's surface [46, 67]. A good quality back coating requires film formation of both coating materials and bonding to the substrate [35, 39].

2.3.24 Washing

Textile washing is a process that removes some of the dyes used in the dyeing and printing processes by washing them in hot baths between 40 and 100 °C with wetting agents and detergents. This operation is classified into mechanical or dry washing for delicate fabrics and chemical or wet washing for conventional fabrics [15, 51].

2.3.25 Dry Cleaning

There are two types of dry washing: washing with sanded stones and washing by microponing, which is more commonly used for jeans (denim).

2.3.26 *Washing with Sandstone*

It is used to make soft and appealing fabrics by striking them with sanded stones and/or volcanic rocks in large washing machines. The latter produce an additional effect of a faded or worn appearance by sanding the surface of the jeans, removing certain particles from the dyes on the wire surfaces [9–11, 39, 42].

2.3.27 *Washing by Microsanding*

It is used to soften sanded and hairless fabric surfaces. To give fabrics a soft appearance, a series of horizontally arranged cylindrical rollers covered with either abrasive paper or an abrasive chemical are used before or after dyeing. Consider the case of jeans being pulled across the face of sanding rollers to create a raised surface finish with some color reduction [23].

2.3.28 *Wet Wash*

Enzymatic washing employs enzymes with auxiliaries, whereas acid washing employs surfactants such as nonyphenol and its ethoxylated derivatives (NP-NPE) [9]. Water used for textile washing usually needs to be slightly softened and heated.

2.3.29 *Enzymatic Washing*

It is a green washing method that employs organic enzymes to eat away at the cellulose fibers of the fabric surface (denim) and remove them during the washing process [3]. A certain amount of dye and fiber are removed from the fabric surface during enzymatic washing. When the desired color is obtained, the enzyme can be stopped by adjusting the alkalinity or temperature of the bath.

2.3.30 *Acidic Washing*

It is accomplished by sanding stones that have been previously soaked in a solution of sodium hypochlorite or potassium permanganate to achieve localized whitening, resulting in a uniform bright blue and white contrast. An optical brightener can be used as part of this wash to improve the color contrast of the denim fabric. This process saves water because it does not require the addition of water [23].

2.3.31 *Drying*

It is a unitary operation that consists of completely or partially removing the impregnating liquid from a textile material by supplying thermal energy. It significantly reduces the mass and volume of textile fabrics while facilitating their transport, storage, and handling [65]. The particularly energy-intensive textile drying process consists of drying the textile material mechanically and thermally [7].

It is a unitary operation that involves supplying thermal energy to completely or partially remove the impregnating liquid from a textile material. It significantly reduces the mass and volume of textile fabrics while making transportation, storage, and handling easier [65]. The particularly energy-intensive textile drying process involves mechanical and thermal drying of the textile material [7].

2.4 Conclusion

In light of this chapter, garment-textile industries represent the fastest-growing industry-wide sector, with mechanical and chemical processes in the textile industry diversified to manufacture yarns and fabrics. This diversity depends on the nature of the textile object (natural, synthetic, and semi-synthetic) to be manufactured, thus its ennoblement using chemical compounds such as acids, bases, additives, pigments, optical brighteners, enzymes, and dyes in the stages of pretreatment, dyeing, printing, and finishing.

References

1. Akbari, A., Desclaux, S., Rouch, J. C., & Remigy, J. C. (2007). Application of nanofiltration hollow fibre membranes, developed by photografting, to treatment of anionic dye solutions. *Journal of Membrane Science, 297*(1–2), 243–252.
2. Al-Etaibi, A. M., Alnassar, H. S., & El-Apasery, M. A. (2016). Dyeing of polyester with disperse dyes: Part 2. Synthesis and dyeing characteristics of some azo disperse dyes for polyester fabrics. *Molecules, 21*(7), 855.
3. Allaoui, M., Berradi, M., Bensalah, J., Es-sahbany, H., Dagdag, O., & Ahmed, S. I. (2021). Study of the adsorption of nickel ions on the sea shells of Mehdia: Kinetic and thermodynamic study and mathematical modelling of experimental data. *Materials Today Proceeding, 45*, 7494–7500.
4. Aniş, P., Davulcu, A., & Eren, H. A. (2008). Enzymatic pre-treatment of cotton. Part 1: Desizing and glucose generation in desizing liquor. *Fibres & Textiles in Eastern Europe, 16*(4), 100–103.
5. Araujo, R., Casal, M., & Cavaco-Paulo, A. (2008). Application of enzymes for textile fibres processing. *Biocatalysis and Biotransformation, 26*(5), 332–349.
6. Aziz, S. H., & Ansell, M. P. (2004). The effect of alkalization and fibre alignment on the mechanical and thermal properties of kenaf and hemp bast fibre composites: Part 1—polyester resin matrix. *Composites Science and Technology, 64*(9), 1219–1230.
7. Babu, B. R., Parande, A. K., Raghu, S., & Kumar, T. P. (2007). Cotton textile processing: Waste generation and effluent treatment. *The Journal of Cotton Science, 11*, 141–153.

8. Bensalah, J., Berradi, M., Habsaoui, A., Allaoui, M., Essebaai, H., El Khattabi, O., Lebkiri, A., & Rifi, E. H. (2021). Kinetic and thermodynamic study of the adsorption of cationic dyes by the cationic artificial resin Amberlite®IRC50. *Materials Today: Proceedings, 45*, 7468–7472.
9. Allaoui, M., Berradi, M., Taouil, H., Es-sahbany, H., Kadiri, L., Ouass, A., Bensalah, J., & Ahmed, S. I. (2019). Adsorption of heavy metals (nickel) by the shell powder of the coast of Mehdia-Kenitra (Morocco). *Analytical & Bioanalytical Electrochemistry, 11*(11), 1547–1558.
10. Berradi, M., Hsissou, R., Khudhair, M., Assouag, M., Cherkaoui, O., El Bachiri, A., & El Harfi, A. (2019). Textile finishing dyes and their impact on aquatic environs. *Heliyon, 5*(11), Article e02711.
11. Berradi, M., Hsissou, R., Khudhair, M., Lakdioui, T., Bekhta, A., El Gouri, M., Rafik, M., El Bachiri, A., & El Harfi, A. (2019). Ultrafiltration of wastewater-models loaded with indigo blue by membranes composed of organic polymers at different percentages: Comparative study. *Moroccan Journal of Chemistry, 7*(2), 230–235.
12. Berradi, M., & El Harfi, A. (2016). Review of the pollution of surface water by effluents from the textile finishing industry. *Journal of Water Sciences & Environment Technologies, 01*, 76–82.
13. Berradi, M., Essamri, A., & El Harfi, A. (2016). Discoloration of water loaded with vat dyes by the membrane process of ultrafiltration. *Journal of Materials and Environmental Science, 7*(4), 1098–1106.
14. Berradi, M., & El Harfi, A. (2016). Performance hybridized process of coagulation-ultrafiltrationin discoloration of water charged by the textile finishing dyes. *Moroccan Journal of Chemistry, 4*, 33–39.
15. Berradi, M., & El Harfi, A. (2017). Discoloration of charged models wastewater with reactive and dispersed dyes by the combined process of coagulation-ultrafiltration. *Journal of Materials and Environmental Science, 8*(5), 1762–1769.
16. Berradi, M., Berradi, O., Chellouli, M., Hsissou, R., El Bouchti, M., El Gouri, M., Sallek, B., El Bachiri, A., & El Harfi, A. (2020). Optimization of the synthesis of ultrafiltration asymmetric membranes based on organic polymers. *Results in Engineering, 6*, Article 100116.
17. Boryo, D. E. A., Bello, K. A., Ibrahim, A. Q., Gin, N. S., Dauda, T. M., & Elabo, V. O. (2013). Effect of alternative scouring agents on dyeing properties of cotton/polyester blend fabric. *Journal of Applied Chemistry, 5*, 11–21.
18. Brameshuber, W. (2016). *Manufacturing methods for textile-reinforced concrete.* In *Textile Fibre Composites in Civil Engineering* (pp. 45–59). Elsevier.
19. Bunday, B. D., & El-Badri, W. K. (1984). A model for a textile winding process. *European Journal of Operational Research, 15*(1), 55–62.
20. Chakraborty, J. N., & Jaruhar, P. (2014). Dyeing of cotton with sulphur dyes using alkaline catalase as reduction catalyst. *Indian Journal of Fibre & Textile Research* 39(3), 303–309.
21. Chen, X., Taylor, L. W., & Tsai, L. J. (2016). Three-dimensional fabric structures. Part 1–An overview on fabrication of three-dimensional woven textile preforms for composites. In *Handbook of technical textiles* (pp. 285–304).
22. Cie, C. (2015). Fixing ink jet printed textiles. In *Ink jet textile printing* (pp. 99–110). Elsevier.
23. Conway, R. (2016). Technical textile finishing. In *Handbook of technical textiles* (pp. 189–210). Elsevier.
24. Elhawary, I. A. (2010). Spin finishes for textiles. In *Advances in yarn spinning technology* (pp. 416–430). Elsevier.
25. Es-sahbany, H., Berradi, M., Belfakir, M., & El Youbi, M. S. (2016). Kinetic study of copper adsorption onto Moroccan Illitic and kaolinite clay. *Moroccan Journal of Chemistry, 4*(2), 352–360.
26. Es-sahbany, H., Berradi, M., Nkhili, S., Bassir, D., Belfaquir, M., & El Youbi, M. S. (2018). Valorization of Moroccan clay: Application to the adsorption of cobalt ions contained in wastewater synthesized. *Moroccan Journal of Chemistry, 6*(1), 1–6.
27. Es-Sahbany, H., Berradi, M., Nkhili, S., Hsissou, R., Allaoui, M., Loutfi, M., Bassir, D., Belfaquir, M., & El Youbi, M. S. (2019). Removal of heavy metals (nickel) contained in wastewater-models by the adsorption technique on natural clay. *Materials Today: Proceedings, 13*, 866–875.

28. Fu, K., & Lu, D. (2014). Reaction kinetics study of α-amylase in the hydrolysis of starch size on cotton fabrics. *The Journal of The Textile Institute, 105*(2), 203–208.
29. Gacen, J., & Cayuela, D. (2000). Comparison of wool bleaching with hydrogen peroxide in alkaline and acidic media. *Coloration Technology, 116*(1), 13–15.
30. Hao, O. J., Kim, H., & Chiang, P. C. (2000). Decolorization of wastewater. *Critical Reviews in Environmental Science and Technology, 30*(4), 449–505.
31. Hmamouchi, S., El Yacoubi, A., Massit, A., Berradi, M., El Hezzat, M., Sallek, B., & El Idrissi, B. C. (2023). Photocatalytic decomposition of methylene blue dye using sand-graphite composite under visible light irradiation. *Materials Today: Proceedings, 72*, 3677–3685.
32. Holkar, C. R., Jadhav, A. J., Pinjari, D. V., Mahamuni, N. M., & Pandit, A. B. (2016). A critical review on textile wastewater treatments: Possible approaches. *Journal of Environmental Management, 182*, 351–366.
33. Holt, R. R. D. (1975). Introduction to superwash wool. *Journal of the Society of Dyers and Colourists, 91*(2), 38–44.
34. Horrocks, A. R., & Anand, S. C. (2000). *Handbook of technical textiles*. Elsevier.
35. Jarmany, A. (2007). Décoloration des rejets liquides de textile par électrocoagulation. https://toubkal.imist.ma/handle/123456789/6071
36. Joshi, M., Bansal, R., & Purwar, R. (2004). Colour removal from textile effluents. *Indian Journal of Fibre & Textile Research* 29(2), 239–259.
37. Joshi, M., & Butola, B. S. (2013). Application technologies for coating, lamination and finishing of technical textiles. In *Advances in the dyeing and finishing of technical textiles* (pp. 355–411). Elsevier.
38. Karmakar, S. R. (1999). *Chemical technology in the pre-treatment processes of textiles*. Elsevier.
39. Khalil, E. (2016). Study on properties analysis of knitwear after acid wash. *AASCIT Communications, 3*(2), 102–106.
40. Khatri, A., Peerzada, M. H., Mohsin, M., & White, M. (2015). A review on developments in dyeing cotton fabrics with reactive dyes for reducing effluent pollution. *Journal of Cleaner Production, 87*, 50–57.
41. Kumar, R. S., & Sundaresan, S. (2013). Mechanical finishing techniques for technical textiles. In *Advances in the dyeing and finishing of technical textiles* (pp. 135–153).
42. Kyzas, G. Z., & Lazaridis, N. K. (2009). Reactive and basic dyes removal by sorption onto chitosan derivatives. *Journal of Colloid and Interface Science, 331*(1), 32–39.
43. Liang, T., & Wang, L. (2015). An environmentally safe and nondestructive process for bleaching birch veneer with peracetic acid. *Journal of Cleaner Production, 92*, 37–43.
44. Liberski, A., Ayad, N., Wojciechowska, D., Kot, R., Vo, D. M., Aibibu, D., Hoffmann, G., Cherif, C., Grobelny-Mayer, K., Snycerski, M., & Goldmann, H. (2017). Weaving for heart valve tissue engineering. *Biotechnology Advances, 35*(6), 633–656.
45. Liu, X., Hurren, C. J., & Wang, X. (2003). Comparative analysis of two selective bleaching methods on alpaca fibers. *Fibers and Polymers, 4*, 124–128.
46. Liu, Y., Pan, Y. T., Wang, X., Acuña, P., Zhu, P., Wagenknecht, U., Heinrich, G., Zhang, X. Q., Wang, R., & Wang, D. Y. (2016). Effect of phosphorus-containing inorganic–organic hybrid coating on the flammability of cotton fabrics: Synthesis, characterization and flammability. *Chemical Engineering Journal, 294*, 167–175.
47. Lockett, A. P. (2003). Mechanical finishing—Traditional and modern. In *Textile finishing* (pp. 114–134).
48. Lucas, J., Berticat, P., Chabert, B., Nemoz, G., Bardy, M., Exbrayat, P. E., & Schutz, R. A. (1978). Etude des copolymères acryliques utilisés dans l'encollage de fils de polyester II. Caractéristiques physicochimiques des copolymères et des systèmes composites polyester-colles. *Die Angewandte Makromolekulare Chemie: Applied Macromolecular Chemistry and Physics, 71*(1), 1–16.
49. Mansour, H., Boughzala, O., Dridi, D., Barillier, D., Chekir-Ghedira, L., & Mosrati, R. (2011). Textiles dyes as a source of wastewater contamination: Screening of the toxicity and treatment methods. *Journal of Water Science, 24*(3), 209–238.
50. Mao, N. (2017). Nonwoven fabric filters. In *Fibrous filter media* (pp. 133–171). Elsevier.

51. Meirowitz, R. (2010). Microencapsulation technology for coating and lamination of textiles. In *Smart textile coatings and laminates* (pp. 125–154). Elsevier.
52. Mesogitis, T. S., Skordos, A. A., & Long, A. C. (2014). Uncertainty in the manufacturing of fibrous thermosetting composites: A review. *Composites Part A: Applied Science and Manufacturing, 57*, 67–75.
53. Mokrejs, P., Krejci, O., & Svoboda, P. (2011). Producing keratin hydrolysates from sheep wool. *Oriental Journal of Chemistry, 27*(4), 1303.
54. Nousiainen, P., Turunen, O. T., & Taurio, R. (1989). Dyeing properties and textile processing of viscose, modal, and cellulosecarbamate-spun fibres. In *International Man-Made Fibres Congress, Dornbirn, Austria, 20–22 September 1989.*
55. Oh, J. M., Biswick, T. T., & Choy, J. H. (2009). Layered nanomaterials for green materials. *Journal of Materials Chemistry, 19*(17), 2553–2563.
56. Poon, C. S., Huang, Q., & Fung, P. C. (1999). Degradation kinetics of cuprophenyl yellow RL by UV/H_2O_2/ultrasonication (US) process in aqueous solution. *Chemosphere, 38*(5), 1005–1014.
57. Priyanka, S., Arvind Bharani, R. S., & John, A. (2023). Biocompatible green technology principles for the fabrication of food packaging material with noteworthy mechanical and antimicrobial properties—A sustainable developmental goal towards the effective, safe food preservation strategy. *Chemosphere*, 139240.
58. Rajamohan, N., & Rajasimman, M. (2013). Kinetic modeling of dye effluent biodegradation by Pseudomonas stutzeri. *Engineering, Technology & Applied Science Research, 3*(2), 387–390.
59. Rayung, M., Ibrahim, N. A., Zainuddin, N., Saad, W. Z., Razak, N. I. A., & Chieng, B. W. (2014). The effect of fiber bleaching treatment on the properties of poly (lactic acid)/oil palm empty fruit bunch fiber composites. *International Journal of Molecular Sciences, 15*(8), 14728–14742.
60. Rouf, A., Rasel, S., & Khalil, E. (2015). Investigation of different washing effects on physical and mechanical properties of cotton knitted garments. *Journal of Textile Science and Technology, 1*(3), 101–109.
61. Royer, B., Cardoso, N. F., Lima, E. C., Ruiz, V. S., Macedo, T. R., & Airoldi, C. (2009). Organofunctionalized kenyaite for dye removal from aqueous solution. *Journal of Colloid and Interface Science, 336*(2), 398–405.
62. Royer, B., Cardoso, N. F., Lima, E. C., Macedo, T. R., & Airoldi, C. (2010). A useful organofunctionalized layered silicate for textile dye removal. *Journal of Hazardous Materials, 181*(1–3), 366–374.
63. Sakoma, K. J., Bello, K. A., & Yakubu, M. K. (2012). Synthesis of some azo disperse dyes from 1-substituted 2-hydroxy-6-pyridone derivatives and their colour assessment on polyester fabric. *Open Journal of Applied Sciences, 2*(1), 54.
64. Santhi, P., & Moses, J. J. (2010). Study on different reducing agents for effective vat dyeing on cotton fabric. *Indian Journal of Fibre & Textile Research,* 35(4), 349–352.
65. Sarkar, J., & Khalil, E. (2014). Effect of industrial bleach wash and softening on the physical, mechanical and color properties of denim garments. *IOSR Journal of Polymer and Textile Engineering, 1*(3), 46–49.
66. Schindler, W. D., & Hauser, P. J. (2004). *Chemical finishing of textiles*. Elsevier.
67. Shim, E. (2013). Bonding requirements in coating and laminating of textiles. In *Joining textiles* (pp. 309–351). Elsevier.
68. Shore, J. (1996). Advances in direct dyes. *Indian Journal of Fibers and Textile Research*, 21, 1–29.
69. Singh, J. P., & Verma, S. (2016). *Woven terry fabrics: Manufacturing and quality management*. Woodhead Publishing.
70. Singh, S. V., & Purohit, M. C. (2012). Applications of eco-friendly natural dye on wool fibers using combination of natural and chemical mordants. *Universal Journal of Environmental Research & Technology, 2*(2), 48–55.
71. Singha, K. (2012). A review on coating & lamination in textiles: Processes and applications. *American Journal of Polymer Science, 2*(3), 39–49.
72. Surikanti, G. R., Bajaj, P., & Sunkara, M. V. (2019). G-C3N4-mediated synthesis of Cu_2O To obtain porous composites with improved visible light photocatalytic degradation of organic dyes. *ACS Omega, 4*(17), 17301–17316.

73. Udakhe, J., Honade, S., & Shrivastava, N. (2011). Recent advances in shrinkproofing of wool. *Journal of the Textile Association, 72*, 171–179.
74. Uddin, K., & Hossain, S. (2010). A comparative study on silk dyeing with acid dye and reactive dye. *International Journal of Engineering & Technology, 10*(6), 1–6.
75. Vasseur, P., Ferard, J. F., Vial, J., & Larbaigt, G. (1984). Comparaison des tests microtox et daphnie pour l'evaluation de la toxicité aigue d'effluents industriels. *Environmental Pollution Series A, Ecological and Biological, 34*(3), 225–235.
76. Verma, A. K., Dash, R. R., & Bhunia, P. (2012). A review on chemical coagulation/flocculation technologies for removal of colour from textile wastewaters. *Journal of Environmental Management, 93*(1), 154–168.
77. Voloshyn, K. (2014). Problématique de la gestion des eaux usées industrielles au niveau municipal dans les régions de l'Estrie et de la Montérégie. https://belsp.uqtr.ca/id/eprint/669/
78. Xu, X. R., Li, H. B., Wang, W. H., & Gu, J. D. (2005). Decolorization of dyes and textile wastewater by potassium permanganate. *Chemosphere, 59*(6), 893–898.
79. Zia, A. W., Birkett, M., Badshah, M. A., & Iqbal, M. (2021). Progress in-situ synthesis of graphitic carbon nanoparticles with physical vapour deposition. *Progress in Crystal Growth and Characterization of Materials, 67*(3), Article 100534.

Chapter 3
Sustainable Yarn Sizing Process

Sanat Kumar Sahoo, Bibhu Prasad Dash, and Asimananda Khandual

Abstract A fine yarn's weavability is greatly influenced by its yarn size. The manufacture of fabrics and ecological load is significantly impacted by the primary and secondary sizing agents, as seen from the environmental conservation perspective. In actuality, textile sizing and desizing processes generate between 50 and 80% of the COD (chemical oxygen demand) load in textile effluent. Modern sizing procedures are centered on many criteria in the current environment, including low COD size options, sophisticated control systems for reproducibility, ideal quality, and ease of handling. The intrinsic physical qualities of the sized yarn may be enhanced, and the quantity of sizing material used may be substantially reduced by pre-wetting technology, precise measurement systems, and individual drives. To comply with environmental protection regulations, green size must meet the functional needs of warp sizing and attenuate air and water pollutants by sizing with low wastewater, pollution, and toxicity levels within specified limits. This chapter covers the evolution of yarn sizing techniques and newly available sustainable substitutes.

Keywords Eco-norms · Size · BOD · COD · Sustainability

3.1 Introduction

The textile industry, one of the world's oldest and largest manufacturing sectors, has a significant impact on the economy, society, and the environment. Consequently, sustainable practices must be implemented throughout various processes,

---

S. K. Sahoo · B. P. Dash (✉) · A. Khandual
Textile Engineering Department, Odisha University of Technology and Research, Bhubaneswar, Odisha, India
e-mail: bpdash@outr.ac.in

S. K. Sahoo
e-mail: sanatkumarsahoo89@gmail.com

A. Khandual
e-mail: asimte@outr.ac.in

M. Shahid et al. (eds.), *Advancements in Textile Finishing*,
https://doi.org/10.1007/978-981-96-6385-9_3

including the critical sizing step. The industry's considerable ecological footprint has been exacerbated by the traditional dependence on synthetic sizing agents, high water and energy consumption, and significant waste generated during the sizing process, which improves fabric strength, durability, and performance [1]. Nevertheless, the industry has been compelled to reassess its measurement processes and implement eco-friendly alternatives in response to the increasing demand for sustainable textiles, which is being driven by consumer awareness, regulatory requirements, and environmental concerns. Sustainable sizing processes are designed to ensure the quality and performance of fabrics while minimizing the environmental impact [2]. This is achieved through eco-friendly sizing agents, optimized application methods, and efficient equipment. The advantages of sustainable sizing processes are not limited to environmental sustainability; they also include improved safety and health for workers, reduced water and energy consumption, reduced chemical usage and waste generation, increased customer satisfaction, and a stronger brand reputation [3]. Despite these advantages, sustainable sizing processes encounter numerous obstacles, such as integrating fabric performance requirements with sustainability, developing cost-effective, eco-friendly sizing agents, enhancing worker training and awareness, and improving process efficiency and productivity. Innovative solutions, including biodegradable sizing agents, nanotechnology, and 3D printing, have been developed by the industry to confront these challenges. These solutions present opportunities for growth and differentiation. Water and energy management, waste reduction and recycling, worker safety and training, quality control and assurance, sizing agent selection and optimization, application method and equipment selection, and raw material selection and sourcing are all components of a comprehensive framework for sustainable sizing processes [4]. This book chapter comprehensively examines sustainable sizing processes, including the principles and advantages of sustainable sizing, traditional and eco-friendly sizing agents, application methods and apparatus, case studies of successful implementations, and future directions.

3.2 Essential Elements for Enhancing Sustainable Sizing

3.2.1 Lifecycle of Product and Process Development

Products and services are secure and environmentally beneficial throughout their life cycle when engineered for durability, reparability, ease of recycling, compostability, or biodegradability; they are also manufactured and packaged using little material and energy [5]. Processes are developed and executed to mitigate or eradicate chemical, ergonomic, and physical risks; they also remove chemical agents, physical components, and circumstances that threaten human health and the environment; they optimize energy and material conservation.

3.2.2 General Considerations Regarding the Impacts

The following primary environmental considerations are addressed in fiber production and processing, spinning, weaving, and chemical processing:

- *Hazardous chemicals* (Table 3.1).
- *Algal Blooming*: The starch used in size causes a rise in algal bloom in river water, which raises the water bodies' need for chemical and biological oxygen [7].
- *Carcinogenic compounds*: When aromatic amines are reduced, insoluble amine groups are produced. These groups then land on the skin and bladder, affecting the body's metabolism and finally causing cancer.
- *Waste*: Reducing waste generation along the textile supply chain and using the identified waste resources to produce textiles (Table 3.2).
- *VOCs*: Volatile Organic Compounds.
- *Energy*: To lower the total carbon footprint of the supply chain. Using insulation, surface reduction, and exhaust air heat recovery, the best available technology (BAT) for equipment size aims to reduce thermal energy consumption for heating [9]. BAT for sizing technology also includes size recovery technology, application technology, and a degradable sizing agent.

Table 3.1 The influence of chemicals on sizing and desizing processes [6]

Process	Chemical used	Hazards cause
Sizing starch paste	Pentachlorophenol (PCP) derived from phenolic and chlorinated chemicals utilized as preservatives	Skin impact and algal blooms are carcinogenic
Desizing starches	Starches preservatives: PCP pyridine-derived reactive softeners	Algal bloom carcinogenic

Table 3.2 Waste produced during sizing and desizing [8]

Procedure	Discharges	Wastewater	Solid waste
Sizing	VOCs	COD, BOD, and metals	Unused starch-based sizing, fiber lint, yarn waste, packaging waste, and cleaning waste
Desizing	Glycol ethers' VOCs	Biocides, anti-static substances, and BOD from sizes	Fiber lint, packaging debris, and supplies for upkeep and cleaning

3.3 Viewpoints on Sustainable Sizing in the Textile Sector

3.3.1 *Purpose of the Sizing*

Warp yarns are utilized as a single set of threads on looms to create cloth. These warp threads frequently experience a variety of pressures, including abrasion from various metal loom components, flexing during unwinding and winding, and cyclic tension during shedding. These inevitable strains result in warp breakage, which leads to loom stoppages; ultimately, fabric flaws and efficiency loss are taken into consideration. The warp yarn's size can help to minimize these issues. The primary goal of sizing is to improve the warp yarn's weavability by using various fibers, their mixes, size materials, application techniques, process management, and novel techniques [10]. Using glue to connect yarn fibers and reduce fluff or hairiness is a crucial step in the warp sizing process. The strands can attain such strength that they endure the mechanical stress experienced during weaving, enhancing weaving efficiency. The objective of sizing is to provide the yarn with a protective coating and to lubricate its surface, enabling it to withstand the abrasive forces encountered during weaving. These data demonstrate how crucial the sizing procedure is to producing effectively woven fabrics [11, 12].

3.3.2 *Basic Ingredients*

3.3.2.1 Fibres

An excellent renewable resource that has been reinvigorated by nature and human innovation is the source of natural fibers. Because they absorb the same amount of carbon dioxide as they produce, these fibers are carbon neutral. They mostly produce organic wastes during processing, with the leftover materials being used to make eco-friendly building materials or electricity. Natural fibers include a range of substances, from biodegradable to permanent ones like heavy metals, and can produce significant amounts of water pollution [13]. New research shows that three tons of carbon dioxide are released into the atmosphere for every ton of manufactured polypropylene. This plastic is often used in rope, containers, and packaging. A major contributor to the warming of the planet is the greenhouse gas. Jute and other natural fibers absorb around 2.4 tons of carbon every ton of dry fiber [14]. Fungi and bacteria feed on natural fibers that have absorbed water, hastening their degradation. Composting materials made of natural fibers can enhance soil structure and produce no harmful byproducts when burned. However, groundwater and soil can be contaminated by heavy metals and other chemicals released by synthetic fibers.

3.3.2.2 Size Materials

Even though cloth size has replaced polyvinyl alcohol in research, there were still certain issues. For example, starch derivatives used today cannot fully replace polyvinyl alcohol. Modified starch research should aim to create grafted starch that performs exceptionally well [15]. The limited market share and high cost of grafted starch made the production process difficult. Acrylic sizing agents were costly and still had issues with re-viscosity following moisture absorption. Consequently, research on acrylic sizing agents has to use the designable property of copolymer size and resolve current issues by optimizing their ratio. Creating a scaling agent similar to solid acrylic would be a developing trend because it was inexpensive to transport and simple to utilize [16].

Furthermore, no substitute could match the superiority of size film performance because of polyvinyl alcohol's appropriate film-forming characteristic [17]. It would also be quite important to figure out how to find a suitable solution for the polyvinyl alcohol refractory difficulties. Research in this field has mostly concentrated on two areas: first, the modification of polyvinyl alcohol to give it the ability to degrade naturally. The second was to create strains capable of breaking down polyvinyl alcohol. Continuous effort, improved quality, additional functionalities, reduced compositions, little or no usage of polyvinyl alcohol, and biodegradable polyvinyl alcohol are necessary for creating truly eco-friendly textiles in practicable sizes [18].

3.3.2.3 Size Paste Concentration [12]

The weight of dry sizing agents divided by the weight of the size paste is the definition of this ratio.

$$\text{Percentage of concentration} = A/B * 100$$

A is the oven-dry weight of the specified materials.
B is the size paste's weight.

3.3.2.4 Impact of Size Add-On

Multiple factors, such as machine velocity, compression pressure, paste viscosity, and temperature, influence the size add-on. The size add-on increases with increasing machine speed and size paste concentration and decreases with increasing squeezing pressure [12]. Numerous tests have demonstrated that sizing increases yarn strength because it results in a larger add-on. The yarn becomes less flexible and more rigid at very high add-on levels because of a thick layer of size that penetrates deeper into the yarn. As the percentage of size increases, the warp breakage rate decreases [18].

$$\text{Size add-on } (\%) = A/C * 100$$

A represents the oven-dry weight of the sized materials.

C is Oven-dry mass of unprocessed yarn.

Oversizing leads to heightened expenses due to increased material consumption, bigger liquid volumes, more energy demands for drying, and elevated waste discharge levels.

3.3.2.5 Wet Pick-Up [12]

It is the size paste picked up by the un-sized yarn.

$$\text{Wet Pick-up} = B/C \times 100$$

B = Weight of size paste, C = Oven-dry mass of unprocessed yarn.

Consequently, Wet Pick-up = (Add-on (%))/(Concentration (%)).

3.3.2.6 Desizing

Hydrolysis breaks down starch, whereas oxidation turns it into molecules soluble in water. Approximately half of the water contamination is caused by wastewater from desizing, which contains a high biological oxygen demand (BOD) that renders it unsuitable [19]. The wastewater produced during the desizing process exhibits a significant concentration of contaminants, characterized by a chemical oxygen demand (COD) of 20,000 mg/l [20]. Additionally, it is highly alkaline, with a pH level exceeding 12. As a result, discharging this effluent into the river is expensive. Before being released into the river, it will need to undergo neutralization. An active sludge polishing phase was included in the treatment stage of a 250 m^3 mesophilic up-flow anaerobic sludge blanket reactor (UASB) treatment stage constructed and put into operation in August 2003 [21]. The wastewater treatment plant has achieved a significant reduction of Chemical Oxygen Demand (COD) by up to 80% and has successfully neutralized the pH from a very alkaline pH of 12 to a more neutral pH range of 7–8 [22]. This leads to optimal wastewater treatment with minimal operating expenses, specifically low energy consumption and reduced reliance on chemicals. The desizing waste exhibits a higher biological oxygen demand within the 300–450 ppm range and a pH level of 4–5 [23]. Hydrogen peroxide can be employed to catalyze the conversion of starch into carbon dioxide (CO_2) and water (H_2O) during the process of starch degradation [24]. Moreover, the enzymatic activity might be facilitated by turning starch into ethanol. Following the process of distillation, ethanol can be utilized as a fuel.

3.3.2.7 Eco-norms

Some chemicals used in the sizing process are hazardous to the environment and human health. A starch paste and PCP made from phenolic and chlorinated compounds—used as preservatives—are among the ingredients [25]. These substances have been shown to irritate skin, encourage algae growth in river water, and raise the water's chemical and biological oxygen demands. In addition, they are known to cause cancer in people. To comply with ecological standards, it is imperative that the use of pesticides like DDT, Dieldrin, and Aldrine in the production of cotton, as well as the use of PCP as a preservative for sizing and Amylase for desizing in the process of textile size, be avoided [26].

3.4 An Eco-friendly Method for the Traditional Yarn Sizing Process

3.4.1 Essential Components

3.4.1.1 Organic Adhesive

The natural starches accessible are sago, potato (farina), maize, tapioca, wheat, and rice [27]. All of these starches are available. The physical characteristics of starches vary; they exhibit different degrees of viscosity, adhesion, and other characteristics. Sago and maize are the two most frequently used alternatives. When working with heavier yarns, it is essential to make use of thin boiling starches to achieve penetration and binding power. The core of their function is to provide the yarn with a denser coating or to give it greater bulk. Compared to traditional starches, they need less boiling time to stabilize their viscosity [28].

Guar gum is a natural plant gum that is produced from the endosperm of guar beans cultivated in India and Pakistan. It is a macromolecular substance with high water solubility. It exhibits exceptional adherence to both pure cotton and polyester-cotton blends [29]. When combined with starch and CMC, the size film showed reduced fracture strength. Nonetheless, it demonstrated outstanding wearability, 1043 times buckling resistance, and a breaking tensile rate of 18.59% [16]. Compared to other possibilities, soy protein performs better when used as an adhesive [30].

3.4.1.2 Synthetic Adhesive

It is possible to apply sodium carboxymethyl cellulose to spun viscose and acrylic fibers by mixing it with PVA [31]. When mixed with soluble starches, polyvinyl alcohol, a synthetic resin composed of polymers, may be used to make spun yarns. Hydrophobic fibers and their mixes, such as nylon, acrylics, acetate, polyester, and others, can be coated with polyacrylic acid, which is a very good adhesive for spun nylon yarns [32]. According to research [33], sizing agents such as PVA, CMC, and polyacrylates have the potential to be recycled. CMC, being a natural substance, is capable of being broken down by biological processes. However, it has a far smaller impact on the amount of oxygen required by organisms than starch. The weavability of yarn that has been sized with polyester resin is superior to that of yarn sized with PVA and polyacrylamide, according to a study [34]. The textile industry used polyvinyl alcohol (PVA) as a sizing agent to solve this problem. In contrast to starch, PVA has a low biological oxygen demand and can be retrieved using the ultrafiltration (UF) technique [35]. The limited ability of PVA to biodegrade has led to its restriction or prohibition in the scaling process in some European nations. Research has indicated that approximately 15–25% of polyvinyl alcohol (PVA) underwent degradation in anaerobic sludge over approximately 300 days [36]. Polyvinyl alcohol (PVA) contributes for 40–80% of COD in textile wastewater and 45% of BOD in textile effluent treatment facilities [37]. Thermally stable membranes remove polyvinyl alcohol and hot water from textile effluent in the investigation. Fabrics typically have a size content ranging from 8 to 15% and are commonly polluted by desizing. The cotton woven fabrics, which are 100% cotton and treated with starch, have a biochemical oxygen demand (BOD) of 45.5 kg and a chemical oxygen demand (COD) of 91 kg for every 1,000 kg of fabric processed [38]. The processing of these fabrics requires approximately 350 L of water per kg of material processed. Acrylic scale exhibits a high capacity to form a film and adhere to surfaces, as well as possessing exceptional strength, flexibility, and resistance to wear. The key cardinal condition for reduced or minimal consumption of PVA in China is outlined in reference [16]. Activated sludge treatment tests demonstrated that TEA-soy biodegraded in a manner comparable to that of modified starch, but it was superior to PVA. This was demonstrated by changes in COD, BOD, and total and ammonia nitrogen [39]. High-speed polyester and polyester/cotton textiles might benefit from using TEA-soy biodegradable sizing agents as an alternative to PVA sizes [40].

3.4.1.3 Softeners and Lubricants

The utilization of lubricants and softeners is essential for the regulation of rigidity and flexibility.

Tallow, spermaceti, and paraffin wax are the most prevalent lubricants. Emulsified tallow and soluble mineral oils are occasionally employed [41].

3.4.1.4 Antiseptics

One additional disadvantage of natural adhesives is their composition, which consists largely of food ingredients. Consequently, there is a risk of microbial growth, especially in warm and humid environments. Subsequently, these microorganisms develop unpleasant odors and discoloration on the warp threads. Industry commonly utilizes zinc chloride, copper sulfate, zinc sulfate, and other similar substances [42].

3.4.2 Secondary Ingredients

Weighting agents, commonly referred to as fillers or filling agents, increase the weight or density of a substance. China clay, often known as kaolin, French chalk, and gypsum, is commonly used in various industries [43]. Due to their high hydrophobicity, synthetic fibers generate static electricity when they come into contact with certain metallic components of the loom and rub against them. Lissapol NX is commonly employed to prevent this from occurring [44]. Brighteners are employed in sizing to enhance the luminosity of grey fabric. Ultramarine blue, a type of acid dye, is commonly employed in sizing [45]. Wetting agents are necessary, particularly during the sizing process, for water-repellent fiber warps that are resistant to being wetted by water-based adhesives. This is especially true for warps that are water-repellent. Cotton or viscose rayon warp size does not necessitate their use. Turkey Red Oil is utilized as a wetting agent [46]. Excessive foam generation occurs when antistatic chemicals or wetting agents are utilized. Excessive foam generation occurs when antistatic chemicals or wetting agents are utilized. To prevent this issue, Pine oil can be utilized as an antifoaming agent [47].

3.4.3 Utilization of Size Paste

3.4.3.1 Preparation of Size Paste

The components are metered out with great precision to create the paste, and cold water is used wherever possible. The preliminary amalgamation of starch with cold water promotes the dissolution of starch granules. Begin by incorporating all the components, excluding the softener, antiseptic, and deliquescent. The addition of gum to the aqueous solution is recommended. It will be introduced 15 min before the boiling cessation [48]. A continual agitator blends the paste while water is introduced to uphold the paste's viscosity. Afterward, the steam is utilized to sustain a boiling temperature while the stirrer speed is set at 50–60 rpm. The cooking duration for maize and tapioca is 1–2 h, whereas for sago, it is 3–4 h [49]. The heating should be adjusted to ensure the boiling point is reached within 20–30 min after the initial steam injection. When using a pressure cooker, the duration of steaming is decreased to 40 min while maintaining a stirrer speed of 40–60 rpm. To preserve the thickness of the mixture when cooking, it is advised not to use cold water. Following the cooking process, the mixture can be preserved or moved to another location using a stirrer that maintains a consistent speed of 15–20 revolutions per minute while maintaining a steady temperature of approximately 90 °C [50].

3.4.4 *The Machines for Sizing*

Figure 3.1 illustrates the division of the size machine into four primary zones. Creel, size box, drying, and head-stock zones are among them [51].

3.4.4.1 Creel Zone

The creel zone contains a significant number of warper beams that can be arranged in various configurations to meet the project's specific requirements. Multiple creel designs are implemented during the procedure. To effectively resolve the challenges encountered in the creel zone, the sizing process employs a variety of creels, including the over-under creel, vertical creel, and inclined creel [52]. Following this, the warp sheets are gathered together, mixed, and then fed by the immersion roller located within the size box. Next, the sheets are sent to the headstock zone and the drying cylinder for further processing. The creel capacity of the modern sizing machine may accommodate a range of 12–20 beams.

3.4.4.2 Size Box Zone

The stainless-steel size box has no sharp edges or inactive areas. The boxes vary in immersion roller positioning, squeeze roller pairs, and auxiliary squeeze rollers. Light-spun, filament, and hard-fiber warps typically use one immersion roller and two squeezing rollers. A double immersion roller and a squeeze roller are utilized for spinning yarns that are either lightweight or heavyweight [53]. To remove the yarn from the creel and add it to the size mixture in a stress-free manner, a dry squeeze roller, also known as a positive dry nip, may be included in each box size. This PDN also eliminates air pockets inside the structure of the yarn, which makes it simple to penetrate the size. Rubber-coated first compressing rollers are another size box improvement. This guides the warp sheet toward the squeezing rollers' nip. Modern machines include 45 °C rubber squeezing rollers. Modern machines use 230–320 kg

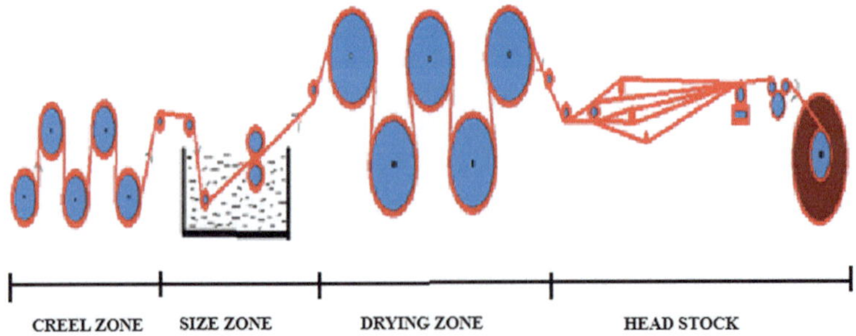

Fig. 3.1 Slasher sizing machine

rubber rollers [54]. The rollers' loading capabilities increase their effective weight to around 450 kg. Growing the weight of the roller from 90 to 227 kg results in a 20% reduction in the amount of size paste consumed in pure sizing [55].

3.4.4.3 Drying Zone

Warp strands maintain 90–140% of their weight in wet-size paste after passing through squeezing rollers. Water is removed while drying, leaving the warp dry [56]. The drying system has three methods. Heat transport involves conduction, convection, and radiation. Traditional-sized machines dry wet yarn sheets by conduction when they touch a hot drum. The yarn sheet and heated surface must be moved synchronously utilizing a PIV gear drive system and Teflon-coated cylinder. It allows the drying surface and the yarn sheet to remain friction-free. With the introduction of multiple-cylinder drying (Fig. 3.2), twin-cylinder drying has been phased out. A PTFE coating is applied to the first two to three drying cylinders [11]. PTFE's low friction and thermal stability allow drying 50–150 m/min. Drying cylinders typically have a 30–34-inch diameter and 60–72-inch working width. Cylinders range from 2 to 32 in the drying chamber [11, 12]. The drying cylinders are heated by 2–2.8 kg/ cm^2 supersaturated steam. Drainage tapes remove condensed steam as water. The damp warp sheet with size paste requires 75–80% of the sizing process's energy to dry (Table 3.3) [57].

The drying loads determine the number of drying cylinders needed to dry yarns [34]. And it is expressed as

$$\text{Drying load} = \frac{\text{Sizing speed} \times \text{total ends} \times \text{WPU} \times (100 - \%\ \text{solids in solution}) \times \text{yarn count}}{1000 \times 1000 \times 100}$$

where, with WPU means Wet pick-up, the sizing speed is in m/min and Yarn count in Tex.

In a drying room, hot air is used to take all the wetness from the warp yarn using the convection method. The air can be heated either through the use of steam or electrical heaters. This method offers the benefit of precise size and prevents yarn

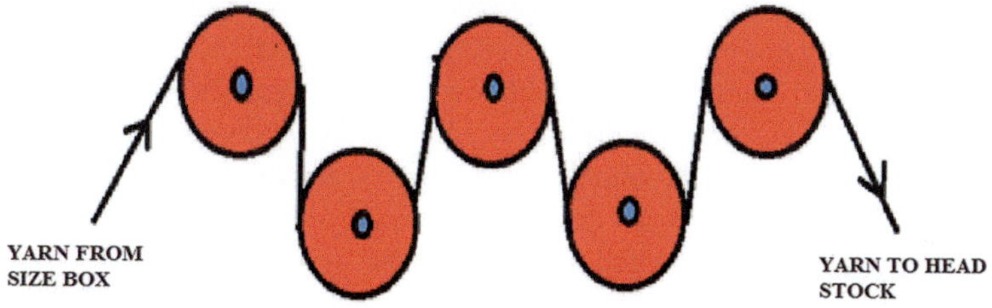

Fig. 3.2 Multi-cylinder drying system

Table 3.3 For spun and filament yarn, cylinder temperature in (°C) [55]

Yarn type	Material	Cylinder number							Min. temp (°C)	Max. temp (°C)
		1	2	3	4	5	6	7		
Spun	Cotton	140	140	140	140	140	140	140	140	140
	Viscose	120	120	120	120	120	100	100	114	120
	P/C blend	110	110	120	120	120	100	100	111	120
	P/V blend	110	110	110	110	110	100	90	106	110
Filament	Polyester	75	95	115	115	115	95	75	98	115
	Viscose	80	90	115	110	110	90	80	96	110
	Acetate	70	80	100	100	100	80	70	86	100
	Nylon	85	95	100	110	110	100	90	99	110

Note Cylinder No. 1 is located adjacent to the size box

from becoming flattened. Additionally, it requires less floor space and creates a nice surrounding atmosphere [58]. The drawback is that the drying process requires a longer duration. Compared to drying in heated cylinders, the energy costs of hot-air drying are at least 40% higher. When hot air dries something, it needs a bigger space and more time to cool down. Thus, the method of conduction utilizing cylinder drying is extensively employed [59]. The final type is radiation. During this procedure, heat is delivered to the surface of the yarn through direct radiation. Electromagnetic radiation is the medium via which energy is conveyed, resulting in significant energy loss during this process. The main benefit is that the yarn surface doesn't get flattened. However, it costs more to set up first than the other two methods [12].

3.4.4.4 Headstock Zone

The yarns in the warp sheet may adhere to one other due to insufficient drying following the application of sizing paste in the drying chamber. Separating the yarns is necessary for the next step; hence it is crucial to individualize the warp yarns. To proceed with the following step, it is necessary to separate the yarns individually. This requires breaking the coat of size film at multiple locations. Intermediary yarns are dragged forward between beam yarn sheets to the machine headstock. Currently, lease yarns are substituted with lease rods [60]. Warp yarns are separated with lease bands in a traditional arrangement. In the contemporary technique, stainless steel rods with a length greater than the warp width are utilized. Figure 3.3 [12] shows these rods 1.5–2 times the lease rod diameter spaced between. Modern sizing involves two forms of splits to produce high-quality sized yarn. Wet splitting refers to the process after the size box and before the drying chamber. Another stage in the process is referred to as dry splitting, which occurs after the drying chamber. Dry splitting offers superior working conditions compared to wet splitting. However, wet splitting offers the highest level of individualization quality [61]. During the process of dry

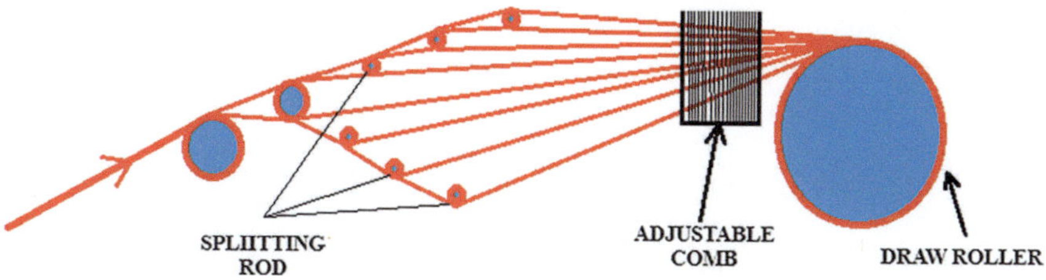

Fig. 3.3 Splitting of yarn sheet

splitting, there is a possibility that some material may flake off as trash. However, this occurrence is less likely when using the wet splitting method.

The threads, customized in size, are split in the warp sheet and then threaded onto the weaver's beams of the specified width and length. Typically, a single sizing machine supplies multiple weaves of different lengths. The headstock of a sizing machine must possess the adaptability to accommodate both the broadest and narrowest weaver beam while also adjusting to changes in color selvage yarns [62]. The weaver's beam is regulated by a guide roller outfitted with a spring-loaded bearing in machines of numerous sizes. It aids in regulating the tension fluctuation between the drag roller and the ultimate dehydrating cylinder. For the warp sheet to be wound around the required beam, its width must be equivalent to its girth. As a result, the warp sheet of a particular dimension is initially inserted through a reed/comb that can be adjusted to adjust the spacing between the threads to the suitable measurement [12]. A moisture meter detects the moisture content of the textile strand after it has passed through the expanded comb. The desired moisture content in the weaver's beam is achieved by adjusting the pace of the entire machine to either increase (when the moisture level is below the lower limit) or decrease (when the moisture level is beyond the upper limit) in response to the type of error.

3.4.5 *Modern Sizing Machine Controls*

The following categories of control systems are included in a contemporary scaling machine.

3.4.5.1 Stretch Control

A net elongation of approximately 1.2–1.5% is typical for a cotton warp sheet. The sizing machine utilizes accessories such as ball bearings, PDN arrangements, and PIV gears to regulate the tension in various areas. The maximum allowable stretch in the creel zone should not exceed 0.5%. The moisture content should be minimized in the wet section and not exceed 1.5% in the drying and splitting area [63]. The effective regulation of tension in the drying region directly influences the elastic characteristics of the scaled warp. The winding area must not exhibit any elongation.

3.4.5.2 Moisture Levels Control

Maintaining a moisture level of 7% is necessary during the process of cotton sizing [64]. Over-drying will result in brittle and harsh yarns, while under-drying will cause the yarns to be sticky. Elevated moisture levels increase.

3.4.5.3 Pressure Control

The wet pick-up is influenced by the pressure exerted by the squeezing rollers, which occurs when the nip of the rollers is pressed together. This, in turn, impacts the cost of the drying process. The pressure level needed is also influenced by the type of sizing substance, the density of the warp yarn, and the desired percentage of size add-on [18].

3.4.5.4 Control of Temperature

The temperature in the sow box influences the viscosity of the sizing substance, as per the research. Consequently, the temperature must be maintained at 92 ± 2 °C. The passage of steam into the sow cage is regulated by the automatic temperature controller. The steam is reduced if the temperature is excessively high and added to sustain the desired temperature if it is too low [18].

3.4.5.5 Level Control

The quantity of size absorbed and added to the material is significantly influenced by the concentration of the size mixture in the size box [11]. A floating roller or sensor can be employed to ascertain the level of the measurement box, thereby enabling it to be managed.

3.4.6 Energy Conservation Methodology

3.4.6.1 Energy Demands

The typical slasher functions at a velocity of 83–120 m per minute, expending around 1500 kilocalories every kilogram of warp size. This energy is essential for preparing the size paste, maintaining it at elevated temperatures, heating it, and drying the sized yarn [65].

3.4.6.2 Procedures for Energy Conservation

The energy needed for size is 0.0031 kWh, which is 5% waste. Reducing the temperature during the size preparation from 100 to 80 °C saves a significant amount of energy [9]. This lower temperature specifically leads to a 25% drop in energy use, equal to 8% of the total energy needed for size. It takes 10% of the total energy to keep the size at a high temperature. This can be reduced by organizing brief storage periods and utilizing non-degradable sizes (storing at low temperatures). Utilizing a single squeeze roller instead of a pair of rollers can save 3–7% of the overall energy consumption. Applying a more forceful compression motion can reduce energy by 5–10%. The high-pressure roll technology reduces energy consumption for drying by 20–50%. The process of stirring a viscosity size mixture during cooking and storage demands a greater amount of mechanical energy compared to low-viscosity sizes. According to reference [66], mixing requires more energy than heating in the preparation of size.

3.4.6.3 Size and Energy Index [9]

$$\mathrm{EIS} = \mathrm{ES}/(\mathrm{FP} \times \mathrm{DW})$$

where EIS = energy indicators for sizing, ES = Size Energy, FP = Fabric manufacture (g/m^2), and DW = Warp density expressed in number per cm.

This index goes to 0 without any sizing agents used.

The energy index and CODs from the sizing agent help convey this. The sizing index of unsized cloth is 0.

3.4.6.4 Sizing Index [9]

$$\mathrm{SI} = \frac{\mathrm{CODs}}{\mathrm{FP}}$$

where SI represents the sizing index and CODs are the chemical oxygen demands or sizing agents in kg/a (a = m^2/kWh). Fabric production (g/m^2) is denoted by FP.

3.5 Applications and Future Trends

These methods were recently developed and are intended for a specific form of yarn and the sizing procedure. It provides effective preservation of water, energy, and materials, resulting in a significant impact on BOD and COD.

3.5.1 Foam Sizing

The add-on can be achieved with minimum pick-up due to the high concentration of size paste. The 13.3% concentration (with a wet pick-up of 90%) was replaced with a 25% concentration (with a wet pick-up of 48%). Approximately 54% of the energy used for drying was saved. Polyester materials of various dimensions create voluminous foam but are prohibitively expensive [67]. Adhesives such as CMC or PVA can be readily transformed into foam. The concentration of the component should be high because less water is utilized in foam sizing. The analysis focused on the benefits of energy conservation and enhanced production speed, which reduced warp bridging and decreased warp hairiness [68]. Foam sizing (Figs. 3.4 and 3.5) produces a more consistent and stable film size on the warped surface while reducing hairiness [69–73]. In 2014, a study was conducted on the effects of combined foam-sizing and warp pre-wetting techniques on foam sizing for industrial applications. The combined process is more advantageous than the standard process since it conserves gas and reduces the need for desizing wastewater and desizing agents [74, 75]. Furthermore, the combined process has a minimal impact on the weaving efficiency while the warp strength is improved, the elongation is reduced, the wear resistance is increased, and the hairiness is decreased. It is important to acknowledge that an increase in the viscosity of the liquor results in more stable froth and delayed drainage of the liquor (Table 3.4) [76].

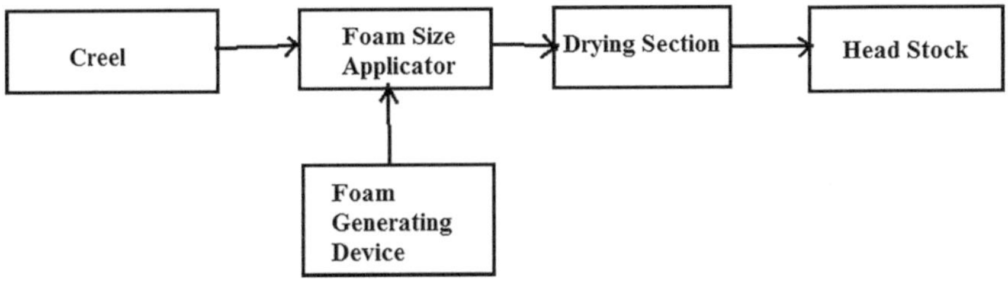

Fig. 3.4 Foam sizing arrangement [12]

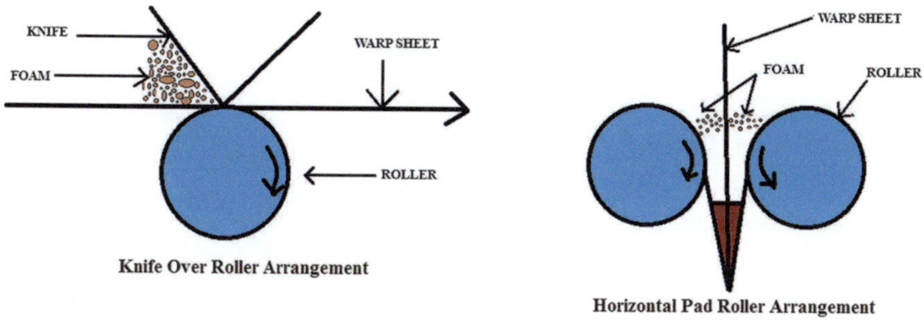

Fig. 3.5 Foam sizing process

Table 3.4 Evaluation of foam sizing in comparison to traditional methods [12]

Parameter/particular	Conventional sizing	Foam sizing
Size add-on % desired	12.0	12.0
Wet pick-up obtainable %	100.0	40.0
Size % in sizing paste %	12.0	30.0
Water to be evaporated kg/kg of yarn	0.88	0.25

3.5.2 High Pressure Squeezing (Fig. 3.6)

The application of high pressure reduces the amount of energy needed for the drying process. The concept involves utilizing elevated concentrations of size mixture and increased squeezing pressure to achieve a typical size add-on [77]. The heat energy (HE) needed for the drying process is significantly greater than the mechanical energy (ME) needed for pressing using a squeeze roll. Low squeezing pressure requires high hydraulic energy (HE) and low mechanical energy (ME). The optimal squeezing effect and energy efficiency for drying cotton yarn can be achieved by using slow-speed running and high squeezing pressure [78]. The high-pressure, high-speed method was more cost-effective than the polyester yarn or poly/cotton blend yarn in terms of the combined sum of heat exchange (HE) and mechanical energy (ME).

Benefits of high-pressure squeezing (Table 3.5):

- According to Ref. [79], cotton exhibits a 6% enhancement in breaking strength.
- In cotton, increased penetration in high-pressure-sized yarn enhances the binding between fibers and enhances abrasion resistance [79].
- HP size for cotton results in a higher packing density or compactness compared to low-pressure sizing [79].
- The breaking extension percentage somewhat reduces in cotton, P/C, and P/V blends [80].
- According to a study, weaving efficiency falls in P/C (polyester/cotton) fabric when using thin boiling starch. However, it increases in P/V (polyester/viscose) blend fabric and cotton fabric [80].

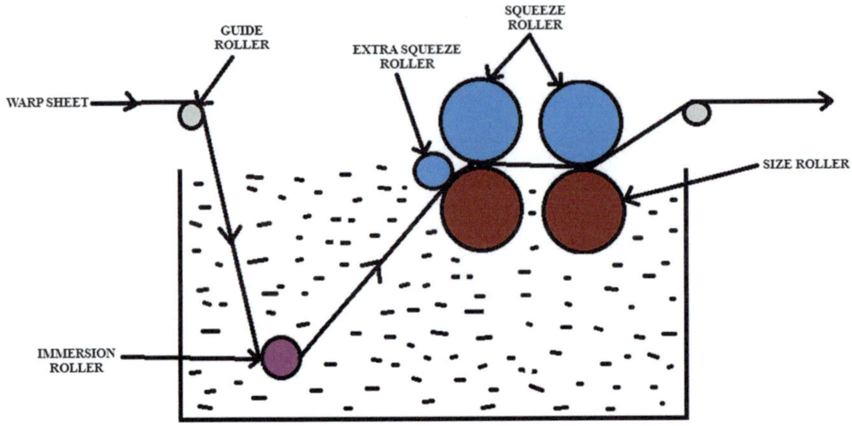

Fig. 3.6 High-pressure squeezing process

Table 3.5 Comparison of standard and elevated-pressure squeezing

Parameters	Normal squeezing	High pressure squeezing
Size concentration, %	8	14
Wet size pick-up, %	120	68.5
Dry size add-on, %	9.6	9.6
Moisture content in the wet-sized yarn, %	110.4	58.9
Moisture in the dry-sized yarn, %	4.5	4.5
Percentage of water evaporated during drying	105.9	54.4

- The skewness of cotton yarn is the opposite of negative. Positive skewness is exhibited by unsized polyester/cotton (p/c) and polyester/viscose (p/v) yarn, but negative skewness is exhibited by sized yarn. Through the use of high pressure during the sizing process, the degree of skewness can be reduced [81].

3.5.3 Combining Dyes and Sizes (Fig. 3.7)

Some dyes that may be colored with sizing include azoic, vats, reactive, pigments, dispersion, and direct dyes. BTRA has successfully sized and dyed azoic mixes using reactive dyes at low temperatures and P.V. alcohol [66]. In this case, the reactive dye maintains its fastness qualities, but the P.V. alcohol size also exhibits increased durability. The technique employs Padazoic colors, which are combinations of azoic colors (such as Rapid Fast colors), and simultaneously uses two slashers. The initial step involves immersing the solution in a solution of the Padazoic dye. A solution is created by combining color, caustic soda, methanol, and a small amount of gum to inhibit color migration during drying. Subsequently, the yarns undergo a drying

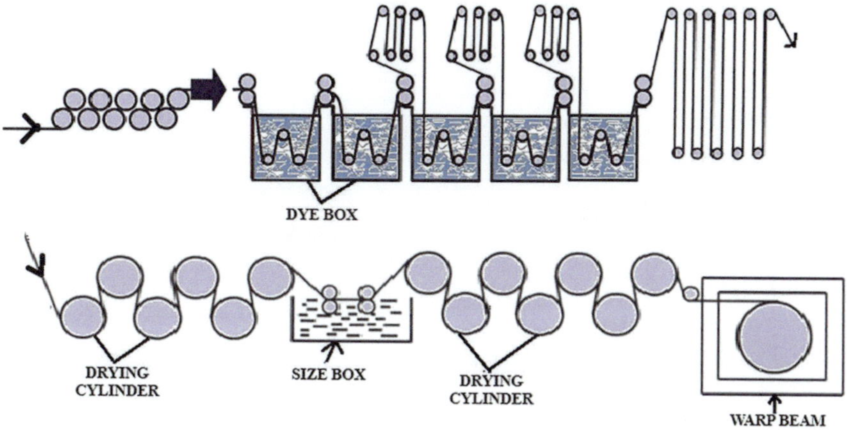

Fig. 3.7 Combined dyeing and sizing process [18]

process. The second box contains a blend of starch, tallow, and a buffering agent, such as a combination of sodium acetate and acetic acid. When utilizing CMC, the sole requirement is acetic acid at a temperature of 96 °C, after which the material is subjected to drying. Occasionally, a reactive dye is applied to cotton yarns along with sizing at a temperature of 130 °C. The curing occurs by the convection process, which is a somewhat gradual process [82].

3.5.4 Sizes for a Single-End (Fig. 3.8)

The yarns are placed into drying cylinders as part of the regular technique for sizing. This placement causes the sized yarns to bind with one another and form a single sheet after they are removed from the drying cylinders. By moving the sticky yarns over and below lease rods, or even with the assistance of an expanded comb, it is possible to separate the sticky yarns. The separation of the threads results in the extraction of fibers from the yarn's surface, yielding a yarn that is fuzzy and not precisely spherical. A certain quantity of size paste has been compromised, and a method was devised to utilize a significant quantity of warp yarns in a dual decorative size container. A separate drying room and dividing setup exist. Single-end sizing is a technique that entails the individual size and drying of each yarn [83]. In single-end sizing, the application of size is executed similarly to conventional sizing. However, the pressure exerted during the squeezing process is significantly higher, resulting in reduced energy requirements for drying the yarn [84].

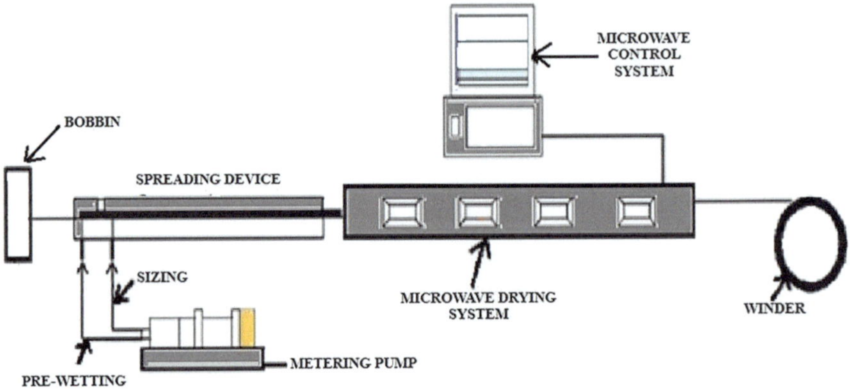

Fig. 3.8 Single-end sizing process [18]

3.5.5 Silica Dispersion

The silica dispersion process improves the ability of the size film to withstand wear and tear, as well as increases the film's softness and flexibility [85]. The idea came from finishing technology, which is used to make cotton, rayon, and jute materials smoother and more useful. The binding power of the rubber latex film is very high. The auxiliary makes up 4–6% of the size blend, based on the weight of the glue in it. Not only does the warp experience an increase in its advantage, but the surface qualities of the yarn also undergo improvement, resulting in a reduction in hairiness and the application of lubrication. The breakage rate on cotton warp is projected to decrease by 25–33% [86]. Even though it was just added, the size blend's price has decreased a little.

3.5.6 Treatment with Heat Shock

It works better to run the warp yarn through a hot roller before putting it through the size [66]. The benefits of this treatment include enhanced yarn strength, a 29% reduction in warp breakages, a 6% increase in loom productivity, and a little decrease in elongation.

3.5.7 The Integration of the Yarn Surface Fibers

This method is utilized for thermoplastic fibers, including nylon and polyester.

It is advisable to subject the yarn to a thermal treatment at a temperature exceeding the melting point of the fiber for a limited duration. This results in the melting and fusion of the surface fibers, forming a compact and resilient surface layer [66]. The benefits of this treatment include enhanced abrasion resistance and strength without any negative impact on dyeing qualities.

3.5.8 Hot-Melt Sizing (Fig. 3.9)

During the warping process, the warp yarn is stretched to a specific size while being wound onto a section beam at a speed of 550 m per minute. It is then rewound onto a weaver's beam. A heated grooved metal roller at 205 °C is needed. From the creel to the warper beam, the grooved chamber spins 10 revolutions per minute in the warp sheet's direction. One warp end passes through each groove. The size paste is compressed by a strong block against this roller when it reaches a melting temperature of 125–155 °C. The grooved surface is uniformly heated to around 150 °C to dry the yarn. Upon contact with the hot roller, it melts and produces a coating on the warp end. Molten size demonstrates low viscosity, enabling it to readily infiltrate yarn fibers [87]. Size rapidly solidifies when the warp is separated from a roller. A minimum of three distinct grooved rollers, ranging from 12 to 120 tex, is necessary. When warp threads exit the groove region, chilled air is directed onto the yarn's surface. It facilitates the efficient and rapid application of paste onto a yarn surface by allowing for easy size adjustment. It was discovered by Hiroshi Y in Japan that the employment of vinyl acetate copolymer heat melt sizing agents during the process of sizing polyester and cotton fibers produced outstanding results [10]. However, it is only relevant to a specific category of yarns.

3.5.9 Semi-dry Sizing

Compared to the usual sizing procedure, the semi-dry sizing approach is designed to reduce the amount of energy consumed during the drying process of wet-sized yarn [88]. An important benefit of this is the decrease in energy use, reduced water usage, and increased productivity compared to the conventional slasher-size machine.

3.5.10 Solvent Sizing (Fig. 3.10)

When it comes to the size method, solvent sizing aims to promote sustainability while simultaneously lowering the amount of energy consumed. This procedure makes use of chlorinated hydrocarbon as a solvent rather than water throughout the entirety of the process [89]. Polystyrene resin functions as the primary material for sizing. The solvent consists of chlorinated hydrocarbons, namely trichloroethanes, perchloroethylene, and trichloroethylene. Chlorinated hydrocarbons have a lower

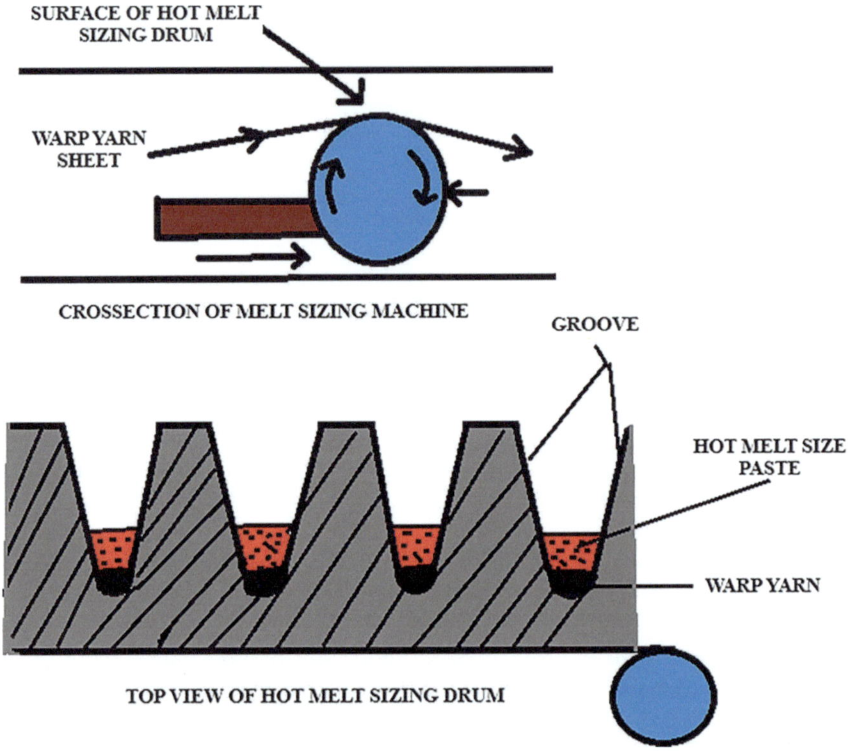

Fig. 3.9 Hot-melt sizing process

heat need for evaporation, resulting in energy savings by not utilizing water. In the solvent procedure, the size is recycled and reused.

The benefits of solvent sizing are as follows: Firstly, the energy needed to evaporate the chlorinated hydrocarbon is reduced as water is not used. Secondly, there is no

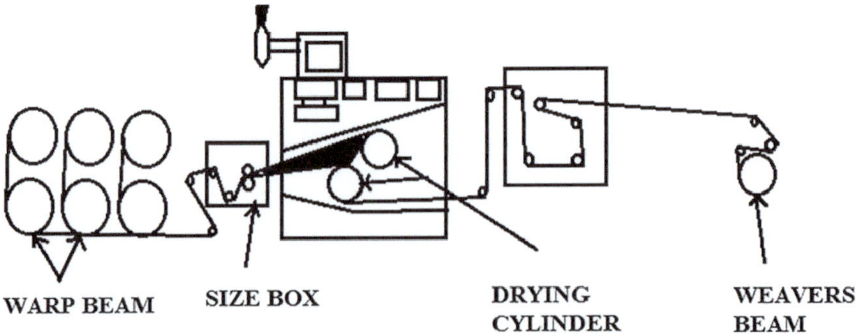

Fig. 3.10 Solvent sizing machine

pollution as the solvent is evaporated and recycled within a closed system. Lastly, the warp sheets dry quickly, allowing for a high production rate [33]. The drawbacks of this procedure are the high toxicity of chlorinated solvents, which can lead to the development of carcinogenic diseases, and the substantial initial investment cost.

3.5.11 After Waxing

A softening agent is applied to the sized yarn using a method more suited to the standard process. When the softener is used in the process, it diminishes the potency and stretchiness of the yarn. However, it can be prevented by employing the post-waxing technique [10]. This technique can be applied to pure cotton, polyester/cotton blend, and synthetic yarns, significantly reducing end breaks during the weaving process. This approach is primarily designed for synthetic and mixed yarns. It aids in flattening the projecting fiber. Mutton tallow cakes and polyethylene glycols, both of which have a melting temperature of around 55–60 °C, are examples of agents that might be used. The recommended quantity of softener is approximately 0.5% of the total weight of the yarn [90].

3.5.12 Cold Sizing (Fig. 3.11)

The cold sizing technique has been specifically designed to accommodate heat-sensitive textiles like wool. Due to the potential degradation of fiber properties, drying this particular fiber requires less heat than other fibers. Polyvinyl alcohol (PVA) is the primary substance used for sizing and as an anti-static agent. It should be dissolved in water [91–93]. A waxing system is employed to apply the cold sizing agent, which is situated between the beaming machine and the warping disk of a sectional warping machine. The two-sided inclined applicator indirectly applies the measure. The motion differential between the warp sheet and the applicator roll can be adjusted to alter the pick-up. A radiation-based drying technique is employed to enhance the outcome of this process [94, 95].

This drying process has several benefits, including reduced energy use, fast and even drying, smoother yarns due to lower heat, and eliminating the need for sizing paste cooking.

3.5.13 Integrated Warping and Sizing (Fig. 3.12)

Since it combines the duties of sizing and warping, this machine is perfect for working with a wide variety of yarns and modest orders. Weaving lines that need fine-count and high-density materials, such as cotton, colored yarns, mixed fabrics, linen, and

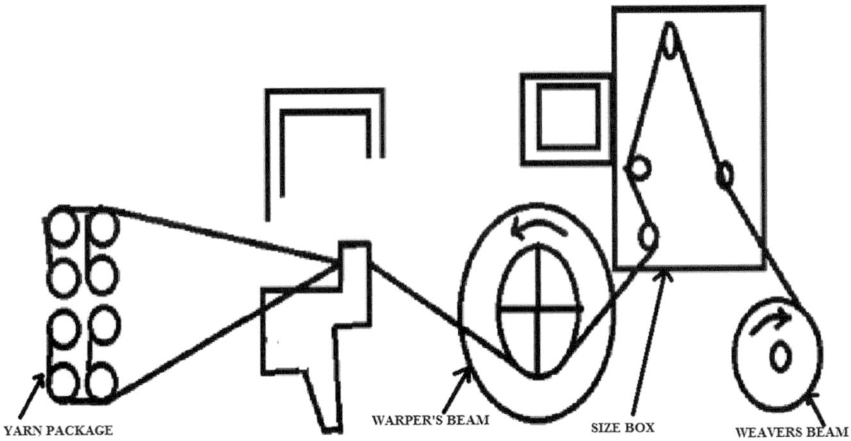

Fig. 3.11 Cold sizing machine

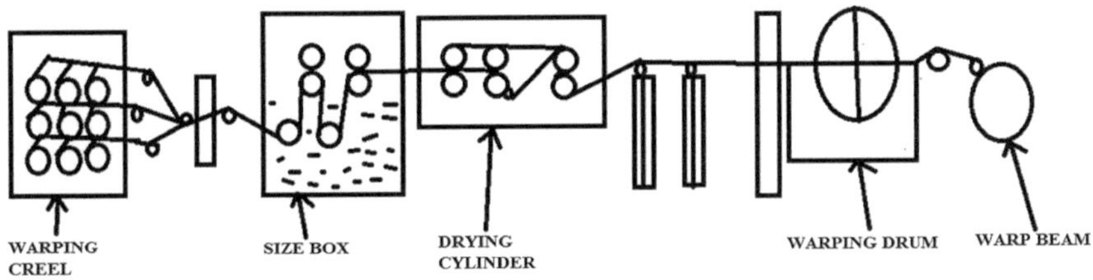

Fig. 3.12 Combined warping-sizing machine

industrial fabrics, are the ones that are meant to use this product. It is particularly well-suited for small orders of yarn-dyed fabrics [51]. Beams with a diameter of up to 10 m and a working width of 23–36 m may be produced by this machine. Both beam speeds of 0–100 m per minute and size speeds ranging from 10–80 m per minute are attainable with this equipment. The brakes work well, and the accuracy is less than 2 m. With a travel speed of 0.001 to 99.999 mm/rev, it can move through endless sections. It can cut down on section placement errors by up to 0.01%. The distance between the drum and the movable reed stays the same. This machine has an automatic rental feature. This machine has six drums that dry things and the boiling force is 0.35 MPa. There is room for 80 L in the size box [18].

3.6 Conclusion

In summary, the sustainable sizing process is an essential element of textile manufacturing, as it allows the industry to reduce its environmental impact while maintaining the quality of the fabric. Textile manufacturers can substantially decrease waste generation, chemical usage, and water and energy consumption by incorporating eco-friendly sizing agents, optimizing application methods, and utilizing efficient equipment. In addition to environmental stewardship, sustainable sizing processes improve fabric durability, enhance worker safety and health, and increase customer satisfaction. The textile industry has made substantial progress in adopting sustainable sizing practices despite the challenges of harmonizing sustainability with fabric performance requirements, devising cost-effective, eco-friendly sizing agents, and improving process efficiency. Collaborative efforts among suppliers, manufacturers, and stakeholders have facilitated the development of innovative sizing technologies and transparent supply chain practices. Using sustainable sizing methods, the textile sector has the potential to lessen its impact on the environment, enhance its reputation for social responsibility, and create a more sustainable future.

References

1. Chen, L., Reddy, N., & Yang, Y. (2013). Soy proteins as environmentally friendly sizing agents to replace poly (vinyl alcohol). *Environmental Science and Pollution Research, 20*, 6085–6095.
2. Niinimäki, K., & Hassi, L. (2011). Emerging design strategies in sustainable production and consumption of textiles and clothing. *Journal of Cleaner Production, 19*(16), 1876–1883.
3. Charter, M., & Tischner, U. (Eds.) (2017). *Sustainable solutions: Developing products and services for the future*. Routledge.
4. Zorpas, A. A. (2016). Sustainable waste management through end-of-waste criteria development. *Environmental Science and Pollution Research, 23*(8), 7376–7389.
5. Principles of Sustainable Production, Adapted from Lowell Center for Sustainable Production. Retrieved January 15, 2014, from http://www.sustainablemeasures.com/node/35
6. Kumari, P., Saroj, S., & Rose, J. S. N. M. (2013). Eco-textiles: For sustainable development. *International Journal of Scientific & Engineering Research, 4*(4), 1379–1390.
7. Ibrahim, N. H., et al. (2022). A review on the biological, physical and chemical mitigation of harmful algal bloom. *Squalen Bulletin of Marine and Fisheries Postharvest and Biotechnology, 17*(2), 95–110.
8. Curteza, A. Sustainable textiles. https://www.scribd.com/document/368281014/Sustainable-Textiles
9. Tobler-Rohr, M. I. (2011). *Handbook of sustainable textile production*. Elsevier.
10. Drexler, P. G., & Tesoro, G. C. (2018). Materials and processes for textile warp sizing. In *Handbook of fiber science and technology* (Vol. 1, pp. 1–89). Routledge.
11. Ajgaonkar, D. B., Talukar, M. K., & Wadekar, V. R. (1982). *Sizing: Materials*. Textile Trade Press.
12. Muthusamy, R., & Kathirvelu. S. (2003). Technology of fabric manufacture—1: Yarn preparation and shuttle weaving, Komarapalayam SSMITT Students' Co-Operative Stores.
13. Udayakumar, G. P., et al. (2021). Ecofriendly biopolymers and composites: Preparation and their applications in water-treatment. *Biotechnology Advances, 52*, 107815.
14. A Sustainable Choice. (2009). *International year of natural fibers production*. http://www.naturalfibres2009.org/en/iynf/sustainable.html
15. Tharanathan, R. N. (2005). Starch—value addition by modification. *Critical Reviews in Food Science and Nutrition, 45*(5), 371–384.
16. Xiao, H., & Zhang, W. (2009). Current situation of environment protection sizing agent and paste. *Journal of Sustainable Development, 2*(3), 172–175.
17. Vatanpour, V., et al. (2023). Polyvinyl alcohol-based separation membranes: A comprehensive review on fabrication techniques, applications and future prospective. *Materials Today Chemistry, 28*, Article 101381.

18. Singh, M. K. (2014). *Industrial practices in weaving preparatory*. Woodhead publishing India Pvt. Ltd.
19. Rabbi, M. A., et al. (2018). Investigation of waste water quality parameters discharged from textile manufacturing industries of Bangladesh. *Current World Environment, 13*(2).
20. Lei, L., et al. (2000). Wet air oxidation of desizing wastewater from the textile industry. *Industrial & Engineering Chemistry Research, 39*(8), 2896–2901.
21. Patel, B. B., Gundaliya, P. J., & Amin, D. D. (2021). Optimized hybrid upflow anaerobic sludge blanket with post treatment processes for wastewater. *Current World Environment, 16*, 282–303.
22. Ramesh Kumar, M., et al. (2022). pH neutralization of textile industry wastewater for effective recycling. *Pigment & Resin Technology, 51*(1), 80–90.
23. Patil, S. T., et al. (2022). Wastewater reuse and recycling in the textile industry. International Journal of Science Research and Engineering Trends, 8(1), 2395–2566.
24. Tolvanen, P. (2013). *Development of an environmentally friendly method of starch oxidation by hydrogen peroxide and a complex water-soluble iron catalyst*. Åbo Akademi-Åbo Akademi University.
25. Söderhjelm, L., & Sipiläinen-Malm, T. (1996). Paper and board. In *Migration from food contact materials* (pp. 159–180). Springer US.
26. Rashid, A. (2011). *Investigations on organochlorine pesticide residues in soil from cotton growing areas of Pakistan*. Department of Plant Sciences, Quaid–i–Azam University Islamabad–Pakistan (Ph. D. Dissertation).
27. Herbert, E. L., & Kugler, J. (1967). *Starch and related products in selected countries* (Vol. 41). US Government Printing Office.
28. Karim, A. A., et al. (2008). Starch from the sago (Metroxylon sagu) palm tree—Properties, prospects, and challenges as a new industrial source for food and other uses. *Comprehensive Reviews in Food Science and Food Safety, 7*(3), 215–228.
29. Thombare, N., et al. (2016). Guar gum as a promising starting material for diverse applications: A review. *International Journal of Biological Macromolecules, 88*, 361–372.
30. Zhao, Y., Xu, H., & Yang, Y. (2013). *Textile size from soyprotein for high speed weaving*. University of Nebraska.
31. Ibrahim, S., & Hamouda, T. (2024). Spinning techniques of poly (vinyl alcohol) fibers for various textile applications. *Egyptian Journal of Chemistry, 67*(1), 447–466.
32. Sen, A. K. (2008). Coated textiles. In *Principles and applications* (2nd ed., pp. 72–73).
33. Var, C., & Palamutcu, S. (2024). Sustainable approaches in textile-sizing process. In *Sustainable manufacturing practices in the textiles and fashion sector* (pp. 55–74). Springer Nature Switzerland.
34. Behera, B. K., & Pakhira, A. (1998). Evaluation of the weavability of polyester-fibre multifilament sized yarn. *Journal of the Textile Institute, 89*(3), 522–531.
35. Sarkar, A., et al. (2012) Recovery of polyvinyl alcohol from desizing wastewater using a novel high-shear ultrafiltration module. *CLEAN—Soil, Air, Water, 40*(8), 830–837.
36. Chiellini, E., et al. (2003). Biodegradation of poly (vinyl alcohol) based materials. *Progress in Polymer Science, 28*(6), 963–1014.
37. Kishor, R., et al. (2021). Ecotoxicological and health concerns of persistent coloring pollutants of textile industry wastewater and treatment approaches for environmental safety. *Journal of Environmental Chemical Engineering, 9*(2), 105012.
38. Panda, S. K. B. C., Sen, K., Mukhopadhyay, S. (2021). Sustainable pretreatments in textile wet processing. *Journal of Cleaner Production, 329*, 129725.
39. Yu, N. Y., Shi, W., Zhang, B. B., Su, G. Y., Feng, J. F., Zhang, X. W., Wei, S., & Yu, H. X. (2013). Occurrence of perfluoroalkyl acids including perfluorooctane sulfonate isomers in Huai River Basin and Taihu Lake in Jiangsu Province, China. *Environmental Science & Technology, 47*(2), 710–717.
40. Zhao, Y., Zhao, Y., Xu, H., & Yang, Y. (2015). A sustainable slashing industry using biodegradable sizes from modified soy protein to replace petro-based poly (vinyl alcohol). *Environmental Science & Technology, 49*(4), 2391–2397.

41. Myddleton, W. W. (2016). Chemistry of oils, fats and waxes. In *Handbook of cosmetic science: An introduction to principles and applications* (p. 59).
42. Madhav, S., et al. (2018). A review of textile industry: Wet processing, environmental impacts, and effluent treatment methods. *Environmental Quality Management, 27*(3), 31–41.
43. Ciullo, P. A. (1996). *Industrial minerals and their uses: A handbook and formulary.* William Andrew.
44. Kalinowski, S. (1966). The origin, prevention, and removal of oil stains. *Journal of the Society of Dyers and Colourists, 82*(8), 299–303.
45. Liman, L. R., et al. (2017). A comparative study on the application of optical brightening agent on cotton knitted fabric in one bath and two bath exhaust method. *International Journal of Textile Science, 6*(5), 135–142.
46. Caupin, H.-J. (2018). Products from castor oil: past, present, and future. In *Lipid technologies and applications* (pp. 787–795). Routledge.
47. Achaw, O.-W., & Danso-Boateng, E. (2021). *Chemical and process industries.* Springer International Publishing.
48. Sultana, S., Haque, M. Z., & Nur, H. P. (2014). Preparation and application of different size materials on the cotton yarn and investigating the effect of sizing on the tensile properties of cotton yarn. *Bangladesh Journal of Scientific and Industrial Research, 49*(1), 25–30.
49. Yusnita, H., Joling, T., Wan Zaliha, W. S. (2017). Effect of selected sago starch modifications on its properties and quality of instant 'keropok lekor'. *Malaysian Applied Biology, 46*(1).
50. Brennan, J. G. (2006). Mixing, emulsification and size reduction. In *Food processing handbook* (p. 513).
51. Drean, J.-Y., & Decrette, M. (2022). Weaving preparation. In *Advanced weaving technology* (pp. 3–80). Springer International Publishing.
52. Rozelle, W. N. (1994). Environmental concerns of warp sizing. Textile World, 144(11), 50–57.
53. Choudhury, A. K. R. (2017). *Principles of textile finishing.* Woodhead Publishing.
54. Capelle, G. (2018). Calendering technology. In *Rubber products manufacturing technology* (pp. 179–265). Routledge.
55. Gandhi, K. L. (2020). Yarn preparation for weaving: Sizing. In *Woven textiles* (pp. 119–166). Woodhead Publishing.
56. Malik, T., Roy, P. K., & Singh, H. K. Development in sizing machine. https://static.fibre2fashion.com/articleresources/PdfFiles/58/5747.pdf
57. Hacking, H. (1980). *The sizing of spun yarns.* Scientific Era Publications.
58. Carr, W. W., Stephen Lee, H., & Ok, H. (2006) Drying of textile products. In *Handbook of Industrial Drying*, Taylor & Francis Group, pp. 803–813.
59. Jones, E. H. (1964). In J. B. Smith (Ed.), *Technology of warp sizing.* Columbine Press.
60. Landau, A. L. (1941). *Warping and slashing, sizing part 1.* International Text Book Co.
61. Luo, C. J., et al. (2012). Electrospinning versus fibre production methods: From specifics to technological convergence. *Chemical Society Reviews, 41*(13), 4708–4735.
62. Thakkar, A. I. (2020). *Developing new system for yarn warping.* Diss. Maharaja Sayajirao University of Baroda (India).
63. Bergman, R. (2021). Drying and control of moisture content and dimensional changes. *Chapter 13 in FPL-GTR-282*, 13-1.
64. Ormerod, A., & Sondhelm, W. S. (1995). *Weaving—Technology & operations.* The Textile Institute.
65. Chakrabarti, S. K., et al. (2016). Energy efficient novel sizing of jute yarns. *Journal of Natural Fibers, 13*(2), 178–191.
66. Sherin, P. (2018). Modern developments in sizing, VJTI, Mumbai, India. https://textilelearner.blogspot.com
67. Prindle, B. (1934). The microbiology of textile fibres: I. Cotton fibre. *Textile Research, 4*(10), 463–478.
68. Namboodri, C. G. (1986). Foam sizing of cotton and blend yarns: Slashing trials. *Textile Research Journal, 56*(2), 87–92.

69. Hristova, E., Khristov, K., & Exerowa, D. (2005). Foam film characterization of commercial surfactants and their mixtures. *Tenside Surfactants Detergents, 42*, 217–225; https://doi.org/10.3139/113.100262.
70. Song, S. L., Hu Z. G., Dong J. F., & Zhang G. Y. (2005). Surface and foaming properties of alkylamine oxide: Protonation and Counterion effects. *Tenside Surfactants Detergents, 42*, 232–236. https://doi.org/10.3139/113.100266
71. Vernkar. (1992). Foam sizing-perspectives and limitations. *Man-Made Textil in India, 35*, 51–52.
72. Shah, G. (1990). Technique of foam sizing. *Journal of Industrial Textiles, 10*, 184–185.
73. Beck. (1989). Foam technology in yarn sizing. In *Textil praxis international* (p. 623).
74. Gao, W. D., Fan, X. R., Liu, J. L., Zhang, J. X., & Du, L. X. (2014). Technology study of foam sizing. *Cotton Textile Technology, 42*(9), 1–5.
75. Lu, Y. Z., Zhang, J. X., Liu, J. L., & Gao, W. D. (2014). Sizing effects of foam-sizing and yarn pre-wetting combined process. *Journal of Textile Research, 12*(35), 47–51.
76. Zhu, B., Cao, X., Liu, J., & Gao, W. (2017). Effects of foamers and stabilizers on foaming ability and stability of oxidized starch/polyacrylic acid size liquor in foam sizing. *Journal of Engineered Fibers and Fabrics, 12*(4), 155892501701200420.
77. McMahon, J. F., & Van der Walt, G. H. J. (1985). *A review of recent developments in warp sizing*. http://hdl.handle.net/10204/13119
78. Wang, K., Wang, J., & Gao, W. (2024). Enhancing warp sizing effect and quality: A comprehensive review of the squeezing process and future research. *Textile Research Journal,* 94, 19–20.
79. Hari, P. K., Behera, B. K., Prakash, J., & Dhawan, K. (1989). High pressure squeezing in sizing: Performance of cotton yarn. *Textile Research Journal, 59*(10), 597–600.
80. Behera, B. K., & Hari, P. K. (1993). Weaving performance of polyester blended sized yarns: Roles of size recipe and high squeeze pressure. *Indian Journal of Fibre and Textile Research,* 18, 66–68.
81. Behera, B. K., & Hari, P. K. (1993). Mechanism of yarn break for high-pressure squeezed sized yarn. *Indian Journal of Fibre and Textile Research,* 18, 165–169.
82. Castro Cabado, M. M. (2016). Development of a bio-based binder for its application in the mineral wool insulation material. https://hdl.handle.net/10016/23843
83. Ok, H., Carr, W. W., & Park, H. (2007). Single-end sizing of yarn using a slot applicator. *Textile Research Journal, 77*(9), 686–695.
84. Ok, H. (2001). *Single end sizing of yarn using a slot applicator.* (Doctoral dissertation, Georgia Institute of Technology).
85. Wang, H., Jin, X., & Cui, J. W. (2013). Study of organic/inorganic hybrid nanoparticles in sizing agents for mechanism and effect of fiber adhesion. *Advanced Materials Research, 821*, 511–517.
86. Islam, Md. A. (1987). *The damage suffered by the warp yarn during weaving*. The University of Manchester (United Kingdom).
87. Bowman, L. E., et al. (1996). Sizing and desizing polyester/cotton blend yarns using liquid carbon dioxide. *Textile Research Journal, 66*(12), 795–802.
88. Dalton, R. G., Lasham, A., & Savidge, G. F. (1988). A new rapid semi-dry blotting technique for multimeric sizing of von Willebrand factor. *Thrombosis Research, 50*(2), 345–349.
89. Drexler, P. G., Tesoro, G. C. (1984). Materials and processes for textile warp sizing. In *Handbook of fiber science and technology*, (pp. 1-89) Routledge.
90. Brnada, S., Kovačević, S., & Sabljak, B. (2013). Prevention of faults on wool fabrics by application of cold sizing. *Textiles and Light Industrial Science and Technology,* 2(4), 189–193.
91. Sampathkumar, J. P. (1994). Studies on cold sizing. *JTA, 54*, 167–170.
92. Parikh, H. B. (2012). Innovation in sizing—Starch free cold sizing. *JTA, 72*, 400–401.
93. Bauer, H. (2007). Practical experience with cold sizing in worsted and carded yarn weaving mills. *Melliand International, 13*, 44–45.
94. Palliwal, M. C., & Kimothi, P. D. (1983). *Process control in weaving.* Ahmedabad Textile Industry's Research Association.
95. Chavan, H. V., & Patil, S. S. (2014). Yarn performance with cold sizing. *Melliand International, 20*(2).

Chapter 4
Wellness and Cosmeto Textiles: Sustainable Finishing Techniques

Akhtarul Islam Amjad and Bharti Pahuja

Abstract Textile use in wellness originated in history when medicinal clothing (Ayurvastra) was utilized as a treatment to promote health and cure ailments. It saw a decline due to the advent of industrialization and poor promotion. However, in the last few decades, people have become more demanding for comfort and functionality in textiles, and they expect novelty in apparel. Wellness and cosmeto textiles are produced using modern technological methods to enhance well-being through wearable products. The finishing process is crucial in achieving these attributes in the fabric. These finishes aim to release adequate medical or cosmetic ingredients onto the wearer's skin for healing effects. These finishes contribute to the wearer's aroma, health, and hygiene by rejuvenating the skin, protecting against microbes, and providing a pleasant fragrance. Essential oils and biologically active compounds are preferably used for wellness finishing. The cosmeto and well-being finishes can be achieved by infusing or embedding essential oils or active ingredients while manufacturing fibers, yarns, and fabrics. These are done by microencapsulation (vitamins, fragrances, and drugs), coating of biologically active compounds (hydrogels or polymer metrics), and plasma treatments of textiles. This chapter will explore the various aroma, health, and hygiene finishes on various textile substrates. Additionally, sustainable embedded techniques for achieving wellness finishes will also be discussed.

Keywords Skincare · Microencapsulation · Essential oils · Plasma technology · Wellness textiles · Ayurvastra · Cosmetotextiles · Sustainable textile finishes

A. I. Amjad (✉)
Department of Fashion Technology, National Institute of Fashion Technology, Panchkula, India
e-mail: akh.textile@gmail.com

B. Pahuja
Department of Textile Design, National Institute of Fashion Technology, Panchkula, India
e-mail: bharti.pahuja@nift.ac.in

M. Shahid et al. (eds.), *Advancements in Textile Finishing*,
https://doi.org/10.1007/978-981-96-6385-9_4

4.1 Introduction

Well-being denotes the condition of being comfortable, healthy, and pleasant, both psychologically and physically. Textiles are crucial to well-being by directly impacting comfort, health, and emotional state [1]. They contribute to well-being through softness, breathability, moisture-wicking, antimicrobial properties, UV protection, and sustainability. These properties in the fabric can be imparted by the finishing process. Conventional textile finishes primarily focus on modifying the basic feel of the fabric, enhancing surface softness or stiffness. On the other hand, functional finishes go beyond mere tactile qualities, offering a range of performance properties such as moisture management, flame retardancy, oil and water repellency, sun protection, and hygiene enhancement. By acknowledging the intricate nature of human skin chemistry and its interaction with environmental factors such as humidity, microbes, pH, temperature, and wind, functional finished clothing can be engineered to provide well-being solutions. These clothing materials have the potential to deliver multiple functionalities in protective clothing, cosmeto textiles, aroma textiles, and health and hygiene textiles, which are part of well-being. The physical finishing process consists of three techniques: impregnation, padding, and coating. The weak bonding force between the finishing agent and textiles is a major limitation. The chemical finishing method entails grafting a functional monomer onto a polymer substrate to create a new functional textile with long-lasting functionality. Biological, ecological finishing is a newer method that utilizes biological enzymes with biological activity for textile finishing [2].

The well-being finishes resulted in various value-added products that provide comfort and various benefits, including anti-acne, antibacterial, aroma, anti-inflammatory, sedative, UV protection, and calming characteristics. Aroma finish is a process in which textile materials are treated with bioactive agents, such as chitosan or a mixture of essential oils. This treatment provides the fabric with multifunctional properties, encompassing therapeutic effects and a sensation of well-being and freshness for the wearer. The focus of hygiene finishes lies in managing infection and undesired bacteria, specifically preventing dust mites. These finishes are utilized on various surfaces, such as bedding, mattresses, carpets, upholstery, and protective clothing. While certain bacteria may not be harmful on their own, they can contribute to unpleasant odors. On the other hand, some bacteria are pathogenic, and controlling their presence can aid in minimizing the spread of infection or cross-infection in shared spaces and medical facilities. Cosmetotextile is a new term coined in the last decade in the field of textiles. This includes aroma, health, and hygiene of the wearer. Cosmetotextile pertains to textiles or fabrics that contain cosmetic ingredients or properties intended to deliver skincare advantages when worn or applied to the skin. These fabrics are specially made to dispense cosmetic substances when in contact with the skin, delivering a range of skincare benefits, including hydration, tightening, anti-aging, and shielding against environmental elements such as UV radiation [3].

4.2 Finishing Technique for Functionalization of the Textile for Aroma, Health, and Hygiene

Wellness substances are incorporated into textile material during different manufacturing steps to get a cosmetic effect through various methods. The choice of methods relies on the characteristics of the cosmetic ingredient, the properties of the fabrics, and the desired quantity of cosmetic substances to be incorporated. Integrating wellness ingredients into a textile can be categorized as follows.

4.2.1 Active Agents During the Dope Formation of Textile Fiber

Binding the active agents to textiles can be done in various ways, depending on the required result and the application technique. By incorporating an active agent into the polymers during the master batch preparation phase, active compounds can be impregnated into different synthetic fibers. The fiber extrusion in standard production systems is made possible by the design of the doped master batch. If required, the fibers may be texturized and chopped into short staples or created in filament form [4]. Incorporating active agents early in the textile manufacturing stage allows for the use of various fibers without changing procedures or equipment, resulting in quick and easy production of fabrics with added value. Copper oxide was successfully embedded into the polymeric structure, which was then used to produce pillowcases, demonstrating the potential for lifting and whitening facial skin after a night's sleep [3, 4].

4.2.2 Surface Polymer Grafting for Functionalization

Grafting is a method used to introduce various functional groups onto textile polymers, allowing for the modification of their surface morphology to enhance specific qualities or introduce new features. This can broaden the potential applications of textiles by including novel sites. By grafting various active ingredients onto fibers, yarns, and fabrics, it is possible to achieve the desired effects. Cyclodextrins have an exceptional ability to form inclusion complexes with a variety of guest compounds. These compounds can include those with antibacterial and insecticidal activities, as well as fragrances. Methods have been developed for grafting cyclodextrin to cellulose or cellulose-based textiles to add functionality and to use it as a carrier for ingredient delivery [5]. Chitosan has also been explored for grafting onto cotton fabrics through UV-curing to provide sustainable and effective solutions for enhancing antibacterial properties that are maintained even after repeated washings. However, this process ensures minimal impact on fabric characteristics while presenting economic advantages, making it a promising alternative treatment option [6].

4.2.3 Functionalization by Microencapsulation

Textiles can be functionalized for aroma, health, and hygiene by including microcapsules that release the active component continuously with the body's natural movement. These capsules contain a specific amount of an active substance and are attached to the textile. Micro-encapsulation can be achieved through various methods like interfacial polymerization, in situ polymerization, sol–gel, coacervation, and phase separation. The active ingredient is released from the capsules through heat, biodegradation, friction, or pressure caused by the body and fabric during utilization. This process breaks the capsules into fragments, which liberates the encapsulated active ingredients [4, 7]. Microcapsules can be applied to textiles through stamping, exhaustion dyeing, impregnation, padding, coating, or spraying. Microcapsules can be absorbed directly into textile material without altering its handle, texture, or color. A binder such as starch, acrylic, polyurethane, or silicone is essential for these processes to secure the capsules onto the cloth. They secure the capsules in place even after washing and wear. In the padding technique, the surplus solution from the cloth is removed using squeeze rolls after being transferred to a bath containing microcapsules. The printing technique involves preparing a printing paste by compounding microcapsules with a binder. The paste is then printed onto the fabric's surface using a rotary screen attached to a rotating roll. Coating involves heating microcapsules deposited on the cloth's surface. An alternative technological procedure includes applying microcapsules on fabric surfaces within a specific space using a spray nozzle, which can be further stabilized using traditional high-temperature heating (130–170 °C). In sustainable textiles, the idea of using microcapsules made using biopolymers such as chitosan, gelatin, alginate, and cellulose instead of synthetic polymers due to their eco-friendliness, lower production costs, and wider availability have also been explored [3, 7].

Microcapsules can be added to silk, cotton, synthetic fibers, and other materials to provide a diverse array of functional textiles for application in sports, medicine, fragrance, Electronic textiles, UV protection, thermochromism, dyes, phase change materials, and other applications.

An advantage of this technology is its controlled release properties, which can increase efficiency and reduce environmental damage [8].

4.3 Wellbeing Finishing in the Form of Cosmetotextile

The wellness and health-promoting benefits of textile finishes have gained prominence in the twenty-first century. Wellness is regarded as favorable conditions devoid of illness and characterized by a harmonious equilibrium of body and mind, often representing a desire for youthful vitality amidst aging. Textiles infused with extracts of natural products and essential oils not only promote healing but also impart a sense of freshness and vigor to the wearer. In applications like wound dressings requiring controlled drug release, natural compounds like chitosan, hyaluronic acid, and alginates are used to achieve this purpose effectively. The concept of Natural Moisturizing Factor (NMF) serves as a measure of skin's moisture levels, with optimal conditions (around 20% moisture) maintaining skin's luster and elasticity [9]. The convergence of cosmetics and textiles opens new avenues for the development of cosmetotextiles, bridging these traditionally distinct sectors to explore innovative applications. The concept of cosmetotextiles emerged from the combination of the cosmetics and textile industries, employing techniques like microencapsulation [10]. As per the European Cosmetic Directive's definition, "cosmetotextiles refer to textile products that incorporate substances or preparations that gradually release onto various superficial areas of the human body, particularly the skin." These products possess specific functions, including cleansing, perfuming, altering appearance, providing protection, maintaining good condition, or addressing body smells. In these types of finished fabrics, bioactive compounds are released upon contact with the skin [9, 10]. Substances used in the cosmo-textile may be induced directly into the textile material, or textile functionalization can be accomplished by advanced techniques such as cyclodextrins, microencapsulation, or nanotechnology. Cosmetotextiles are capable of imparting skincare properties, reducing aging, and improving the feeling of wellness or well-being. The integration of active compounds in cosmetotextiles enhances application methodologies, as these systems function as reservoirs facilitating sustained release [11]. Oxidative stress caused by free radicals increases skin aging and is associated with dermatological conditions such as atopic dermatitis. Ultraviolet light stimulates the production of free radicals in cells; hence, the use of topical antioxidants is advised [10]. The fabrication of textiles with antioxidant capacity involves adding resveratrol, a stilbenoid produced by several plants. This process shows a remarkable effect on the successive release of resveratrol into the skin layers. Innovations aimed at integrating cosmetic ingredients into textiles began to surface in the late 1980s, with Japan leading the initial efforts. The European Union expanded on this work in 1995, exemplified by Hermes' launch of a scented scarf containing encapsulated Calèche, an eighteenth-century women's accessory. French company Euracli made significant strides in cosmetotextile innovations, setting the stage for multinational companies like Cognis (2001) and Invista (2003) to introduce their branded solutions, such as Skintex and Lycra Body Care. In 2003, the French cosmetotextile brand Lytess experienced notable growth in European market sales, further propelling the field's development and market visibility [12].

4.4 Sustainable Cosmetic Ingredients for Wellbeing Finishes

4.4.1 Inorganic Compounds

Various inorganic and synthetic compounds are employed to enhance textiles for well-being purposes. For instance, zinc oxide, titanium oxide, iron oxide, carbon black, and other compounds are utilized to shield the skin against UV radiation. Additionally, acetyl-glucosamine and D-glucosamide are used to impart deodorant properties to textiles, while copper oxide promotes healing and antimicrobial functionality in textile applications. These compounds serve specific functions such as UV protection and antimicrobial activity, enhancing the performance and appeal of textiles [13, 14].

4.4.2 Animal Derivatives

Chitosan, squalene, and sericin are animal-derived substances sourced from shrimp exoskeletons, shark livers, crabs, and the degumming fluid of silk cocoons. These compounds are used for their antibacterial, wound healing, deodorization, nourishment, moisture level balancing, and ability to promote cell regeneration. Chitosan is a versatile animal-derived compound widely utilized in cosmetic applications for its healing, antibacterial, and deodorant properties. Derived from the N-deacetylated derivative of chitin, a polysaccharide found in shellfish exoskeletons, chitosan is renowned for enhancing skin texture and moisture levels and stimulating cell regrowth. The abundance of amine groups renders chitosan a biocompatible polymeric material suitable for encapsulation. It forms a protective film on the skin, contributing to its overall nourishment and protection against environmental stressors [14]. Another noteworthy compound is squalene, a fatty substance sourced from shark liver or vegetable oils like palm and olive oil. Squalene acts as a natural antioxidant and is converted into squalane, which resembles the skin's natural lipid structure. This similarity allows squalane to deeply penetrate the skin, supporting natural hydration and aiding in the absorption of other beneficial ingredients. Squalane, when combined with vitamins like E and C, ascorbyl phosphate, and hyaluronic acid, helps shield the skin from photoaging and the development of age spots. Additionally, it softens the skin, reducing the appearance of fine lines and wrinkles [10, 11]. Sericin, derived from silk protein, offers significant potential in cosmetotextiles due to its antibacterial, oxidative resistance, and moisturizing properties. Recovered from silk waste during processing, sericin has been used to enhance textile functionality by increasing hydrophilicity and moisture retention. Coating fabrics with sericin not only improves comfort but also reduces formaldehyde content and minimizes skin irritation, making it ideal for applications such as diapers to enhance skin integrity and moisturization levels. Sericin's compatibility with other materials allows for innovative product development, improving functional properties across various textile applications. The fabric can be enhanced with microcapsules to provide various benefits such as moisturizing, cooling, energizing, relaxing, anti-heavy legs, and mosquito-repellent properties. The active compounds are released through mild friction between the microcapsules and the skin or the biodegradation of the chitosan membrane by skin enzymes. However, the limited aqueous solubility of chitosan restricts its use in delivering lipophilic compounds to the skin [6, 13].

4.4.3 Plant Derivatives

Medicinal plants have been utilized for ages as treatments for human diseases. In recent years, they have gained significant interest in textile finishes. This is attributable to their low toxicity, pharmacological properties, and economic viability. There is a significant shift from chemical and nonsustainable products to natural alternatives. These products are non-toxic, biodegradable, and confer health and wellness advantages.

Cosmetic substances like essential oils and extracts are extracted from plants, herbs, and fruits through processes of cold pressing and distillation of blossoms and leaves. Natural additives from plants may consist of compounds, groups of compounds, or essential oils that show multiple benefits, including antiviral, antifungal, antibacterial, antioxidant, antiparasitic, insecticidal, radical-scavenging properties, anti-inflammatory, antiseptic, germicidal, healing, and emollient effects. They are derived from specific sources such as the leaves of the Aloe Vera plant, brown algae, wheat germ oil, and various essential oils like innone (violet), cedaroil (lilac), hydroxy citronellol (lily), alpha-hexylcinnamaldehyde (jasmine), citral (lemon scent), allyl caproate (rose scent), anillin (apple scent), cinnamaldehyde (pineapple), prenyl acetate (banana), heliotropin (cherry), peppermint, lavender, thyme, sage, eucalyptus, and chamomile oil. These derivatives offer numerous benefits, including antimicroorganism and antimycotic properties, wound healing and inflammatory reduction, maintenance of skin elasticity and firmness, antioxidant qualities, moisture-binding capacity, and providing fragrance for relaxation and refreshment to the wearer.

Algae, when combined with textile fibers, release minerals, proteins, and vitamins when worn. Algae are widely used in cosmetics for their thickening, water-binding, and antioxidant properties. Rich in vitamins and minerals, algae nourish, hydrate, rejuvenate, and detoxify the skin, replenishing essential elements like proteins, vitamins (A and B), iron, sodium, phosphorus, magnesium, copper, and calcium. Different types of algae help firm and moisturize the skin, reduce puffiness, and protect against skin-damaging agents. Phytomer has identified coben, an active element in algae, as crucial in regulating melanin production, leading to the development of whitening products. Aloe vera, known for its nutritional value, comprises more than 75 nutrients and 200 active constituents. It encompasses minerals, amino acids, and vitamins. Careful processing is essential to retain these beneficial components, as overprocessing can diminish the potency of aloe vera products. Scientific research supports the use of aloe vera in textiles, offering antibacterial, wound-healing, and anti-inflammatory effects. Ginseng, valued for its adaptogenic properties, can be used to protect skin from cancer and inflammation through microencapsulation techniques. Various fruit oils provide refreshing and relaxing aromas to textiles, enhancing the wearer's experience with scents like lemon, rose, apple, pineapple, banana, and cherry [14].

Certain flowers, like violet, lilac, lily, and jasmine, yield specific chemicals used in wellness products through extraction techniques. These extracts, rich in well-being properties, can be applied to textiles using microencapsulation for diverse cosmetic benefits. Marigold flower petal extract, known as Calendula, possesses antibacterial, antioxidant, and anti-inflammatory qualities, making it effective in healing and soothing skin conditions like eczema. Calendula also benefits scalp health, repairs damage, and hydrates the skin, especially during winter. Its anti-inflammatory properties aid in healing wounds, sores, acne, ulcers, and other skin issues, similar to aloe vera. Vegetable glycerin, derived from plant oils, acts as a moisturizer due to its ability to form hydrogen bonds with water. It reduces the movement and provides a syrup-like texture. It is used extensively in cosmetics and has potential applications in cosmetotextiles due to its skin-absorbing properties. Padina pavonica, extracted from Mediterranean brown algae, enhances skin firmness and elasticity. Collaborations like Cosmetil's Hydrabra combine P. pavonica extracts with textiles for moisturizing and firming effects. Hinokitiol, a natural wood oil, exhibits antibacterial properties against various microorganisms and offers a relaxing aroma. Vitamin E, a lipid-soluble antioxidant, protects cells from oxidative stress, benefiting skin health and overall well-being in cosmetics and skin care products [15].

4.5 Applications of Various Sustainable Well-Being Finishes

4.5.1 Application in the Form of Aroma Finishes

The aroma finishing of textiles enhances product value by adding unique scents, catering to changing market demands. This process involves incorporating various scents into fabrics and creating scented textiles.

The incorporation of scents into textiles has been practiced for many years through the use of fabric conditioners during washing or drying to impart a fresh scent to the cloth. The impact is transient, enduring only one or two wash cycles. Microencapsulation enables perfumes to be preserved on textiles for an extended duration, allowing for retention through approximately 30 washes without total aroma loss [16]. The dimensions of capsules vary from 1 to 20 μm. Larger capsules emit more aroma upon breaking. Smaller capsules result in enhanced product coverage and prolonged scent release, as they require more time to be ruptured by physical pressure. Hako-Werke GmbH has developed a coated floorcloth containing microencapsulated scents, which may be utilized on many textile substrates, including hosiery [11, 15, 16].

The grafting of cyclodextrin (CD) derivatives onto textile fibers presents an opportunity to either absorb undesirable odors, such as those caused by perspiration, or to facilitate the regulated release of fragrance. Studies show that using vanillin or citronella oil as fragrance agents on polyester, wool, or cotton in conjunction with CDs improves scent retention and release characteristics. A direct correlation between the quantity of fixed CDs and the persistence of the impact was discovered. Moreover, γ-CD appeared to be the most adaptable cyclodextrin concerning the array of substrate molecules. It was assumed that γ-CD remains more accessible to the substrates despite the crosslinking reaction occurring during grafting, as it possesses the largest open cavity. Furthermore, it can be demonstrated that textile finishing with CDs enhances the endurance of perfumes during laundering [17, 18].

4.5.2 Applications in the Form of Odour-Resistant Finishes

The human body odor is produced by the microbial decomposition of sweat constituents. The breakdown of long-chain fatty acids into butyric acid and other chemical components is responsible for the odor. Other chemical compounds, such as allyl methyl sulfide from garlic, cocaine, and other chemicals that function as mosquito attractants, can also be found in sweat. These chemicals, among others, may be found in human perspiration and produce a distinct odor based on the combination of chemical components [19].

Textiles possessing odor-reducing characteristics are implemented in two distinct manners. One inhibits the decomposition of perspiration, while the other neutralizes or complexes the chemical produced during this process. Antibacterial agents emitted from the textile eradicate skin microorganisms, but using polymer coatings with various antibacterial substances can harm the natural skin microbiome. This creates a risk as more resistant microorganisms may take over. A safer alternative involves the chemical breakdown of sweat components directly on the fiber surface. This is achieved through the amalgamation of titanium dioxide with silver chloride [20]. Cyclodextrins and Calixarenes can form complexes with various organic molecules, including those present in sweat. By incorporating cyclodextrins into textile fibers or coatings, they can trap and immobilize odor-causing molecules, preventing them from being released into the air. This helps to keep the fabric smelling fresher for longer periods [17]. Calixarenes are another class of molecules that can be used for binding odor substances. Similar to cyclodextrins, calixarenes have a unique structure that allows them to encapsulate or trap specific molecules within their cavity [5, 20].

4.5.3 Application in the Form of Health and Hygiene Finishes

Maintaining proper hygiene is crucial in healthcare and medicine to prevent the spread of diseases and infections and ensure good health. These finishes may be used for various clothing and textile products, such as military uniforms for soldiers in harsh environments, intimate clothing like undergarments and socks, sportswear for athletes, surgical wound dressings in medical textiles, and home furnishing fabrics like bedding, mattresses, carpets, and upholstery. Cotton, linen, wool, and silk are particularly susceptible to bacterial and fungal growth, especially in warm and humid conditions. Even synthetic fibers which are known to be inherently resistant can become susceptible because of textile auxiliaries like starches and spin finishes. Microbial growth is also facilitated by sweat, oil, blood, dust, dirt, and food residue on fabrics, which can lead to foul odors.

Some bacteria may not cause disease, but they can cause bad smells that affect the comfort of textiles. Other bacteria can cause diseases and controlling their growth can help prevent the spread of infections in shared spaces and medical facilities. Various possibilities can be used for deodorization and antimicrobial properties of textile materials.

The basic approach is the use of

- **Adsorbents**, like activated carbon and zeolites, can function by absorbing odor and perspiration. A study conducted by Ahmad et al. revealed that fabrics containing activated carbon can reduce the odor that was picked up by the fabrics [21]
- **Active** compounds designed to interact with odors possess distinct chemical functional groups. Examples of such actives include biguanide polymers, which form complexes with organic molecules having organically attached nitrogen and/or sulfur atoms, and fatty alcohol esters of methyl methacrylic acid, which react with thiols, amines, and aldehydes. Successful application of amine-containing polymers such as polyethyleneimine and its derivatives, polyhexamethylene biguanide, and polyacrylate has been described in the literature [22].
- **Antimicrobial agents** can stop the growth of bacteria, thereby preventing the degradation of sweat into odoriferous compounds and preventing the growth of microorganisms. A variety of textile products utilize synthetic anti-microbial chemicals like triclosan, metals, and their salts, organometallics, phenols, quaternary ammonium compounds silver, etc.
- **Supramolecular compounds** such as cyclodextrins can form stable inclusion complexes with odors and can also be used to add suitable actives to the fabric. Wacker Chemie AG provides a registered cyclodextrin-based treatment known as Cavatex. It enables finished textiles to absorb undesirable odors while selectively releasing beneficial active compounds, including scents and conditioning agents, upon exposure to moisture.

Researchers have also been studying the effectiveness of natural substances like chitosan, neem extract, tulsi, aloe vera, tea tree oil, sericin, clove oil, onion skin and pulp extract, and eucalyptus oil in promoting hygiene. This is because of the increasing awareness among individuals about the environment and the demand for sustainable alternatives to synthetic hygiene products. This has led to the development of safe and non-toxic products that are friendly to both the skin and the environment. Thilagavati et al. [23] discovered the effective antimicrobial components in neem, pomegranate, and prickly chaffflower that can inhibit microbial growth. Neem leaves (Azadirachta indica), prickly chaff flower (Achyranthus aspera), tulsi leaves (Ocimum basilicum), and pomegranate rind (Punica granatum) demonstrated antibacterial efficacy against Staphylococcus S. aureus and E. coli pathogens [23]. The study conducted by Benltoufa et al. [24] showed that Chitosan hydrogel, when applied to cellulosic fabric, had antimicrobial properties for E. coli, Listeria monocytogene, and Staphylococcus aureus bacteria. Wijesirigunawardana & Perera [25] reported on the encapsulation of synthetic lime oil using chitosan and Arabic gum. The microcapsules were attached to cotton fabric through a grafting method with succinic acid binder, and their antibacterial effectiveness was evaluated against E. coli, B. cereus, S.typhimurium, and S. aureus. It was determined that the antibacterial activity of the lime oil could be preserved within the microcapsule [25]. The use of β-Cyclodextrin/Grafted Chitosan and lavender oil has demonstrated notable enhancements in fragrance longevity, performance characteristics like tensile strength and bending length, and antibacterial properties, surpassing those achieved by conventional approaches as reported by Singh et al. [11]. Nanoparticles derived from the shade-dried gel of A. vera were generated using the top-down procedure of ball milling. These nanoparticles were then applied as a coating on cotton fabrics, and they demonstrated antibacterial properties for aureus and E. coli. [26]. Basak et al. [27], developed a scent pack by utilizing cotton nonwoven, paper, and natural citronella essential oil. The product emits a revitalizing and invigorating scent that lasts for a duration of seven to eight days in the surrounding environment once the container is opened. Furthermore, it could deter mosquitoes for a duration of four to five days. This product is versatile and may be used in a variety of settings, including offices, washrooms, bedrooms, living rooms, and cars.

4.5.4 Application in the Form of Vitalizing, Refreshing, and Relaxing Finishes

A finishing effect that provides a refreshing and relaxing sensation. Two main methods for promoting relaxation and refreshment include deploying calming fragrance oils that are released actively and utilizing phase change materials to regulate body temperature. In the 1980s, the US National Aeronautics and Space Administration (NASA) explored the notion of employing microencapsulated PCMs to regulate skin temperature in space suits. The primary characteristic of these materials is their ability to alter their phase within a limited temperature (from solid to liquid and vice versa), absorbing or expelling heat. Thus, PCMs can be employed to give a cooling or heating impact in the garment microclimate based on the ambient temperature. Adaptive Comfort technology uses patented microencapsulated PCMs (Thermocules) in textile applications, including gloves, socks, hats, outdoor wear (vests, thermals, parkas, snowsuits, and trousers), and household textiles (blankets, duvets, mattresses, pillowcases). Numerous fruit oils like ginger, menthol, orange, or rosemary, are utilized to impart a refreshing and calming aroma to the individual wearing them. Aloe vera essential oil which includes several vitamins, minerals, and other active ingredients was able to provide the wearer with pleasant feelings while also increasing energy levels. It has also been utilized for its antibacterial and anti-inflammatory effects. Ajinomoto, an amino acid producer, collaborated with Mizuno Corp to create the 'Amino Veil' brand for sports products. Amino acids in Amino Veil tennis and golf clothing absorb moisture and maintain skin pH equilibrium. The product contains arginine, an amino acid known to promote skin regeneration. Yonex, a sports clothing company, has introduced a product line using xylitol, a sweetener typically found in chewing gum. When in contact with water, it absorbs heat and provides cooling benefits. Xylitol-enhanced jeans can help keep the user cool while they sweat [28]. Yonex's Xylitol-impregnated Very Cool Polo Shirts reduce body heat by 3 degrees for added comfort during play. These shirts are worn by tennis stars and top badminton players. Bombay Dyeing's Aroma Rich bed linen collection features soft cotton satin with a high thread count. It is infused with a natural fragrance that promotes relaxation and sleep. The smoothness and scent derive from the microencapsulated smell, which emits a subtle aroma upon contact with the skin. The Aroma Rich bed linen collection, which includes Lavender, Musk, Jasmine, and Rose, not only encourages sound sleep but also prevents odor and keeps the bed linen fresh after several washing. The Tradekey company's Ayurvastram eye cushion offers mild support and coolness to the area around the eyes. This eye cushion is made of pure silk. The eye pillow contains Ushira (Vetiveria zizanioides), as well as a variety of other Ayurvedic herbs and flaxseed. According to Ayurveda, Ushira is a plant that calms, enhances mood, and relieves tension. When mixed with essential oils, this herb has excellent cooling qualities. Similar type of active ingredients and uses are summerised in Table 4.1.

Table 4.1 Active ingredients for relaxing aromatherapy

Active ingredients	Uses
Valerian	For relaxing properties
Amber tree resin oil	Their bark yields a soothing, aromatic oil with therapeutic benefits
Chamomile	For relaxing and sleep-inducing effects
Jasmine	For relaxing, balancing, confidence enhancing, and aphrodisiac qualities
Lavender	For harmonizing and soothing
Passion fruit oil	For its skin soothing and smoothing attribute

4.5.5 Application in the Form of UV Protection Finishes

Extended exposure to UV light can cause sunburn, allergies, early aging of the skin, and even skin cancer. Based on wavelength, UV radiation is divided into three zones: (i) UVC (200–280 nm), which is primarily absorbed in the upper atmosphere's ozone layer; (ii) UVB (280–320 nm), which penetrates the skin's outer layer and is more likely to cause sunburn, skin aging, and other related conditions; and (iii) UVA (320–400 nm), which may also cause skin aging. Therefore, it is necessary to make sure that the textile effectively blocks the transmission of both UVA and UVB rays to design and produce an efficient UV skin care textile. Fiber type and its chemical composition, Fabric cover factor, type of dye applied on the surface, and final finishes play a significant role in protecting against harmful rays, however, they are not enough for prolonged exposure. The UV protective performance of a textile is measured by the Ultraviolet Protection Factor (UPF), indicating how long a person can be in the sun before skin reddening occurs. Zinc oxide and titanium dioxide at the nanoscale are the two most effective substances for absorbing and scattering UV light, offering superior UV protection over materials of standard sizes. Several plant compounds (extract) and natural or biopolymers have also been used recently in the creation of environmentally friendly UV-protective textiles. Enzyme-treated fabric dyed with Acacia catechu contains saponins, flavonoids, glycosides, phenols, and terpenoids. Phytochemical analysis showed excellent UPF protection, displaying the potential use of Acacia catechu as a dye for sun-protective textiles [29, 30]. Lemongrass oil coated on wool fabric using chitosan as a cross-linking agent showed significant improvement in the UV protection factor. The uncoated cloth had the lowest UPF rating of all the tests, measuring 43.9. The UPF was observed to rise with the concentration of lemongrass oil, with a maximum of 122 [31].

4.5.6 Application in the Form of Moisturizing Finishes

When the temperature is arid in autumn and winter, human skin often becomes dehydrated. Researchers have discovered that applying moisturizing finishes to fabrics can effectively moisturize the skin, enhancing the comfort of wearing these fabrics. Moisturizing ingredients like vitamin E, vitamin C, chitosan, aloe vera, and hyaluronic acid can transfer from the fibers to hydrate the outer layer of the skin (stratum corneum). These natural ingredients can be encapsulated using methods like spray drying, sonication, and complex coacervation. Fabrics treated with these finishes are soft, smooth, and pleasant to the touch, providing excellent skin moisturization and beautification effects. They are also safe, eco-friendly, and free from toxic substances, meeting the standards for ecological textiles [22].

Researchers have emphasized the importance of moisturization in treating eczema, as dry skin can result in inflammation that compromises the skin barrier. They developed patches using a combination of electrospun hydrophobic polystyrene (PS) and hydrophilic nylon 6 (PA6) infused with oils to effectively moisturize atopic skin. Tests confirmed the patches' potential through oil release on the skin and their moisturizing effects. Chaheh and colleagues demonstrated that wearing textiles treated with protease can stimulate the production of active ingredients like vitamin E on the skin, resulting in moisturizing effects [32]. Additionally, squalene, known for its humectant properties, can reduce wrinkles and fine lines without leaving oily residues on the skin. Integration of TiO_2 into textiles enhances moisture absorption through a photocatalytic process. For instance, a thin film of TiO_2 can be deposited on polyethylene fibers using the layer-by-layer deposition technique to create quick-drying fabric suitable for indoor or outdoor clothing [16].

4.5.7 Application in the Form of Energizing Finishes

Energizing finishes for well-being are designed to boost human energy levels through textile articles. One key ingredient used in these finishes is Coenzyme Q10 (CoQ10), a natural antioxidant that cells use to enhance movement and energy. While scientific evidence may be lacking, CoQ10 is considered a promising substance for developing energizing textiles. Perfuming finishes for well-being are textile treatments that absorb unpleasant odors and emit pleasing fragrances. Various synthetic and natural products like chitin, chitosan, essential oils (such as clove, jasmine, lavender, etc.), and acetyl-glucosamine are used to imbue textiles with fragrant properties. Manufacturers like Matsui Shikiso Chemical Co. in Japan have developed techniques to encapsulate essential oils like musk, civet, ambergris, pine, and citrus and integrate these microcapsules into fabric fibers using polymerization methods. Encapsulation of fragrances helps protect volatile constituents from evaporation, oxidation, and contamination, enhancing the longevity of the perfuming effect. These fragrances can also be incorporated into textile binders and softeners to improve durability against washing and handling while maintaining a pleasant scent [33].

4.6 Commercial Availability of Wellbeing Finishes

The commercial availability of well-being finishes in textiles has seen significant growth driven by consumer demand for functional and value-added products. Companies worldwide, including those in the UK, Germany, and France, are actively developing and marketing textiles treated with microencapsulated formulations containing vitamins, aloe vera, chitosan, and other beneficial agents. These finishes offer diverse benefits such as skin care, moisturization, cooling effects, and anti-cellulite properties across a range of products including bedding, underwear, T-shirts, stockings, socks, and other apparel items. This provides consumers with innovative options to enhance comfort, health, and overall well-being through their everyday clothing choices. Notable brands like Microban with their Scentry™ and Scentry Revive specialize in fabric treatments that provide long-lasting freshness and odor control using encapsulated fragrances. Devan's BI-OME® technology allows fabrics to stay fresh longer, reducing the need for frequent washing and saving energy and water. Similarly, Odour Breakdown® addresses the issue of odors in polyester garments by weakening bacteria fixation and neutralizing malodors, enabling easier cleaning even at lower temperatures [34]. Additionally, Cognis developed Skintex Supercool, utilizing high-tech microcapsules with natural ingredients like menthol for a cooling effect. Other notable innovations include the use of sensory perception technology by Wool Development International Limited (WDI) to impart aroma, moisturizers, and anti-bacterial properties in various textile products. Furthermore, Indian and international brands such as RIL, Arvind Limited, and Welspun offer textiles treated with technologies like Intellifresh™ and HygroCotton® to enhance well-being through odor control and moisture management. Collaborative efforts between global companies like Lanxess, Lytess, and Lenzing have resulted in the development of specialized fibers and treatments like Tencel® C and Bayscent® Aromatherapy, targeting skin health and comfort. Legends & Heroes introduced Denim Spa Therapy for Legs under the Ript Skinz brand, providing specialized skincare benefits through denim wear. Pulcra Chemicals developed Cyclofresh Plus, offering textile brands, resistance from body odor and fragrances through a dual mechanism using cyclodextrins and silver ions [35]. Italian company Solidea innovated with micro massage magic clothing in shorts and socks, aiming to improve skin health and reduce cellulite on the buttocks and legs. Lytess, a French company, offers specific products like collants that drain and slim legs, alongside other textiles like denim and shirts designed to release cosmetic substances. Additionally, Camangi Corporation introduced UMORFIL, a line of cosmetotextiles based on nylon and polyester infused with ocean collagen peptide amino acids for enhanced skin comfort and flexibility. These innovations highlight the growing trend of integrating skincare and wellness features into textile products, catering to consumer demands for functional and beneficial clothing options. Several manufacturers offer a range of well-being finishing products with distinctive brand names and innovative technologies. Dogi International Fabrics incorporates aloe vera into their fabrics for added comfort and skincare benefits. Hefel Textil GmbH from Austria introduces "LYOSILK®" and "SEACELL®ACTIVE", featuring advanced textile technologies aimed at promoting skin health and well-being [36]. Additionally, Hydrabra and other brands contribute to the diverse landscape of well-being textiles, showcasing the industry's commitment to innovation and consumer wellness. The use of specific engineering techniques in textiles can enhance wellness benefits. For instance, Invista International in Switzerland developed Lycra leg care stockings with graduated compression, which reduce fatigue and swelling while enhancing athletic performance. Skintex technology by incorporating microencapsulated active ingredients into textiles offers sustained benefits like skin hydration and protection [8]. Solidea, Italy, patented a fabric design for cellulite-reducing garments, and Swiss Federal Laboratories developed lightweight cooling garments for multiple sclerosis patients. Cognis, Germany, introduced baby diapers with phase change materials for improved comfort. These innovations bridge the gap between cosmetics and fashion, strengthening the concept of cosmetotextiles. Spanx, Inc. specializes in body-shaping undergarments, and INTERFILIERE Hong Kong underscores Hong Kong's importance in intimate and swimwear fashion [11, 37].

4.7 Future Scope and Challenges for the Sustainable Wellbeing Finish

The future potential of sustainable well-being finishes is bright, as there is a growing demand from consumers for environmentally friendly products. This involves the creation of new and innovative materials sourced from renewable sources, such as bamboo, organic cotton, and recycled fibers, that mitigate environmental effects. The focus of sustainable well-being finishes will be on biodegradability and recyclability, achieved through using natural polymers and bio-based substances to minimize pollution at the end of their life cycle. Furthermore, advancements in low-impact processing techniques like waterless dyeing and chemical-free finishes will further enhance sustainability. By incorporating natural active ingredients like aloe vera and essential oils, these finishes will provide benefits for the skin and ensure responsible sourcing practices. Although numerous challenges exist to the well-being finishes and finishing industry [38]. There is a pressing need for innovative material development to create textiles that can effectively integrate cosmetic and skincare components while ensuring comfort, durability, and sustainability. Another challenge is maintaining consistent and long-lasting performance of well-being finishes, such as moisturization and anti-aging benefits, even after frequent washes or extended use. Additionally, adhering to stringent regulatory standards for textiles, cosmetics, and medical devices across different regions is crucial for market compliance and consumer well-being. Educating consumers about the advantages and proper usage of well-being finishes while addressing concerns regarding safety, effectiveness, and environmental impact is essential for broad acceptance and adoption. Furthermore, developing techniques for customizing and personalizing well-being finishes based on individual skin types and preferences will be vital to optimize effectiveness and enhance consumer satisfaction in this evolving industry.

4.8 Conclusion

The integration of wellness-enhancing technologies into textiles represents a promising frontier in the intersection of fashion, cosmetics, and healthcare. Sustainable well-being finishes in textiles are characterized by a concerted effort to integrate eco-friendly materials, low-impact processing techniques, and natural active ingredients to minimize environmental impact while enhancing consumer well-being. Various companies and research institutes worldwide are pioneering the development of textiles that offer tangible wellness benefits, ranging from reducing fatigue and muscle soreness to promoting skin health and comfort. Innovations like microencapsulation of active ingredients graduated compression in garments, and the use of specialized fibers and materials demonstrate the potential of cosmetotextiles to revolutionize traditional apparel. As consumer demand for multifunctional and performance-driven textiles grows, collaborations between industries such as textiles, cosmetics, herbal medicine, and pharmaceuticals will play a crucial role in further advancing the field of cosmetotextiles.

Declaration The authors declare that they have no known competing financial interests or personal relationships that could have appeared to influence the work reported in this chapter. Generative AI and AI-assisted technologies have been used in the writing process to improve the readability and language of the work only. After using this tool, the authors carefully reviewed and revised the content to ensure accuracy and quality. The authors take full responsibility for the final version of this publication.

References

1. Trippeer, B., Stark, J., Gam, H., Ellis, N., Morgan, B., & Park, J. G. P. (2022). Sustainable textiles for health and well-being: An investigation of curricular opportunities for fashion design and interior design student collaboration. *Journal of Textile Design Research and Practice, 10*(2), 214–238.
2. Zhang, Y., Xia, X., Ma, K., Xia, G., Wu, M., Cheung, Y. H., Yu, H., Zou, B., Zhang, X., Farha, O. K., & Xin, J. H. (2023). Functional textiles with smart properties: Their fabrications and sustainable applications. *Advanced Functional Materials, 33*(33), 2301607.
3. Peng, X., Umer, M., Pervez, M. N., Hasan, K. F., Habib, M. A., Islam, M. S., & Cai, Y. (2023). Biopolymers-based microencapsulation technology for sustainable textiles development: A short review. *Case Studies in Chemical and Environmental Engineering* , 100349.
4. Borkow, G. (2016). Cosmetotextiles-sometimes the simple things work. *Journal of Cosmetology & Trichology, 2*(1), 1000e103.
5. Cova, T. F., Murtinho, D., Pais, A. A., & Valente, A. J. (2018). Combining cellulose and cyclodextrins: Fascinating designs for materials and pharmaceutics. *Frontiers in Chemistry, 6*, 271.
6. Ferrero, F., Periolatto, M., & Ferrario, S. (2015). Sustainable antimicrobial finishing of cotton fabrics by chitosan UV-grafting: From laboratory experiments to semi-industrial scale-up. *Journal of Cleaner Production, 96*, 244–252.
7. Borkow, G., & Gabbay, J. (2004). Putting copper into action: Copper-impregnated products with potent biocidal activities. *The FASEB Journal, 18*(14), 1728–1730.
8. Jamal, Z., & Rani, S. (2018). Cosmetotextiles: A wearable skin care. *International Journal of Home Science, 4*, 31–35.
9. Mathis, R., & Mehling, A. (2011). Textiles with cosmetic effects. In *Handbook of Medical Textiles* (pp. 153–172). Woodhead Publishing.
10. Cheng, S. Y., Yuen, C. W. M., Kan, C. W., & Cheuk, K. K. L. (2008). Development of cosmetic textiles using microencapsulation technology. *Research Journal of Textile and Apparel, 12*(4), 41–51.
11. Singh, N., Yadav, M., Khanna, S., & Sahu, O. (2017). Sustainable fragrance cum antimicrobial finishing on cotton: Indigenous essential oil. *Sustainable Chemistry and Pharmacy, 5*, 22–29.
12. Islam, J. M., Akter, T., Mokbul, M., Afroz, S., & Mondal, M. I. H. (2022). Textiles in cosmetics and personal care. In *Medical Textiles from Natural Resources* (pp. 457–497). Woodhead Publishing.

13. Matijević, I., Bischof, S., & Pušić, T. (2016). Cosmetic preparations on textiles: Cosmetotextiles. *Journal of Textile and Apparel, Technology and Management, 65.*
14. Singh, M. K., Singh, A., & Morris, H. V. (2023). Cosmeto-textiles. *Textile Progress, 55*(3), 109–163.
15. Alonso, C., Martí, M., Martínez, V., Rubio, L., Parra, J. L., & Coderch, L. (2013). Antioxidant cosmeto-textiles: Skin assessment. *European Journal of Pharmaceutics and Biopharmaceutics, 84*(1), 192–199.
16. Yarkent, Ç., & Oncel, S. S. (2022). Recent progress in microalgal squalene production and its cosmetic application. *Biotechnology and Bioprocess Engineering, 27*(3), 295–305.
17. Bezerra, F. M., Lis, M. J., Firmino, H. B., Dias da Silva, J. G., Curto Valle, R. D. C. S., Borges Valle, J. A., Scacchetti, F. A. P., & Tessaro, A. L. (2020). The role of β-cyclodextrin in the textile industry. *Molecules, 25*(16), 3624.
18. Martel, B., Le Thuaut, P., Bertini, S., Crini, G., Bacquet, M., Torri, G., & Morcellet, M. (2002). Grafting of cyclodextrins onto polypropylene nonwoven fabrics for the manufacture of reactive filters. III. Study of the sorption properties. *Journal of Applied Polymer Science, 85*(8), 1771–1778.
19. Broadhead, CRC Press, R., Craeye, L., & Callewaert, C. (2022). Odour control or inhibition using antimicrobial finishing. In *Odour in Textiles* (pp. 95–116). CRC Press.
20. Hemamalini, T., & Dev, V. R. G. (2022). Fragrance finishing on textile substrate for odour control. In *Odour in Textiles* (pp. 143–166). CRC Press.
21. Ahmad, W. Y. W., Omar, K., & Ahmad, M. N. (2012). Activated carbon as anti-odour coated and pigment printed fabric. In *Presented at Conference of Business Engineering and Industrial Applications* (ISBEIA) (pp. 690–695).
22. Muñoz, V., Gonzalez, J. S., Martínez, M. A., & Alvarez, V. A. (2017). Functional textiles for skin care active substance encapsulation. *Journal of Textile Engineering and Fashion Technology, 2*(6), 538–545.
23. Thilagavati, G., Rajendrakumar, K., & Rajendran, R. (2005). Development of eco-friendly antimicrobial textile finishes using herbs. *Indian Journal of Fibre, 30*, 431–436.
24. Benltoufa, S., Miled, W., Trad, M., Slama, R. B., & Fayala, F. (2020). Chitosan hydrogel-coated cellulosic fabric for medical end-use: Antibacterial properties, basic mechanical and comfort properties. *Carbohydrate Polymers, 227*, Article 115352.
25. Wijesirigunawardana, P. B., & Perera, B. G. K. (2018). Development of a cotton smart textile with medicinal properties using lime oil microcapsules. *Acta Chimica Slovenica, 65*(1), 150–159.
26. Subramani, K., Kolathupalayam Shanmugam, B., Rangaraj, S., Palanisamy, M., Periasamy, P., & Venkatachalam, R. (2018). Screening the UV-blocking and antimicrobial properties of herbal nanoparticles prepared from aloe vera leaves for textile applications. *IET Nanobiotechnology, 12*(4), 459–465.
27. Basak, S., Saxena, S., Raja, A., Patil, P., Krishnaprasad, G., Narkar, R., & Kambli, N. (2021). Cotton-based fragrance packet for well-being applications. *Journal of Cotton Research, 10*(1).
28. Sun, G. (Ed.). (2016). *Antimicrobial textiles.* Woodhead Publishing.
29. Muthu, S. S. (Ed). (2015). *Handbook of sustainable apparel production.* CRC Press.
30. Samant, L., Jose, S., Rose, N. M., & Shakyawar, D. B. (2022). Antimicrobial and UV protection properties of cotton fabric using enzymatic pretreatment and dyeing with Acacia catechu. *Journal of Natural Fibers, 19*(6), 2243–2253.
31. Gogoi, M., Kadam, V., Jose, S., Shakyawar, D. B., & Kalita, B. (2022). Multifunctional finishing of woolens with lemongrass oil. *Journal of Natural Fibers, 19*(4), 1353–1365.
32. Wang, Q., Lu, J., Jin, Z., Chen, K., Zhao, M., & Sun, Y. (2022). Study on the structure and skin moisturizing properties of hyaluronic acid viscose fiber seamless knitted fabric for autumn and winter. *Materials, 15*(5), 1806.
33. Julaeha, E., Nurzaman, M., Wahyudi, T., Nurjanah, S., Permadi, N., & Anshori, J. A. (2022). Development of antibacterial microcapsules of citrus essential oil for cosmetotextile application: A review. *Molecules, 27*(22), 8090.

34. Odour control technology-Antimicrobial-Bi-Ome-Devan. Retrieved May 15, 2024, from https://devan.net/bi-ome/
35. CYCLOFRESH–Pulcra Active Freshness Technology (cyclofresh-pulcra.com). Retrieved May 15, 2024, from https://cyclofresh-pulcra.com/
36. Amjad, A. I., & Kumar, R. (2023). Effect of eco-friendly bamboo rayon, lyocell, and Seacell fibers on the properties of blended knitted melange fabrics. *Fibers and Polymers, 24*(5), 1845–1858.
37. Live it up-Intellifresh. Retrieved May 15, 2024, from https://intellifresh.com/
38. Mondal, M. I. H., Saha, J., & Rahman, M. A. (2021). Functional applications of Aloe vera on textiles: A review. *Journal of Polymers and the Environment, 29*, 993–1009.

Chapter 5
Enhancing Comfort Through Functional Finishes in Outdoor Textiles

Radostina A. Angelova

Abstract Outdoor textiles provide protection, comfort, and functionality during various outdoor activities. This chapter explores the importance of functional treatments in enhancing outdoor textiles' performance to meet consumers' evolving demands and the challenges of modern lifestyles. Functional treatments such as moisture-wicking, thermoregulation, UV protection, antimicrobial properties, and durable water-repellent (DWR) coatings are crucial for creating high-performance fabrics that offer environmental protection, improved comfort, and enhanced user experience. With advancements in technology, innovative approaches like nanotechnology, smart materials, and sustainable manufacturing practices are transforming traditional textiles into versatile, high-tech solutions. This chapter also addresses the growing consumer interest in quality, functionality, and sustainability in outdoor textile products, examining how these trends shape the industry's research, design, and production processes. Key mechanisms of functional treatments and their practical applications in outdoor textiles are analyzed, highlighting their role in optimizing comfort, wearability, and performance. The chapter concludes by emphasizing the balance between innovation, sustainability, and accessibility, presenting strategies for developing cutting-edge outdoor textiles that meet environmental, economic, and user-centered demands.

Keywords Functional treatments · Outdoor textiles · Moisture-wicking · Thermoregulation · Sustainable manufacturing

5.1 Introduction

Outdoor textiles are an important part of our daily lives, providing protection and comfort in various weather conditions. With the advancement of technology, functional treatments of textiles play a key role in enhancing user comfort [78]. These specialized treatments improve the physical properties of the materials while adding

R. A. Angelova (✉)
Technical University of Sofia, Sofia, Bulgaria
e-mail: radost@tu-sofia.bg; joy_angels@abv.bg

M. Shahid et al. (eds.), *Advancements in Textile Finishing*,
https://doi.org/10.1007/978-981-96-6385-9_5

new functionalities that meet the growing demands of modern consumers and their lifestyles. In recent decades, there has been a significant increase in interest in outdoor activities. People increasingly appreciate the time spent in nature through hiking, camping, mountaineering, or simply walking. The rise in outdoor activities has led to a growing demand for high-quality textile products that provide both environmental protection and maximum comfort. Comfort is crucial for a positive user experience [6]. Whether it is professional athletes or nature enthusiasts, the comfort and functionality of clothing can significantly impact performance and enjoyment of the activity [38]. Functional treatments in outdoor textiles are key in achieving this enhanced comfort. By integrating specific chemical and physical treatments, textile materials acquire new properties or improve their existing characteristics [65]. This includes moisture-wicking abilities, thermoregulation, UV protection, antimicrobial properties, etc. Imagine a hiker climbing a steep mountain trail in variable weather conditions. Clothing that effectively wicks sweat away from the body allows the skin to breathe, and at the same time protects against sudden rain can make the difference between a successful hike and an uncomfortable, even dangerous, experience. In this context, functional treatments like DWR (Durable Water Repellent) coatings and moisture-wicking finishes become indispensable [29]. With the advancement of technologies in the textile industry, new possibilities are emerging for creating materials with exceptional properties. Nanotechnology, smart materials, and sustainable manufacturing practices enable the development of textiles that are lighter, stronger, and more functional [78]. These innovations enhance people's comfort while also meeting the growing environmental demands of modern society.

It is particularly important to note that consumers are becoming more informed and demanding about the quality and functionality of textiles and clothing [2]. They are looking for items that not only look good but also provide real benefits in the context of outdoor activities. This challenges manufacturers to integrate the latest technologies and offer products that meet these high expectations [4]. In this sense, the topic of the role of functional treatments in outdoor textiles is crucial for several groups:

- For consumers, to make informed choices when purchasing clothing and equipment that meet their specific needs and preferences.
- For manufacturers and designers, to develop innovative products that stand out in the market and meet the growing demands for quality and functionality.
- For researchers, to advance new technologies and materials that enhance the properties of textiles and offer sustainable solutions.

Inevitably, the progress in the development of outdoor textile production faces the challenges of the modern world. The demands for sustainability and eco-friendliness compel manufacturers to reduce their environmental footprint by using recycled materials, lowering water and chemical usage, and implementing sustainable production practices [40]. Achieving an optimal balance between different functional properties requires a focus on comfort without compromising wearability. Furthermore, the production of high-tech outdoor textiles must be offered at prices accessible to a broad range of consumers [68].

This chapter examines the significance of functional treatments in outdoor textiles and their role in enhancing users' comfort. Treatments such as moisture-wicking, thermoregulation, UV protection, and antimicrobial properties are presented as key technologies that transform traditional fabrics into high-performance products. The chapter analyses various types of functional treatments, their mechanisms of action, and real examples of their application in textile products.

5.2 Individual Comfort

Comfort, by its nature, is a relative and subjective concept. According to Hatch [37], comfort is defined as "the absence of pain and discomfort. It is a state of neutrality." Slater [97] offers another definition, suggesting that comfort is "a pleasant sensation of physiological, psychological, and physical harmony between an individual and their surrounding environment." The factors determining individual comfort and the influence of textile products and clothing are summarised in Table 5.1.

The physical aspects of comfort are closely related to a person's perceptions and their subjective sense of discomfort or pain [46]. The factors that affect physical comfort include the senses: touch, sight, hearing, taste, and smell. In interactions with textile products and clothing, touch is the most important, followed by sight (more frequently), and sometimes smell (in specific cases). From a psychological perspective, comfort reflects an individual's preferences for the materials, colors, and designs of textile products [51]. These factors are key to providing convenience and confidence during various activities. In this context, textiles shape a person's appearance and relationships with others (e.g., creating positive impressions). Moreover, clothing can express individuality and character [109].

Physiological comfort can be divided into the following types [8]:

- Physiological comfort determined by sensory perceptions is known as neurophysiological comfort. It depends on irritations such as pricking, itching, roughness, thermal discomfort (unpleasant sensations of cold or heat), as well as the tendency of textiles to accumulate static electricity.
- Physiological comfort related to movement is determined by the following characteristics of outdoor textiles: stretchability, weight, and pressure on the surface of the body.

Table 5.1 Factors affecting individual comfort

Factors	Physical comfort	Physiological comfort	Psychological comfort
Tactile factors	Touch	Thermal sensations	Material
Visual factors	Visual perceptions	–	Colour and shape
Olfactory factors	Smell	–	–
Dynamic factors	–	Movement	Style and design

- Physiological comfort related to thermal sensations is defined as thermo-physiological comfort. It depends on the balance between the heat produced by the body and the heat dissipated into the environment. When these two values are equal, the body is considered to be in a neutral state.

5.3 Weather-Resistant Finishes

The finishes applied to textile products to enhance their resistance to weather conditions play a key role in providing protection and comfort to users in outdoor environments. From mountain hikers to people working outdoors, these treatments ensure that the user stays dry and protected from the harmful effects of nature [16]. Two of the most important and long-lasting treatments are waterproofing and UV protection. In practice, these treatments have revolutionized textiles for both clothing and technical textiles used outdoors.

5.3.1 Waterproofing

The primary function of waterproofing treatments is to prevent water from penetrating fabrics used as the outermost layer of protective and workwear or technical textiles with specialized applications, such as tents, umbrellas, awnings, and others [58]. The goal of water-repellent finishes is to create a surface that repels water, preventing it from penetrating between the yarns (the macrostructure of the fabric) or between the fibers in the yarns (the mesostructure of the fabric)—Fig. 5.1. These treatments also prevent the fibers from absorbing water. The surface of treated fabrics causes water droplets to break into smaller drops and roll off without being absorbed. This behavior is primarily due to manipulating surface tension and the contact angles between the treated surface and water molecules [3].

Hydrophobic substances are applied to fabrics to increase the contact angle between water and the material, typically above 90° [3]. The higher the contact angle,

Fig. 5.1 The surface of a wet water repellent fabric

the more water will bead up on the surface rather than spread out and be absorbed. This effect can be observed in nature, for instance, on the leaves of the lotus plant, which are naturally hydrophobic and effectively repel water [80]. Inspired by this, scientists have developed various synthetic water-repellent treatments for textiles.

There are several types of water-repellent treatments widely used in the textile industry, each with its own advantages and applications.

5.3.2 Durable Water Repellent (DWR)

Durable water repellent (DWR) is one of the most widespread technologies. A hydrophobic layer is applied to the surface of the textile, protecting it from getting wet [39]. This technology is most commonly used in jackets, raincoats, tents, and other products that need to withstand rain and moisture. For DWR treatments, fluorinated compounds, also known as fluorocarbons (PFCs), are traditionally used [88]. They are particularly effective at repelling water, oils, and dirt. The most used substances in DWR treatments are perfluorocarbons (PFCs), such as PFOS (perfluorooctane sulfonate) and PFOA (perfluorooctanoic acid) [111]. Fluorinated compounds are highly effective because they create a strongly hydrophobic surface. However, there are serious environmental and health concerns associated with the use of fluorocarbons, as they do not break down easily in the environment and can accumulate in the human body and nature [111]. Due to these risks, many textile and clothing manufacturers have started transitioning to fluorine-free alternatives for DWR treatments [88]. Some of the new substances used to achieve a water-repellent effect include hydrocarbon compounds, silicone, and polyurethane compounds. While not as effective, they do not rely on harmful fluorocarbons and still offer a good level of moisture protection. Gore-Tex® is an example of the development of waterproof and breathable fabrics (Fig. 5.2), which often use DWR treatment for the outermost layer of clothing or products [112]. This coating protects the textile surface from getting wet, while the membrane underneath ensures breathability by allowing the transfer of heat, air, and water vapor. Gore-Tex is actively developing and implementing fluorine-free alternatives for its products to reduce its environmental footprint [30].

The North Face also uses DWR technologies for its products, with some transitioning to Futurelight™ technology [104]. This technology combines water-repellent properties with moisture transfer processes from the body outward, causing less environmental harm. They use membranes and layers with DWR treatment to provide maximum protection for users in harsh weather conditions. One of the challenges with DWR is that the effectiveness of the coating can decrease over time, especially after frequent washing or intensive use [39]. To maintain the effectiveness of DWR treatments, users can apply special products for reapplying water-repellent coatings [28].

Fig. 5.2 A jacket with a Gore-Tex® membrane

5.3.3 Membrane Technologies

Membranes combine waterproofing with breathability, preventing water and wind from penetrating from the outside while allowing heat and water vapor to escape from the body to the environment [112]. This makes membrane textile systems ideal for outdoor activities where it is important not only to protect against rain but also to maintain thermophysiological comfort.

Gore-Tex® membranes are among the most well-known technologies for waterproof, windproof, and breathable clothing, footwear, and accessories. This is due to their structure, which contains billions of microscopic pores that are about 20,000 times smaller than a water droplet [64]. This makes it impossible for water to penetrate the membrane, even during heavy rain or snow. At the same time, the membrane is designed to allow water vapor (sweat) to pass through the pores, which are approximately 700 times larger than a water vapor molecule [64]. This prevents excessive sweating and overheating of the body, especially during physical activities.

Gore-Tex® membranes also block wind, thereby protecting the human body from heat loss. They are durable and retain their waterproof and breathable qualities even after prolonged use and multiple washes [13].

There are various technological variations of Gore-Tex:

- Gore-Tex® Pro: This is the most durable and technically advanced version of the membrane, designed for professional mountaineers and athletes. It is a 3-layer fabric with high abrasion resistance and suitable for intensive activities in extreme conditions [34].
- Gore-Tex® Active: Designed for active sports such as running, cycling, and skiing, this 3-layer membrane is thinner and more flexible. It is ideal for high-intensity physical activities where comfort and quick moisture wicking are important [31].
- Gore-Tex® Paclite®: This is a lightweight and compact version of the membrane. The 2.5-layer laminate construction is designed for situations where the weight and bulk of clothing or equipment are crucial [32].
- Gore-Tex® INFINIUM™: This version focuses on comfort and performance, but it is not as waterproof as other Gore-Tex® technologies. It is designed for activities where windproofing and breathability are of critical importance [33].

Gore-Tex® membranes are widely used in clothing from brands such as The North Face and Arc'teryx, as well as in footwear by Salomon. eVent membranes are high-tech waterproof and breathable membranes designed for extreme weather conditions [26]. They feature a unique Direct Venting™ Technology, making them a preferred choice for active sports such as hiking, climbing, and cycling. Traditional waterproof membranes use hydrophilic layers that absorb moisture from the body and release it outward through evaporation [35]. With eVent membranes, this process is unnecessary—the micropores in the membrane directly release water vapor, resulting in faster and more efficient breathability [26]. They also offer excellent protection against water by blocking external moisture while allowing sweat to evaporate. Their waterproof rating, typically over 20,000 mm water column, makes them suitable for extreme weather conditions [26]. eVent membranes are highly resistant to wear and stress, making them suitable for intensive use in harsh outdoor activities. They retain their waterproof and breathable properties even after prolonged use and multiple washes [103]. This is why they are used in the design of various products, including clothing for mountaineering, hiking, and other active sports, mountain hiking and running shoes, backpacks, and accessories where waterproofing and breathability are essential. Brands that use eVent membranes include Montane, Rab, and La Sportiva.

5.3.4 *Fluorine-Free Technologies*

Fluorine-containing compounds (such as PFOS and PFOA) in traditional water-repellent treatments have significant environmental drawbacks. They are toxic and do not break down easily in nature [111]. In search of alternatives to these chemicals, many companies and research teams are developing fluorine-free water-repellent finishes. They use environmentally friendly substances such as silicones, waxes, paraffins, or polyurethane compounds that possess water-repellent properties [93]. These materials create a barrier on the surface of the textile that repels water without blocking breathability, thereby preserving thermophysiological comfort. An example of fluorine-free technologies is Schoeller® Ecorepel, developed by the Swiss company Schoeller, which allows textiles to remain waterproof even after repeated washing [90]. Bionic-Finish® Eco, from the German company Rudolf Group, uses modified wax and silicone molecules [86]. Sympatex® offers water-repellent breathable membranes and textile technologies that are completely fluorocarbon-free, making them popular for the production of active sports and hiking clothing [100]. Although the brand Teflon is associated with fluorine-containing products, Teflon EcoElite™ is a biodegradable and renewable water-repellent finish that contains no fluorocarbons. It is used in sportswear and home textiles [102].

5.3.5 *Applications of Waterproof Textiles*

Waterproof textiles have widespread applications across many industries and activities. They are essential in the production of sports and outdoor clothing for hiking, climbing, and active outdoor sports. Protective clothing for firefighters, rescue teams, and military personnel often uses water-repellent membranes to shield against water and moisture in harsh conditions.

Fabrics used for making tents and canopies are typically treated with DWR or membrane technologies to protect against rain. Many fashion brands also integrate waterproof treatments into everyday outerwear to offer protection during rainfall.

5.4 UV Protection

In recent decades, increasing awareness of the harmful effects that UV rays can have on the skin has made UV protection one of the most important functions of modern textiles. UV protection is achieved through special treatments or by using materials that have the ability to absorb or reflect ultraviolet rays [20]. UV rays are divided into two main categories: UVA (long waves that penetrate deep into the skin) and UVB (short waves that can cause surface skin burns) [73]. Protection from these rays can prevent premature skin aging, skin cancer, and other related diseases. The UV protection of textiles is measured by the UPF (Ultraviolet Protection Factor), which indicates how much UV radiation is blocked by the material [60]. For example, fabric with UPF 50 allows only 1/50th of UV rays to pass through, meaning that 98% of the rays are blocked. Several technologies are used to provide UV protection for outdoor textile products.

5.4.1 Adding UV UV-Absorbing Agents

Adding UV-absorbing agents to textiles is a technology based on integrating special chemical substances that absorb and block UV radiation [72]. Zinc oxide (ZnO) is commonly used in the cosmetics industry, as it is added to sunscreen creams, but it can also be incorporated into textiles [107]. It absorbs both UVA and UVB rays, which are the most harmful spectrums of UV radiation [89]. Zinc oxide does not degrade when exposed to light and provides long-lasting protection. Titanium dioxide (TiO_2) is another widely used mineral in cosmetics and textiles. It is a white, water-insoluble mineral that absorbs UVB rays and reflects UVA rays [45]. UV protective agents can be applied to textile surfaces through impregnation or the application of nanoparticles [105]. Impregnating textiles with a solution containing UV absorbing agents provides deep protection for the fibres. Several companies worldwide specialize in producing UV-protective textiles using technologies such as the addition of UV absorbing agents. Among them are Coolibar, whose products are certified with UPF (Ultraviolet Protection Factor) 50+ [17], Columbia Sportswear, a global leader in outdoor clothing and accessories, The North Face, Patagonia, and others.

5.4.2 Nanotechnologies and UV Protection

The integration of nanoparticles into textile products leads to higher efficiency in UV protection. It becomes durable over time, even after multiple washes and sun exposure. Traditional UV protection materials such as titanium dioxide (TiO_2) and zinc oxide (ZnO) are used [96]. Titanium dioxide nanoparticles can be applied as a coating on fabrics or incorporated directly into the structure of the fibres [7]. They provide a high degree of protection without disrupting the breathability of the material. Similar to titanium dioxide, zinc oxide is integrated into fabrics through nanotechnologies, providing excellent UV protection and durability to the textile without reducing the comfort and elasticity of the clothes [113]. Nike and Under Armour use nanotechnologies in some of their sports lines to provide clothing with high UV protection. Coolibar offers swimsuits and beachwear treated with nanoparticles for maximum UV protection. The North Face and Columbia Sportswear offer jackets and shirts made with nanoparticles for long-lasting UV protection, suitable for extreme conditions. Medical uniforms and workwear exposed to UV radiation also use nanoparticles to protect outdoor workers. Dickies produces workwear with UV protection, utilizing nanoparticles to improve the material's resistance to sunlight. The advantages of nanotechnologies for UV protection lie primarily in their durability [77]. Nanoparticles remain active even after repeated washing and sun exposure. At the same time, they are imperceptible to the user, preserving the comfort and softness of the textile [92]. Nano-treatments offer better protection compared to traditional UV protection methods, as they reflect and absorb both UV-A and UV-B rays [53].

5.4.3 *Use of Fibers with Natural UV Protection*

The use of fibers with natural UV protection does not require the addition of extra chemicals. These fibers have a structure that naturally blocks or reflects UV radiation, thus providing skin protection from prolonged sunlight. Figure 5.3 shows the textile fibers with UV protection, ranked by the degree of protection reduction.

Merino wool is a natural fiber with excellent thermoregulation and moisture management properties and the ability to block UV rays. Its structure contains amino acids that absorb UV radiation [70]. Merino wool can provide UPF protection of up to 40–50, making it highly effective for outdoor activities such as hiking and high-altitude sports [71]. It is used in hiking apparel, sports shirts, and base layers for climbers and mountaineers. Icebreaker, a New Zealand company, specializes in producing merino wool clothing, with many of its products offering UV protection and being designed for outdoor activities. Smartwool, an American brand, offers a wide range of merino wool products, including socks, shirts, and thermal clothing with UV protection. Hemp is a natural fiber with a dense structure that has the ability to block UV rays. It is known for its strength and durability and can provide a high UPF factor (30–50) [83]. Hemp is used in the production of summer clothing and accessories. Patagonia uses hemp in combination with organic cotton and other materials to create garments with natural UV protection. Toad&Co offers sustainable clothing that includes hemp fabrics, combining natural UV protection with breathable and comfortable materials. Polyester is a synthetic fiber that, due to its structure, can reflect and block UV rays. Combined with tight weaving, polyester is often used for clothing that provides high sun protection. Polyester garments can offer UPF

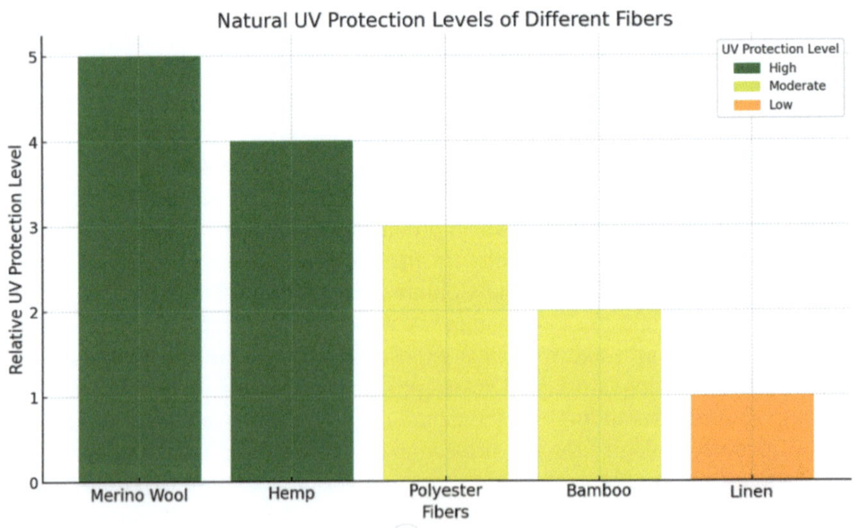

Fig. 5.3 Ultraviolet protection levels of different fibers without any additional treatment

30–50 depending on the density and finish of the fabric [60]. Nike offers sportswear that uses polyester fabrics with built-in UV protection, especially in running and soccer lines. Under Armour also provides sportswear with UV protection made from polyester fabrics, offering comfort and protection during outdoor workouts. Bamboo and linen are additional examples of fibres with natural UV protective properties. They are eco-friendly options for clothing designed to protect against harmful UV rays. Bamboo typically has a UPF range of 20–40, depending on processing, while linen, depending on fibre density, can offer UPF 15–30 [66].

5.5 Durability and Comfort

When it comes to the durability and comfort of outdoor textiles, two of the most important aspects are abrasion resistance and antibacterial coatings. These features not only extend the lifespan of the garments but also make them more comfortable to wear, even in harsh weather conditions.

5.5.1 Abrasion Resistance

Abrasion resistance refers to the fabric's ability to withstand friction, which is crucial for outdoor clothing [36]. When in contact with rough surfaces (such as rocks, trees, backpacks, or even during movement), the fibers, and subsequently the yarns, can wear down. Therefore, special coatings are applied, or fibers and structures (mesostructures and macrostructures) that are naturally resistant to abrasion are used. The enhanced abrasion resistance of outdoor textiles significantly contributes to wearer comfort. When fabrics and garments retain their structure and protective properties over time without becoming coarse, thin, or uncomfortable due to wear, it helps maintain comfort even during intense physical activity.

5.5.2 Abrasion Resistance Coatings

Polyurethane (PU) coatings and polyamide blends play a significant role in producing modern abrasion-resistant textiles for outdoor activities and sports. At the same time, they provide flexibility, water resistance, and durability. Polyurethane is a synthetic material applied as a thin layer on fabrics to make them more resistant to abrasion and wear, as well as to water and wind [94]. These coatings do not compromise the flexibility of the textile, which is particularly important for garments that require freedom of movement. Some more advanced polyurethane coatings are also designed to offer breathability [15]. Polyurethane coatings are commonly found in products such as backpacks and tents, making them more durable and moisture-resistant.

Polyamide is an extremely strong synthetic material known for its durability, abrasion resistance, and lightweight. When blended with other materials (such as polyester or elastane), it produces fabrics that combine flexibility, strength, and resistance to stretching and tearing [52]. This makes them an excellent choice for outdoor and sportswear exposed to intense wear. It is worth noting that polyamide blends result in lightweight garments, which is an advantage for outdoor activities where the weight of the clothing plays a key role in comfort and freedom of movement [67]. Garments from such blends dry quickly, making them convenient for use in rainy conditions or environments with high humidity. Brands like Nike and Adidas often use polyamide blends in their sportswear to ensure durability and lightness. Arc'teryx and Patagonia incorporate polyamide blends in their outdoor lines, providing strength and durability in garments without sacrificing comfort.

5.5.3 *Cordura® Fabrics*

Cordura® is a trademark of Invista for a series of fabrics known for their exceptional strength, durability, and abrasion resistance. It encompasses a range of fabrics made primarily from polyamide or blends of polyamide and other, often synthetic, fibers [18]. The original Cordura® material is nylon 6,6, renowned for its remarkable strength and abrasion resistance. Cordura® fabrics are up to 10 times more resistant to abrasion compared to regular fabrics, making them ideal for products subjected to intense friction or physical impact [27]. Depending on the specific application, Cordura® can also be made from polyester or cotton blends [49]. Blending with polyester increases abrasion resistance and makes garments more resistant to UV rays, while blends with cotton make garments softer and more breathable. Cordura® is also produced from recycled polyamide, such as discarded fishing nets or plastic bottles, helping to reduce the carbon footprint [18]. Cordura® fabrics are characterized by a tightly woven macrostructure, key to their strength and abrasion resistance (Fig. 5.4). The fabric is woven using a plain weave, one of the strongest weaves in textile manufacturing. Additionally, a Ripstop weave is employed, with thicker threads woven at intervals to create reinforced sections in the fabric. This structure is designed to prevent small holes or tears from expanding. Even if the fabric tears, the damage is limited to a small area, thanks to the reinforced threads. To increase the softness of Cordura® fabrics, weaves without reinforcement and with less cross-threading, such as twill and Oxford, are used.

Cordura® Ballistic, for instance, is woven in a plain weave and provides high resistance to punctures and abrasion. It is used in motorcycle jackets and protective gear. Cordura® Classic, on the other hand, features a Ripstop weave for maximum strength and durability while maintaining relatively lightweight properties, making it popular for military and outdoor clothing [18]. Cordura® Denim is based on a cotton blend and uses a twill weave to produce denim fabric with enhanced abrasion resistance, making it suitable for workwear and everyday use. Cordura® Oxford employs an Oxford weave, producing a soft and flexible fabric [18]. It is applied

Fig. 5.4 Cordura® fabric with drops

in producing lightweight but durable outdoor clothing and backpacks. Brands like Osprey and Deuter use Cordura® in their hiking backpacks to ensure durability and long-lasting performance. In fact, many companies utilize Cordura® for their products. Patagonia, Arc'teryx, and The North Face incorporate it in their hiking jackets and trousers, designed for tougher and more challenging conditions. Carhartt uses Cordura® for increased durability and resistance in their workwear, which is intended for industrial work environments.

5.6 Antibacterial Coatings

Antibacterial coatings are important for outdoor textiles when the wearer is exposed to prolonged physical activity, heat, and moisture [99]. This combination creates favourable conditions for the growth of microorganisms. Through additional treatments, the growth of bacteria, fungi, and unpleasant odours is reduced or prevented, thereby enhancing the comfort and hygiene of the clothing.

5.6.1 Silver-Based Coatings

Silver ions (Ag^+) effectively inhibit the growth of bacteria that cause unpleasant odors [11]. The ions bind to the cell walls of bacteria, leading to the destruction of the cell membrane. This disrupts the vital functions of the bacteria, preventing their reproduction. The ions also interfere with the bacteria's ability to produce the enzymes and proteins they need for growth [95]. This makes them highly effective in preventing the accumulation of bacteria on textile fibers. One of the advantages of silver ion coatings is their durability. They remain active after multiple wear and wash cycles [21]. At the same time, they are safe and generally do not cause skin irritation or allergies [55]. The technology is carefully regulated and extensively researched to ensure it is safe for users. It is used in the clothing and products of companies such as Icebreaker, Under Armour, and Smartwool.

The most important methods of applying silver include:

- Impregnation by coating, where silver ions are applied to the textile surface [85]. Silver nanoparticles or compounds are added to a polymer solution or binding agent. The coating is applied using the dipping, spraying, or padding method, followed by thermal treatment. This method is popular in the production of sportswear and socks.
- The implementation of silver nanoparticles within the fibres themselves ensures that the particles do not wash out over time [48]. Silver can be embedded during the extrusion of synthetic fibres, such as polyester or polyamide [24]. In addition to outdoor clothing, this method is used for underwear and medical textiles, where the durability of antibacterial properties is essential.
- Chemical fixation involves the use of chemical binding agents that attach silver ions to the surface of the fibres [108]. This creates a strong bond between the fabric and the antibacterial agent, ensuring it remains active even after washing and prolonged use. The treatment is applied during the finishing process, where the fabrics are passed through chemical solutions containing silver ions. This method is used for fabrics in medical garments, work uniforms, and everyday clothing.
- Nanocoating with silver ions is another option. The coating contains particles smaller than 100 nm, which are applied to the textile fibres [54]. The effectiveness of the nanocoating is very high, as the small size of the particles allows a large surface area to come into contact with bacteria. The nanoparticles are applied via spraying, dipping, or other technological processes, after which the fabric is left to dry or undergoes thermal treatment [79]. These coatings are popular in the sports industry and for outdoor clothing, as they provide lightweight, invisible protection against microbes and odours.
- Electrolytic coating uses an electrochemical process to apply silver ions onto textile fibres. In this process, the fabric is placed in an electrolyte solution containing silver ions, and by applying an electric current, the silver ions bond to the surface of the fibres [12]. This method results in a durable and highly effective antibacterial coating, suitable for military uniforms, protective clothing, and medical textiles.

5.6.2 Zinc-Based Coatings

Zinc oxide (ZnO) coatings are increasingly popular in the textile industry due to their antimicrobial properties and environmental benefits. Zinc oxide prevents the growth of bacteria and other microorganisms on fabrics, offering long-lasting protection [82]. This is achieved by destroying the cell membranes of microorganisms and inhibiting their enzymes, leading to their death. Zinc oxide also blocks the production of toxins that can cause infections or unpleasant odors, especially when sweating or wearing clothes for extended periods [11]. Zinc oxide also has excellent UV protective properties, making it essential in treating outdoor clothing. It absorbs and disperses UVA and UVB rays [56]. Unlike some other chemical antibacterial agents, ZnO is relatively environmentally friendly. It does not release toxic substances into nature and is safe for human skin [42]. This is why it is widely used in sportswear, where odor prevention and hygiene are crucial, and skin contact is inevitable. Fabrics treated with zinc oxide can be worn during intense physical activities for extended periods. The beneficial properties of ZnO treatment are also important for medical uniforms, medical sheets, and other textiles to prevent hospital-acquired infections. The same applies to workwear and protective clothing for firefighters or construction workers, where protection from bacteria and UV radiation provides better hygiene during long-term use.

There are several methods for applying zinc oxide to fabrics:

- Similar to silver ions, ZnO can be incorporated into the fibres themselves during the manufacturing process. This involves adding zinc particles to polymers (such as polyester or polyamide) before the fibres are extruded [61].
- The application of ZnO as a coating on the fabric surface is done by dipping the fabric into a solution containing zinc particles, or by using a nanocoating, which ensures an even distribution of zinc across the fibres [98].

Fabrics treated with zinc oxide retain their antimicrobial properties even after multiple washes. This long-lasting protection and hygiene benefit clothing brands like Smartwool, Patagonia, and many others.

5.6.3 Natural Antibacterial Coatings

Derived from natural sources like chitosan (from the shells of crustaceans) or tea tree oil, these coatings are an eco-friendly alternative to synthetic antibacterial treatments [22]. They provide antimicrobial protection and prevent the growth of odors and microorganisms on fabrics while being safer for the environment and posing a lower risk of skin irritation or allergies [10]. Chitosan is derived from chitin, found in the shells of crustaceans such as shrimp, crabs, and lobsters. This biopolymer inhibits the growth of bacteria, fungi, and viruses by binding its positively charged molecules to the negatively charged cell walls of bacteria [101]. The interaction disrupts the bacterial cell membrane and prevents their reproduction. Chitosan can also absorb moisture from textile surfaces, further reducing the risk of bacterial growth, which thrives in humid environments [25]. This makes it useful in sportswear, socks, and garments exposed to intense sweating, helping to prevent unpleasant odors and maintain hygiene during physical activity. It is also applied in medical textiles (such as bandages and clothing), preventing infections and promoting wound healing. Tea tree oil (Melaleuca alternifolia) is a natural essential oil renowned for its potent antimicrobial and antifungal properties. It contains terpenes, which have the ability to disrupt the cell membranes of bacteria and fungi, inhibiting their growth [19]. It is widely used in personal care and medical products, but it also has applications in the textile industry, such as in the production of socks, sportswear, and outdoor gear. The application of natural antibacterial agents can be achieved through various methods, including impregnation, microencapsulation, nanocoating, spraying, or plasma treatment [99]. Each method has its own advantages, with the choice depending on the type of textile and the specific requirements for durability and effectiveness of the antibacterial action.

5.6.4 Balance Between Antibacterial Properties and Thermophysiological Comfort

The application of antibacterial coatings can reduce the transmission of air and water vapor through fabrics. This reduction in "breathability" is crucial for thermophysiological comfort, as it prevents moisture from evaporating from the skin's surface, hindering the body's ability to stay cool and dry. Modern coatings are designed to balance these two factors, providing fabrics that remain breathable while effectively preventing the growth of bacteria. This ensures that garments offer both hygiene and comfort, especially during intense physical activities.

5.7 Comfort and Performance

5.7.1 Moisture Management

Moisture management is key to ensuring thermophysiological comfort, especially during sports activities or outdoor work. Proper moisture management involves techniques that keep the skin dry and help regulate body temperature [41]. Their goal is to facilitate the movement of moisture (mainly sweat and water vapor) away from the body towards the outer layer of the clothing. From there, the moisture can be quickly released into the environment, mainly through evaporation. Retaining moisture on the surface of the skin can lead to overcooling in cold outdoor environments or create heat stress in hot environments [8]. Additionally, the presence of moisture increases the risk of microorganism growth and the occurrence of unpleasant odors. Excessive moisture accumulation can also lead to friction and skin irritation [110].

5.7.1.1 Moisture Management Through Composition

The use of specific materials, primarily with hydrophobic properties, allows for the rapid transfer of water vapour from the surface of the skin to the layer of clothing in direct contact with the surrounding air. These materials are widely used in sportswear, outdoor clothing, and work clothing. The main types are:

- Polyester. Synthetic polyester fibres are known for their durability, lightness, and ability to wick moisture. They are the most commonly used material for moisture-wicking garments. Polyester does not absorb moisture, allowing sweat to move easily through the fibres and evaporate quickly.
- Polyamide. Polyamide is characterised by high strength and resistance to wear. It is lightweight, flexible, and often used in combination with other fibres. Being hydrophobic, it shares similar moisture management properties with polyester, allowing sweat to be drawn away from the skin and evaporated quickly.
- Polypropylene. Another synthetic fibre, polypropylene has exceptional moisture-wicking properties. It repels moisture and transfers it to the outer layer of the fabric, where it evaporates. This makes it ideal for intense physical activities in extreme conditions.
- Merino wool. This natural fibre is renowned for its excellent thermoregulating and moisture-wicking properties. It can absorb up to 30% of its own weight in moisture without the wearer feeling wet, while also helping to regulate body temperature.
- Bamboo. Bamboo is soft, breathable, and resistant to microbes. Bamboo fabrics absorb moisture and wick it away from the skin, keeping the wearer cool and dry. Additionally, they possess natural antibacterial properties, making them suitable for long-term use.

In addition to these natural and synthetic fibres, special patented fabrics used by various brands in their outdoor clothing and accessories should also be mentioned. Coolmax® by The Lycra Company is one of the most well-known brands of moisture-wicking fabrics, which effectively draw moisture away from the body and help it evaporate quickly [62]. Nike's Dri-FIT technology provides moisture-wicking properties to keep athletes dry and comfortable during intense workouts [74]. Under Armour's HeatGear® technology allows garments to wick moisture and keep the wearer cool during physical activity in hot conditions [106].

5.7.1.2 Moisture Management Through Structure

The macrostructure of fabrics plays a key role in moisture management, with knitted structures being particularly effective for this purpose. Double Knit or Layered Knit fabrics allow the inner layer to wick moisture away from the skin, while the outer layer facilitates its evaporation [47]. The inner layer typically has a smooth or hydrophilic surface, which helps absorb moisture from the skin and transfer it to the outer layer. The outer layer has a hydrophobic structure that enables moisture to spread and evaporate more quickly. Mesh weaves or knits refer to a more open weave or knit with small holes that increase airflow and allow moisture to evaporate faster [9]. Jacquard Knit or Weave incorporates intricate patterns and textures in the fabric, enabling different areas of the fabric to have different properties [84]. Textured Knit or Weave enhances moisture management by creating air pockets between the fabric and the skin [50]. These air pockets allow for better air circulation and quicker removal of moisture from the skin. The textured design also increases the surface area, which accelerates evaporation. Bi-Component knitted or woven fabrics use a combination of hydrophilic (moisture-attracting) and hydrophobic (moisture-repelling) fibres to facilitate moisture movement [43]. The hydrophilic fibres draw moisture away from the skin, while the hydrophobic fibres in the outer layer of the fabric allow moisture to spread and evaporate more quickly.

5.7.1.3 Moisture Management Finishes

These treatments aim to enhance the fabric's ability to wick moisture away from the skin and evaporate it more quickly. Various materials are used, most commonly applied as a film on the fabric's surface. Silicone-based finishes use silicone compounds that prevent fibers from absorbing moisture [76]. This treatment distributes moisture evenly across the fabric surface, allowing it to evaporate quickly. In addition, it enhances the softness of the garment, increasing comfort upon contact. These finishes are applied to sportswear, fitness clothing, and outdoor gear, where rapid sweat-wicking and keeping the skin dry are essential. Similarly, Fluorocarbon Finishes create a water-repellent layer, but they are also used for moisture-wicking fabrics, which aid in faster water evaporation and prevent moisture retention in the fabric. These finishes are commonly used in outdoor and sports clothing, where both sweat-wicking and protection from rain are required. Cationic Finishes use cationic agents (positively charged) that interact with water molecules [87]. They attract water molecules to the fabric and facilitate their movement along the fibers, improving capillary action. These finishes are often used in sports and fitness apparel made from synthetic fibers for intense physical activities to enhance their naturally low ability to manage moisture. Hydrophilic Polymer Finishes are applied to hydrophobic fabrics (e.g., polyester or polyamide) to improve their ability to absorb and wick moisture [63]. They create a surface that attracts moisture from the skin, allowing it to move along the fibers and evaporate quickly. This increases the capillary activity of the fibers and speeds up the evaporation process. Nanotechnologies are used to create ultra-fine coatings that can enhance the moisture-wicking properties of fabrics without significantly altering the texture or feel of the material [81]. Nanoparticles are applied to the fabric, creating fine channels that facilitate moisture movement from the inside of the fabric to its outer surface. These particles also help improve the distribution of moisture for faster evaporation. Nano-finishes are used in sportswear, thermal underwear, and outdoor gear to enhance moisture-wicking properties and long-lasting durability.

5.7.2 *Breathability*

Breathability is crucial for clothing comfort, especially during intense physical or outdoor activities. Clothing must wick moisture and allow air to circulate to prevent the accumulation of heat and moisture inside the garments. Materials, technologies, and finishes that enhance breathability help maintain comfort during wear while also providing protection from weather conditions.

5.7.2.1 Breathability Finishes

Breathability finishes are applied to fabrics to facilitate air transfer, which helps with moisture removal and prevents overheating [14]. Microporous membranes have millions of tiny pores that are small enough to repel water droplets (like rain) but large enough to allow water vapour (sweat) to pass through, ensuring breathability [44]. These membranes are often used in outdoor gear, such as Gore-Tex®. Polytetrafluoroethylene (PTFE) membranes are popular in breathable fabric technologies and are used to create coatings that repel water while allowing water vapor to pass through [57]. They are utilized in high-tech outdoor clothing and professional equipment for extreme conditions. Polyurethane coatings used for moisture management can also be designed to provide breathability [15]. Polyurethane membranes contain micropores that allow sweat to evaporate but block water [59]. Similarly, membranes that combine polyurethane and polyester components perform the same function. Thanks to their multilayer structure, they allow water vapour to pass through the fabric while repelling water droplets.

5.7.2.2 Maintaining Breathability While Adding Protective Functions

It is important to maintain a balance between breathability and protective functions in outdoor and sports apparel, which needs to be waterproof, windproof, or resistant to other weather conditions. One strategy is to combine breathable and waterproof membranes (such as Gore-Tex®, eVent®, and others). Another solution is using ventilation openings, mesh elements, and adjustable vents or zippers. These are placed in heat-generating areas during active movement, such as under the arms or on the back, to support airflow—Fig. 5.5. These areas are covered with flaps or zippers that prevent water or wind from penetrating.

Two-layer or three-layer structures in garments are also a way to combine good protection and breathability [64]. Two-layer systems include an outer waterproof layer and an inner breathable layer. The inner layer may be mesh or microporous to provide comfort and prevent heat buildup. Three-layer systems include an additional protective layer that supports breathability and repels moisture while ensuring resistance to external factors like wind and rain.

Fig. 5.5 A jacket with ventilation openings covered with zippers

5.7.3 Thermophysiological Comfort

When designing outdoor clothing, achieving thermophysiological comfort is crucial. This is especially important under changing weather conditions or prolonged exposure to harsh climates, such as extreme cold or heat. Therefore, clothing must provide comfort in two ways: warming and cooling the body.

5.8 Heat Retention Finishes

Insulating finishes play a vital role in regulating body temperature. They are particularly important for clothing intended for cold weather but can also apply to garments used in variable or extreme conditions [16]. One of the most commonly used solutions is thermal linings. These are applied inside the clothing to create a barrier that traps body heat. Frequently used materials include fleece or down. Essentially, these are mechanical insulation layers that retain heat by creating air pockets within the fabric. The pockets trap body heat and keep it close to the skin, preventing cold air from penetrating. These solutions are highly effective in cold conditions but must be combined with other technologies to prevent overheating. Another option is using chemically treated materials whose properties have been modified to enhance their thermal insulation. This can include fabrics treated with chemicals that improve the heat-retaining capabilities of the fibers. An example of such a solution is 3 M's Thinsulate, a synthetic insulation material [1]. It combines fine fibers with a chemical treatment to retain more heat in thinner layers. Thinsulate is commonly used in gloves, winter jackets, and footwear. Natural insulation materials like goose down (feathers) and wool also remain widely used due to their excellent natural heat retention properties. Down, for example, is extremely lightweight and warm because it traps a large amount of air between the fibers. It is used in many high-quality jackets and sleeping bags. Conversely, wool has the unique ability to retain heat even when wet, making it a preferred material for winter clothing and blankets. With technological advancements, intelligent insulation finishes are being developed which can regulate temperature in real time. These materials adapt to the body or environmental temperature and change their insulating properties as needed. A typical example of this technology is using phase-change materials (PCM). The Outlast technology, in particular, is popular for outdoor clothing and products [75]. It uses materials with different transition temperatures between solid and liquid states, where heat is released or absorbed. Often in microcapsules, PCM absorbs, stores, and releases heat depending on the temperature of the body and the environment [5]. This type of insulation is especially suitable for garments that need to be worn throughout the day in both cold and warm conditions.

5.8.1 Cooling Finishes

Cooling finishes are important in textiles designed for comfort in hot weather [69]. They are intended to lower body temperature by dissipating heat and moisture, providing a cooling sensation. Cooling finishes are used in various types of clothing, particularly in garments for active lifestyles and outdoor activities, such as hiking, running, or sports. Some fabrics are specifically engineered to reflect heat or sunlight and accelerate sweat evaporation to create a cooling effect on the body. Special fibers or fabric structures are used to achieve this. An example of such a material is the synthetic Coolmax, designed to wick moisture away more quickly, keeping the wearer dry and cool [91]. It is primarily used in sportswear and socks for active pursuits. Some textiles are treated with cooling chemicals or substances that create a cooling sensation upon contact with the skin. These finishes can be temporary or long-lasting, depending on the treatment. An example is cooling finishes with microcapsules of menthol, which are activated when they come into contact with sweat, creating a pleasant cooling effect. This effect is often used in outdoor workout clothing. Thermoregulatory technologies allow garments to respond to changes in temperature and humidity, maintaining a constant microclimate close to the skin. These technologies can provide both cooling and insulation depending on the conditions. An example of this is again the phase-change materials (PCM). Technologies like Outlast, previously mentioned in the context of insulating finishes, can also be used for cooling by storing heat and releasing it when necessary. Another method for achieving a cooling effect is using fabrics impregnated with minerals [23]. Ceramic particles or tourmaline can reflect infrared rays while creating a cooling sensation. These minerals absorb body heat and radiate it outward. This type of textile is used in summer clothing and sports gear.

5.8.2 Interaction with Other Functional Finishes

Both heat-retention and cooling finishes must be integrated with other functional coatings to provide comprehensive protection and comfort without compromising the effectiveness of the garments. It is important to balance these finishes with other functions such as waterproofing, wind resistance, UV protection, and breathability to achieve optimal results. Clothing for hot weather, for example, often combines waterproof and cooling finishes to provide protection in adverse weather conditions while keeping the body cool. However, waterproof materials must be sufficiently breathable to prevent excessive sweating. In contrast, clothing for cold weather must balance insulation and breathability. Too much insulation can lead to overheating and sweating, while insufficient insulation may result in feeling cold. To avoid this imbalance, many modern garments combine insulating finishes with breathable materials that allow moisture to evaporate while retaining heat.

5.9 Future Trends

In the future, outdoor textiles will develop with a focus on comfort, as new technologies and sustainable practices will play an important role in this process. Consumers are increasingly looking for materials that offer both protection and convenience. The popularity of eco-based fibers, such as recycled polyester and organic cotton, is growing, which provide comfort to people while ensuring greater care for the environment. The integration of smart textiles and protective finishes will become more widespread. For example, sensors will monitor temperature and physiological parameters while garments change their breathability. Such technologies would improve comfort while providing additional safety during outdoor activities. The advancement of nanotechnology will provide new solutions for moisture management and breathability, which is key to maintaining optimal comfort in variable climatic conditions. The role of UV protective coatings and coatings that can provide healthy operating conditions for garments will also become increasingly significant. Along with technological advancements, ways to improve the accessibility of clothing for outdoor activities must be found, as some of the new functionalities are at unreachably high prices. Besides technical innovations, design trends will focus on versatility and aesthetics, allowing outdoor textiles to fit both extreme conditions and the everyday life of consumers. Ultimately, the future of outdoor textiles will be defined by the pursuit of greater comfort, sustainability, and innovations that will enrich the experiences of outdoor users.

References

1. 3M (n.d.) *Thinsulate™ insulation*. Retrieved October 16, 2024, from https://www.3m.com/3M/en_US/thinsulate-us/
2. Aakko, M., & Niinimäki, K. (2022). Quality matters: Reviewing the connections between perceived quality and clothing use time. *Journal of Fashion Marketing and Management: an International Journal, 26*(1), 107–125.
3. Ahmad, D., Van Den Boogaert, I., Miller, J., Presswell, R., & Jouhara, H. (2018). Hydrophilic and hydrophobic materials and their applications. *Energy Sources, Part a: Recovery, Utilization, and Environmental Effects, 40*(22), 2686–2725.
4. Ahmad, S., Miskon, S., Alabdan, R., & Tlili, I. (2020). Towards sustainable textile and apparel industry: Exploring the role of business intelligence systems in the era of industry 4.0. *Sustainability, 12*(7), 2632.
5. Albdour, S. A., Haddad, Z., Sharaf, O. Z., Alazzam, A., & Abu-Nada, E. (2022). Micro/nano-encapsulated phase-change materials (ePCMs) for solar photothermal absorption and storage: Fundamentals, recent advances, and future directions. *Progress in Energy and Combustion Science, 93*, Article 101037.
6. Amoroso Leslie, C., Lau, Y. Y., Suen, D.W.-S., & Tsang, C.-W. (2024). Revolutionizing the textile and clothing industry: Pioneering sustainability and resilience in a post-COVID era. *Sustainability, 16*(6), 2474.
7. Anaya-Esparza, L. M., Villagrán-de la Mora, Z., Ruvalcaba-Gómez, J. M., Romero-Toledo, R., Sandoval-Contreras, T., Aguilera-Aguirre, S., & Pérez-Larios, A. (2020). Use of titanium dioxide (TiO_2) nanoparticles as reinforcement agent of polysaccharide-based materials. *Processes, 8*(11), 1395.
8. Angelova, R. A. (2015). *Textiles and human thermophysiological comfort in the indoor environment*. CRC Press.
9. Begum, M. S., & Milašius, R. (2022). Factors of weave estimation and the effect of weave structure on fabric properties: A review. *Fibers, 10*(9), 74.
10. Beşen, B. S. (2019). Production of disposable antibacterial textiles via application of tea tree oil encapsulated into different wall materials. *Fibers and Polymers, 20*, 2587–2593.
11. Broadhead, R., Craeye, L., & Callewaert, C. (2022). Odour control or inhibition using antimicrobial finishing. *Odour in textiles* (pp. 95–116). CRC Press.
12. Celik, Y. C., Pulletikurthi, G., & Endres, F. (2016). Electrodeposition of Al, Zn, and Pt on silver-coated textile fibres from ionic liquids. *Journal of Solid State Electrochemistry, 20*, 2781–2790.
13. Chang, Y., & Liu, F. (2023). Review of waterproof breathable membranes: Preparation, performance and applications in the textile field. *Materials, 16*(15), 5339.
14. Chinta, S. K., & Gujar, P. D. (2013). Significance of moisture management for high performance textile fabrics. *International Journal of Innovative Research in Science, Engineering & Technology, 2*(3), 814–819.
15. Choi, M., Kim, Y., Park, S., Ka, D., Kim, T., Lee, S., & Hong, J. (2021). Functionalized polyurethane-coated fabric with high breathability, durability, reusability, and protection, 31(24), 2101511.
16. Cloud, R. M., Cao, W., & Song, G. (2013). Functional finishes to improve the comfort and protection of apparel. *Advances in the dyeing and finishing of technical textiles* (pp. 258–279). Woodhead Publishing.
17. Coolibar. (n.d.). *Fabric innovation*. Retrieved October 16, 2024, from https://www.coolibar.com/pages/fabric-innovation
18. CORDURA. (n.d.). *Find your match*. Retrieved October 16, 2024, from https://cordura.com/find-your-match
19. Cox, S. D., Mann, C. M., Markham, J. L., Bell, H. C., Gustafson, J. E., Warmington, J. R., & Wyllie, S. G. (2000). The mode of antimicrobial action of the essential oil of Melaleuca alternifolia (tea tree oil). *Journal of Applied Microbiology, 88*(1), 170–175.

20. Čuden, A. P., & Urbas, R. (2023). Advances in ultraviolet (UV) ray blocking textiles. *Functional and technical textiles* (pp. 213–273). Woodhead Publishing.
21. Cui, J., Shao, Y., Zhang, H., Zhang, H., & Zhu, J. (2021). Development of a novel silver ions-nanosilver complementary composite as antimicrobial additive for powder coating. *Chemical Engineering Journal, 420*, Article 127633.
22. Das, S., & Natarajan, S. (2022). Antibacterial activity of chitosan in textiles: A review. *Man-Made Textiles in India, 50*(3), 96–100.
23. De Castro Silva, R. M., Zhao, J., de Souza, L. O., Liebscher, M., Curosu, I., Mechtcherine, V., & de Andrade, S. F. (2024). Mechanical behavior of textile reinforced concrete with newly developed mineral-impregnated carbon fabrics submitted to elevated temperatures. *Cement and Concrete Composites, 148*, Article 105467.
24. El-Ola, S. A. (2008). Recent developments in finishing of synthetic fibers for medical applications. *Designed Monomers and Polymers, 11*(6), 483–533.
25. Enescu, D. (2008). Use of chitosan in surface modification of textile materials. *Roumanian Biotechnological Letters, 13*(6), 4037–4048.
26. Event Fabrics. (n.d.) *Outerwear.* Retrieved October 16, 2024, from https://www.eventfabrics.com/products/outerwear/
27. Fayed, A. I. H., El Amaim, Y. A., & Elgohary, D. H. (2023). Enhancing the performance of cordura and ballistic nylon using polyurethane treatment for outer shell of bulletproof vest. *Journal of King Saud University—Engineering Sciences, 35*(4), 239–246.
28. Fleetwood, S. K., Bell, S., Jetter, R., & Foster, E. J. (2023). Plant-based, aqueous, water-repellent sprays for coating textiles. *Soft Matter, 19*(36), 7020–7032.
29. Gao, H., Deaton, A. S., & Barker, R. (2024). Impact of durable water-repellent finishing in thermal liner on firefighter heat stress. *AATCC Journal of Research*, 11(4), 273–281.
30. Gore-Tex. (n.d.). *Reduce chemical impacts.* Retrieved October 16, 2024, from https://www.gore-tex.com/sustainability/protect-the-planet/reduce-chemical-impacts#:~:text=Gore%20Fabrics%20is%20working%20towards,is%20the%20end%20of%202023
31. Gore-Tex. (n.d.). *GORE-TEX® ACTIVE technology.* Retrieved October 16, 2024, from https://www.gore-tex.com/technology/gore-tex-products/active
32. Gore-Tex. (n.d.). *GORE-TEX® PACLITE® technology.* Retrieved October 16, 2024, from https://www.gore-tex.com/technology/gore-tex-products/paclite
33. Gore-Tex. (n.d.). *What exactly is the GORE-TEX® INFINIUM™ product range*? Retrieved October 16, 2024, from https://www.gore-tex.com/en_uk/blog/what-exactly-is-the-gore-tex-infinium-product-range
34. Gore-Tex. (n.d.). *GORE-TEX® PRO technology.* Retrieved October 16, 2024, from https://www.gore-tex.com/technology/gore-tex-products/pro
35. Gorji, M., Mazinani, S., & Gharehaghaji, A. A. (2022). A review on emerging developments in thermal and moisture management by membrane-based clothing systems towards personal comfort. *Journal of Applied Polymer Science, 139*(27), Article e52416.
36. Gupta, B. S. (Ed.). (2008). *Friction in textile materials.* Elsevier.
37. Hatch, K. L. (1993). *Textile science.* West Publishing Co.
38. Havenith, G., Holmér, I., & Parsons, K. (2002). Personal factors in thermal comfort assessment: Clothing properties and metabolic heat production. *Energy and Buildings, 34*(6), 581–591.
39. Holmquist, H., Schellenberger, S., van Der Veen, I., Peters, G. M., Leonards, P. E. G., & Cousins, I. T. (2016). Properties, performance and associated hazards of state-of-the-art durable water repellent (DWR) chemistry for textile finishing. *Environment International, 91*, 251–264.
40. Hossen, S., Mridha, Y., Rahman, A., Ouboucetta, R., & Amin, M. R. (2024). Consumer perceptions and purchasing trends of eco-friendly textile products in the US market. *International Journal of Business and Economics, 1*(2), 20–32.
41. Hu, R., Liu, Y., Shin, S., Huang, S., Ren, X., Shu, W., & Luo, X. (2020). Emerging materials and strategies for personal thermal management. *Advanced Energy Materials, 10*(17), 1903921.

42. Huang, R., Zhang, S., Zhang, W., & Yang, X. (2021). Progress of zinc oxide-based nanocomposites in the textile industry. *IET Collaborative Intelligent Manufacturing, 3*(3), 281–289.
43. Hussain, S., Glombikova, V., Havelka, A., Jamshaid, H., Batool, S. S., & Khan, M. Z. (2017). Moisture transport phenomena of functional underwears. *Transport, 26*, 27.
44. Júnior, H. L. O., Neves, R. M., Monticeli, F. M., & Dall Agnol, L. (2022). Smart fabric textiles: Recent advances and challenges. *Textiles, 2*(4), 582–605.
45. Kale, R. D., & Meena, C. R. (2012). Synthesis of titanium dioxide nanoparticles and application on nylon fabric using layer by layer technique for antimicrobial property. *Advances in Applied Science Research, 3*(5), 3073–3080.
46. Kamalha, E., Zeng, Y., Mwasiagi, J. I., & Kyatuheire, S. (2013). The comfort dimension; A review of perception in clothing. *Journal of Sensory Studies, 28*(6), 423–444.
47. Kandhavadivu, P., Rathinamoorthy, R., & Surjit, R. (2015). Moisture and thermal management properties of woven and knitted tri-layer fabrics. *Indian Journal of Fibre & Textile Research (IJFTR), 40*(3), 243–249.
48. Kelly, F. M., & Johnston, J. H. (2011). Colored and functional silver nanoparticle–wool fiber composites. *ACS Applied Materials & Interfaces, 3*(4), 1083–1092.
49. Kertmen, N. (2021). New trends in fibers used in denim fabric production. *Tekstil Ve Mühendis, 28*(121), 48–59.
50. Kirtay, E. (2007). A study of the thermal properties of textured knitted fabrics. *Fibres & Textiles in Eastern Europe, 15*(1), 60.
51. Kodžoman, D. (2019). The psychology of clothing: Meaning of colors, body image and gender expression in fashion. *Textiles & Leather Review, 2*(2), 90–103.
52. Kothari, V. K. (2008). Polyester and polyamide fibres–apparel applications. *Polyesters and polyamides* (pp. 419–440). Woodhead Publishing.
53. Kuthi, N. A., Basar, N., & Chandren, S. (2022). Nanonutrition- and nanoparticle-based ultraviolet rays protection of skin. *Advances in nanotechnology-based drug delivery systems* (pp. 227–280). Elsevier.
54. Lakshmanan, A., & Chakraborty, S. (2017). Coating of silver nanoparticles on jute fibre by in situ synthesis. *Cellulose, 24*(3), 1563–1577.
55. Lansdown, A. B. (2006). Silver in health care: Antimicrobial effects and safety in use. *Biofunctional Textiles and the Skin, 33*, 17–34.
56. Lee, S. (2009). Developing UV-protective textiles based on electrospun zinc oxide nanocomposite fibers. *Fibers and Polymers, 10*, 295–301.
57. Li, X., Wang, X. X., Yue, T. T., Xu, Y., Zhao, M. L., Yu, M., & Long, Y. Z. (2019). Waterproof-breathable PTFE nano-and microfiber membrane as high efficiency PM2.5 filter. *Polymers, 11*(4), 590.
58. Loghin, C., Ciobanu, L., Ionesi, D., Loghin, E., & Cristian, I. (2018). Introduction to waterproof and water repellent textiles. *Waterproof and water repellent textiles and clothing* (pp. 3–24). Woodhead Publishing.
59. Lomax, G. R. (2007). Breathable polyurethane membranes for textile and related industries. *Journal of Materials Chemistry, 17*(27), 2775–2784.
60. Louris, E., Sfiroera, E., Priniotakis, G., Makris, R., Siemos, H., Efthymiou, C., & Assimakopoulos, M. N. (2018). Evaluating the ultraviolet protection factor (UPF) of various knit fabric structures. In *IOP Conference Series: Materials Science and Engineering* (Vol. 459, p. 012051). IOP Publishing.
61. Ludaš Dujmić, A., Radičić, R., Ercegović Ražić, S., Cingesar, I. K., Glogar, M., Jurov, A., & Krstulović, N. (2024). Characterization of melt-spun recycled PA 6 polymer by adding ZnO nanoparticles during the extrusion process. *Polymers, 16*(13), 1883.
62. LYCRA. (n.d.). *COOLMAX® ECOMADE technology*. Retrieved October 16, 2024, from https://www.lycra.com/en/business/search-technologies/coolmaxr-ecomade-technology
63. Mahltig, B. (2015). Hydrophobic and oleophobic finishes for textiles. *Functional finishes for textiles: Improving comfort, performance and protection* (pp. 387–428). Woodhead Publishing.

64. Maity, S., Chauhan, V., & Pandit, P. (2022). Waterproof breathable fabrics and suits. *Protective textiles from natural resources* (pp. 347–375). Woodhead Publishing.
65. Maity, S., Singha, K., & Pandit, P. (2023). Introduction to functional and technical textiles. *Functional and technical textiles* (pp. 1–30). Woodhead Publishing.
66. Majumdar, A., & Arora, S. (1997). *Bamboo fibres in textile applications* (pp. 285–304). ENVIS Centre on Forestry, National Forest Research Institute.
67. McCann, J. (2008). Sports applications. *Polyesters and polyamides* (pp. 505–524). Woodhead Publishing.
68. McCann, J. (2023). Identification of design requirements for smart clothes and wearable technology. *Smart clothes and wearable technology* (pp. 327–369). Woodhead Publishing.
69. Menezes, E., & Choudhari, M. (2007). Special finishes and effects. *AATCC Review, 7*(3).
70. Millington, K. R. (2000). Comparison of the effects of gamma and ultraviolet radiation on wool keratin. *Coloration Technology, 116*(9), 266–272.
71. Millington, K. R., & Rippon, J. A. (2016). Wool. In *Structure and properties of high-performance fibers* (p. 367). Elsevier.
72. Mondal, S. (2022). Nanomaterials for UV protective textiles. *Journal of Industrial Textiles, 51*(4_suppl), 5592S–5621S.
73. Morganroth, P. A., Lim, H. W., & Burnett, C. T. (2013). Ultraviolet radiation and the skin: An in-depth review. *American Journal of Lifestyle Medicine, 7*(3), 168–181.
74. Nike. (n.d.). *Dri-FIT clothing*. Retrieved October 16, 2024, from https://www.nike.com/bg/w/dri-fit-clothing-32dxrz6ymx6
75. Outlast. (n.d.). *How it works—Temperature regulating fabric*. Retrieved October 16, 2024, from https://www.outlast.com/en/temperature-regulating-fabric/how-it-works
76. Özek, H. Z. (2018). Silicone-based water repellents. *Waterproof and water repellent textiles and clothing* (pp. 153–189). Woodhead Publishing.
77. Patra, K. (2013). Application of nanotechnology in textile engineering: An overview. *Journal of Engineering and Technology Research, 5*(5), 104–111.
78. Paul, R. (Ed.). (2015). *Functional finishes for textiles: Improving comfort, performance, and protection*. Woodhead Publishing.
79. Perera, S., Bhushan, B., Bandara, R., Rajapakse, G., Rajapakse, S., & Bandara, C. (2013). Morphological, antimicrobial, durability, and physical properties of untreated and treated textiles using silver nanoparticles. *Colloids and Surfaces a: Physicochemical and Engineering Aspects, 436*, 975–989.
80. Qu, M., He, J., & Zhang, J. (2010). Superhydrophobicity, learn from the lotus leaf. In *Biomimetics learning from nature*. IntechOpen.
81. Rahaman, M. T., & Khan, M. S. H. (2024). Applications of green nano textile materials for environmental sustainability and functional performance: Past, present and future perspectives. *Nano-Struct Nano-Objects, 40*, Article 101332.
82. Rajendra, R., Balakumar, C., Ahammed, H. A. M., Jayakumar, S., Vaideki, K., & Rajesh, E. (2010). Use of zinc oxide nanoparticles for production of antimicrobial textiles. *International Journal of Engineering, Science and Technology, 2*(1), 202–208.
83. Ray, A., Singha, K., Pandit, P., & Maity, S. (2020). Advanced ultraviolet protective agents for textiles and clothing. *Advances in functional and protective textiles* (pp. 243–260). Woodhead Publishing.
84. Rezk, E. H. A. E., Elwan, N. A. A., Elkelany, M. M. I. M., & Khalaf, A. M. I. (2023). Design and heat setting of 3D-weft jacquard knitted effect on some properties related to thermal insulation. *Egyptian Journal of Chemistry, 66*(13), 1247–1258.
85. Rodrigues, A. G., de Oliveira, R., Gonçalves, P. J., Ottoni, C. A., de Cássia, R. R., Morgano, M. A., de Araújo, W. L., & De Souza, A. O. (2019). Functional textiles impregnated with biogenic silver nanoparticles from Bionectria ochroleuca and its antimicrobial activity. *Biomedical Microdevices, 21*, 1–10.
86. Rudolf. (n.d.). *Bionic Finish Eco*. Retrieved October 16, 2024, from https://rudolf.com/technologies/bionic-finish-eco

87. Sayed, U., & Dabhi, P. (2014). Finishing of textiles with fluorocarbons. *Waterproof and water repellent textiles and clothing* (pp. 139–152). Woodhead Publishing.
88. Schellenberger, S., Gillgard, P., Stare, A., Hanning, A., Levenstam, O., Roos, S., & Cousins, I. T. (2018). Facing the rain after the phase out: Performance evaluation of alternative fluorinated and non-fluorinated durable water repellents for outdoor fabrics. *Chemosphere, 193*, 675–684.
89. Schneider, S. L., & Lim, H. W. (2019). A review of inorganic UV filters zinc oxide and titanium dioxide. *Photodermatology, Photoimmunology and Photomedicine, 35*(6), 442–446.
90. Schoeller Textiles. (n.d.). *Ecorepel*. Retrieved October 16, 2024, from https://www.schoeller-textiles.com/en/technologies/ecorepel
91. Sewport. (n.d.). *Coolmax fabric*. Retrieved October 16, 2024, from https://sewport.com/fabrics-directory/coolmax-fabric
92. Shah, M. A., Pirzada, B. M., Price, G., Shibiru, A. L., & Qurashi, A. (2022). Applications of nanotechnology in smart textile industry: A critical review. *Journal of Advanced Research, 38*, 55–75.
93. Sharif, R., Mohsin, M., Ramzan, N., Ahmad, S. W., & Qutab, H. G. (2022). Synthesis and application of fluorine-free environment-friendly stearic acid-based oil and water repellent for cotton fabric. *Journal of Natural Fibers, 19*(5), 1632–1647.
94. Sikdar, P., Dip, T. M., Dhar, A. K., Bhattacharjee, M., Hoque, M. S., & Ali, S. B. (2022). Polyurethane (PU) based multifunctional materials: Emerging paradigm for functional textiles, smart, and biomedical applications. *Journal of Applied Polymer Science, 139*(38), Article e52832.
95. Silver, S. (1996). Bacterial resistances to toxic metal ions: A review. *Gene, 179*(1), 9–19.
96. Singha, K., Maity, S., & Pandit, P. (2020). UV protection via nanomaterials. *Frontiers of textile materials: Polymers, nanomaterials, enzymes, and advanced modification techniques* (pp. 153–166). Woodhead Publishing.
97. Slater, K. (1986). Discussion paper: The assessment of comfort. *Journal of the Textile Institute, 77*(3), 157–171.
98. Sudrajat, H. (2018). Superior photocatalytic activity of polyester fabrics coated with zinc oxide from waste hot dipping zinc. *Journal of Cleaner Production, 172*, 1722–1729.
99. Sun, G. (Ed.). (2016). *Antimicrobial textiles*. Woodhead Publishing.
100. Sympatex. (n.d.). *Membran*. Retrieved October 16, 2024, from https://www.sympatex.com/en/membran/
101. Takahashi, T., Imai, M., Suzuki, I., & Sawai, J. (2008). Growth inhibitory effect on bacteria of chitosan membranes regulated with deacetylation degree. *Biochemical Engineering Journal, 40*(3), 485–491.
102. Teflon. (n.d.). *How EcoElite works*. Retrieved October 16, 2024, from https://www.teflon.com/en/industries-and-solutions/solutions/consumer-applications/textile-finishes/how-ecoelite-works
103. Tehrani-Bagha, A. R. (2019). Waterproof breathable layers: A review. *Advances in Colloid and Interface Science, 268*, 114–135.
104. The North Face. (n.d.). *FUTURELIGHT™ technology*. Retrieved October 16, 2024, from https://www.thenorthface.com/en-us/about-us/technology-innovation/technology/futurelight
105. Tsuzuki, T., & Wang, X. (2010). Nanoparticle coatings for UV protective textiles. *Research Journal of Textile and Apparel, 14*(2), 9–20.
106. Under Armour. (n.d.). *HeatGear technology*. Retrieved October 16, 2024, from https://www.underarmour.com/en-us/c/technology/heatgear/
107. Verbič, A., Gorjanc, M., & Simončič, B. (2019). Zinc oxide for functional textile coatings: Recent advances. *Coatings, 9*(9), 550.
108. Vilamová, Z., Konvičková, Z., Mikeš, P., Holišová, V., Mančík, P., Dobročka, E., & Seidlerová, J. (2019). Ag-AgCl nanoparticles fixation on electrospun PVA fibres: Technological concept and progress. *Science and Reports, 9*(1), 15520.
109. Vingilyte, J., & Khadaroo, A. (2022). Personal clothing style and self-concept: Embracing the true, the ideal and the creative self. *Fashion, Style & Popular Culture*.

110. Voegeli, D. (2019). Prevention and management of moisture-associated skin damage. *Nursing Standard, 34*(2).
111. Whittaker, M. H., & Heine, L. (2018). Toxicological and environmental issues associated with waterproofing and water repellent formulations. *Waterproof and water repellent textiles and clothing* (pp. 89–120). Woodhead Publishing.
112. Williams, J. T. (Ed.). (2017). *Waterproof and water repellent textiles and clothing*. Woodhead Publishing.
113. Yadav, A., Prasad, V., Kathe, A. A., Raj, S., Yadav, D., Sundaramoorthy, C., & Vigneshwaran, N. (2006). Functional finishing in cotton fabrics using zinc oxide nanoparticles. *Bulletin of Materials Science, 29*, 641–645.

Chapter 6
Advances and Challenges in Application of Polymeric Materials for Textile Finishing

Md. Himel Mahmud, Md. Reazuddin Repon, Daiva Mikučionienė, Md. Tanvir Raihan, Syed Ziadul Huda, and Abdullah Al Fariz

Abstract Polymeric materials play a crucial role in textile finishing, offering multiple functionalities such as mechanical resistance, durability, and comfort. The chapter focuses on natural and synthetic polymers, including chitosan, lignin, pectin, gelatin, alginate, polyurethane, polyvinyl cholride, polyacrylics and silicones and fabrication techniques like coating, exhaust method, pad-dry-cure, padding, drying and curing. It highlights how these polymers greatly improve the practical and aesthetic qualities of textiles. In-depth, the inherent qualities of these polymeric compounds, such as their durability, elasticity, and adaptability, contribute to the development of textiles with outstanding functional attributes. The chapter highlights various applications of polymeric finishing for flame retardant, drug delivery system, wound healing, antimicrobial, water repellent, hydrophobic and antistatic properties. Furthermore, the discussion focuses on relevant difficulties linked to the incorporation of polymeric materials, such as compatibility problems and environmental factors, while suggesting approaches to overcome these barriers. This chapter provides a detailed assessment of the progress and difficulties in using polymeric materials for textile finishing.

Md. H. Mahmud · A. Al Fariz
Department of Textile Engineering, Northern University Bangladesh, Dhaka, Bangladesh
e-mail: himel.mahmud@nub.ac.bd

Md. R. Repon (✉)
Department of Bioproducts and Biosystems, School of Chemical Engineering, Aalto University, Vuorimiehentie 1, 02150, Espoo, Finland

D. Mikučionienė
Department of Production Engineering, Faculty of Mechanical Engineering and Design, Kaunas University of Technology, Kaunas, Lithuania

Md. T. Raihan
Department of Textile Engineering, Ahsanullah University of Science and Technology, Dhaka, Bangladesh

S. Z. Huda
School of Chemical Engineering, University of Adelaide, Adelaide, Australia

M. Shahid et al. (eds.), *Advancements in Textile Finishing*,
https://doi.org/10.1007/978-981-96-6385-9_6

Keywords Polymeric materials · Textile finishing · Nanocomposite · Electrospraying

6.1 Introduction

Textile finishing encompasses post-loom or post-knitting processes to enhance fabric appearance and functionality. It includes chemical or mechanical treatments, altering properties like texture and aesthetics. Wet or dry methods are employed, with finishes applied via machines or impregnation. Finishing aims to enhance fabric appeal and functionality, tailored to fiber type, physical properties, chemical absorbency, and desired end-use properties [1]. Polymer materials in textile finishing stands as a prominent method for altering textile surfaces, aimed at enhancing comfort and modifying appearance and physical traits. This involves the application of both natural and synthetic polymers onto diverse textile substrates for functional applications such as flame retardancy, drug delivery systems, wound healing, antimicrobial, water repellency, and hydrophobic, and antistatic properties functionalities.

Within this book chapter, a comprehensive exploration of polymers is undertaken, coupled with a consideration of the techniques employed for polymer coating. These range from traditional methods like exhaustion, pad-dry-cure, and coating to innovative approaches. Furthermore, the chapter deals with different methods of functionalization achieved through different polymers, offering insights into the diverse applications and advancements in this field [2].

6.2 Types of Polymeric Materials

Various polymeric materials used in textile finishing are shown in Table 6.1.

Table 6.1 Classification of polymers used in textile finishing

Polymeric material	Synthetic	Polyurethane (PU)
		Polyvinyl Chloride (PVC)
		Polyacrylics (PA)
		Silicones
	Natural	Chitosan
		Starch-based polymers
		Lignin
		Pectin
		Gelatin
		Alginate

6.2.1 Synthetic Polymers

In terms of biodegradability and environmental friendliness, natural polymers have been proven one of the best choices in the world of textile finishing for a long time. However, in the face of rapidly changing market and consumer demands, natural polymers have not been the best options as they offer a limited possibility of structural changes, which leads to some enhanced and desired property enhancements like durability, elasticity, etc. This encouraged the evolution of "man-made" or "synthetic polymers", which offered a remarkable degree of structural control leading to an endless variety of enhanced textile properties. Synthetic polymers are also high molecular weight compounds that are synthesized through the process called the "polymerization of monomers", engineered from different petrochemical sources. Synthetic polymers can be classified into 4 types: Polyurethane, Polyvinyl chloride, Poloacrylics, and Silicones [3]. Using the scope of structural reconfiguration of monomers, synthetic polymers not only offer property enhancement of the textile materials, but they signify improvements by imparting crucial characteristics such as antimicrobial activity, UV protection, and flame retardancy (Table 6.2). Polyurethane, Polyvinyl Chloride, Polyacrylic, etc., are some of the most used synthetic polymers in the textile industries for finishing purposes.

6.2.1.1 Polyurethane (PU)

In the world of textile production, polyurethane is one of the vastly used polymers and it even found its way into the finishing section because of its remarkable elasticity, excellent durability, and high abrasion resistance. The systemization of this polymer is done through a reaction called polyaddition that happens between di or polyisocyanates and polyols. The reaction between these two chemicals lead to the formation of long-chain molecules that contain the urethane linkages [4–6].

In the finishing section, coating technique is mostly used for the polyurethane application of textile materials to enhance fabric performance aspects. These coatings of polyurethane can impart water resistance, breathability, and durability on the apparel.

6.2.1.2 Polyvinyl Cholride (PVC)

PVC is a resin-type polymer that is insoluble in water making it one of the most used materials in plastics and pipes to protect them. It has found its way into the textile finishing section due to its exceptional characteristics like durability, resistance to chemical and water repellency. The backbone of PVC is created from the polymerization process of vinyl chloride monomers which are gas in normal temperatures.

Table 6.2 Synthetic polymers employed in textile finishing

Polymers	Characteristics	Application	Refs
Polyurethane (PU)	Remarkable elasticity Excellent durability Resistant to weather and abrasion Available in different forms—solvent, latex, etc	Used in finishing on textile materials like – Waterproof Clothes Life jackets	[4–6]
Polyvinyl Cholride (PVC)	Features resin-type properties Insoluble in water Resistance to chemicals and fire Water repellent	Applied on finishing process for materials such as, Tent covers Tarpaulins Leather	[7–9]
Polyacrylics (PA)	High tensile strength Great thermal stability ie. Fire retardant Ensures longevity	Used in textile finishing for materials such as, Upholstery Carpet	[4, 8, 10]
Silicones	Wide range of application methods Imparts softness on textile materials Increases the aesthetic properties of the material Reduces static cling	In the finishing section, is applied on – Textiles made out of natural, synthetic materials as well as blends	[7, 9, 11, 12]

PVC provides a lot of scope for modifications where things like plasticizers, stabilizers as well as other additives can be added which can lead to enhancement of its flexibility, stability and performance [4, 8, 9].

The application of PVC or its modified forms is also done mostly through the coating technique which is done in the finishing part of the production. Due to its water-repellent nature, it is a vastly used to waterproof a textile material [8]. Recently it has been gathered attention for imparting flame retardant characteristics on textiles [7].

6.2.1.3 Polyacrylics (PA)

Polyacrylics are one of the important members of the synthetic polymer group which is widely used in textile finishing. One of the most used polymers of this group in textile industries is Polyacrylonitrile (PAN). The acrylics polymers are known for their high tensile strength that helps them to ensure longevity and they show thermal stability. The production process of polyacrylics involves the polymerization of acrylic and methacrylic acids as well as their amides, nitriles and salts [4, 10].

Polyacrylics are a popular choice in the world of textiles due to their exceptional affinity to dye particles. This characteristic of polyacrylics leads to materials retaining their colour for a long time even under harsh conditions. This is a big reason the upholstery materials are coated with PA polymers which confirms a long use under the sunlight [4, 8].

6.2.1.4 Silicones

Silicones, a unique part of the synthetic polymer group have become one of the key parts of the textile finishing recently due to having exceptionally useful properties that improve the functionality as well as make it more visually appealing. Furthermore, they are linked to imparting softness on the apparel after it is applied. Fundamentally the core of the silicone structure is composed of silicon which is bonded with oxygen as well as holding on to some organic side group [11, 12]. Due to their softening ability, silicones are currently used in textile finishing as one of the top softener options. It can be applied to textile materials in various ways and proper application leads to the material exhibiting luxurious softness which enhances the material's comfortability as well as the drape properties [11, 12].

6.2.2 Natural Polymers

Balancing potent biocidal properties with safe handling requirements, such as human non-toxicity at standard concentrations and environmental friendliness, is a frequent challenge associated with synthetic polymers. Consumer and organizational awareness of toxicological and environmental concerns, such as skin irritation, sensitization, allergy risks, biodegradability, and bioaccumulation, has grown. Regulations are being introduced to restrict the handling, disposal, and trade of hazardous substances. These factors underscore the significance of utilizing natural polymers [13]. Natural polymers are large molecular-weight compounds derived from plants, microbes, and animals [14]. Chitosan, alginate, guar gum, starch, pectin, gelatin, etc. (Fig. 6.1) are some examples of natural polymers that can be utilized in textile processing and finishing (Table 6.3).

6.2.2.1 Chitosan

Chitosan is a modified biopolymer with unique attributes such as nontoxicity, biocompatibility, biodegradability, ecological safety, and a multiple biological effects, comprising low immunogenicity, anticancer activity, antibacterial activity, and antifungal activity. Chitosan can dissolve in organic and mineral acids, unlike chitin which remains insoluble in most solvents because of low chemical reactivity

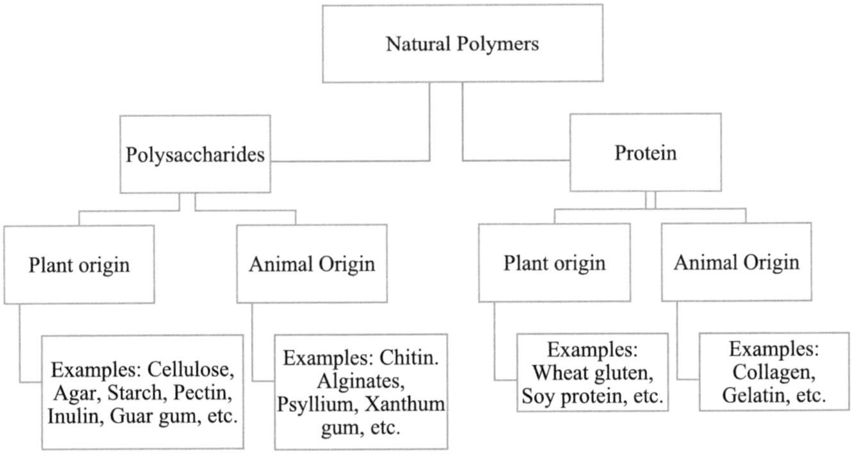

Fig. 6.1 Classification of natural polymers [15]

and solubility. This makes chitosan advantageous for further processing. Microorganisms, terrestrial organisms, and aquatic organisms are the three main sources of chitin and production of chitosan. However, some marine species, such as shrimp, lobsters, and crab shells, are the primary bases of chitin and chitosan. Manufacturing synthetic fibers, textile coloration, long-lasting finishing techniques such as antimicrobial, antistatic, and deodorizing, sizing and printing are some of the textile processes that utilize chitosan [13, 16, 17].

Biological properties of chitosan:

- Biodegradable
- Biocompatible
- Have antimicrobial activity
- Nontoxic
- Hemostatic, fungistatic, spermicidal, antitumor
- Used in wound healing

Chemical properties of chitosan:

- Linear polyamine.
- Has free amino groups which helps it to obtain a more hydrophobic functionalization, can bear anionic charges and can be used in cross-linking.
- Chitosan is soluble in acidic solutions due to the protonation of the reactive amino groups.
- Reactive hydroxyl groups available which allow for various chemical modifications, such as acetylation, alkylation, and crosslinking, to improve its properties and functionality.
- Chelates many transitional metal ions.

Table 6.3 Characteristics and applications of natural polymers

Natural Polymers	Characteristics	Application
Chitosan	• Biodegradable, biocompatible, and non-toxic • Positively charged, enabling interactions with negatives • Antimicrobial properties against bacteria, fungi, viruses • Excellent film-forming and adsorption capabilities	Used in finishing like – • Antibacterial and antifungal • Protective clothing (fire-resistant, chemical-resistant, etc.)
Starch	• White, odorless, and tasteless powder • Insoluble in cold water, gelatinizes when heated with water • Semi-crystalline structure provides resistance to shear and stress • Exhibits some resistance to mild acid and alkali conditions	Widely used as – • Thickener • Binder • Sizing agent • Coating (water-repellency, flame retardancy, or enhanced durability.)
Lignin	• Complex, aromatic, and highly cross-linked polymer • Relatively high thermal resistance • Relatively resistant to biodegradation • Sustainable, thermal, and mechanically resistant fibers	Used in finishing like – • Water repellency • UV protection • Antimicrobial
Pectin	• Exhibits high viscosity in aqueous solutions • Forms gels based on pH, sugar, and calcium • Soluble in water, forming viscous solutions • Relatively stable at high temperatures	• Thickening agent in printing pastes • Coating or finishing agent for properties like water resistance, stiffness • Sizing agent, binder, emulsifier, dispersing agent
Gelatin	• Exhibits varying degrees of viscosity • Soluble in warm water, insoluble in cold water • Biocompatible, and biodegradable	• Enhance dyeability and pretreatment processes like bleaching • Used as antibacterial, antioxidant, and antifungal finishing agent
Sodium Alginate	• Imparts properties like moisture management, water repellency, fire retardancy • Exhibits high viscosity in aqueous solutions • Readily soluble in water, forming viscous solutions • Biocompatible, and biodegradable Ability to undergo ion exchange reactions	• Coating • Thickening agent in printing pastes

- Good thermal stability with a decomposition temperature typically ranging from 200 °C to 400 °C [16, 17].

Due to its intriguing qualities, chitosan has been extensively studied for the many years for different textile applications. In textile finishing, chitosan is primarily used for antimicrobial, anti-odor, blood coagulant, blood anticoagulant, antistatic, wrinkle-resistant, and UV protection finishes, all of which have potential applications in various domains [16].

6.2.2.2 Starch-Based Polymers

Starch-based polymers are biodegradable polymers and derived from starch. Starch is a carbohydrate polymer primarily made up of gliucose units joined together by glucosidic bonds. It consists of amylopectin and amylose [14]. Due to renewability, vast availability, and biodegradability, starch stands out as a highly promising natural polymer. Starch is the second most popular plant-based product after chitosan. Starches are an attractive choice as a cost-effective foundation for developing new biodegradable polymers, given their affordability and compatibility with standard plastic manufacturing equipment. By modifying starch molecules through methods such as blending, chemical modification, or extrusion, starch-based polymers are formed like starch blends, starch composites, starch-based nanocomposites, starch foam, starch films and coatings, and starch adhesives. These modifications improve the mechanical properties, processability, and biodegradability of starch, rendering it suitable for various applications [18].

In textile processing, wheat starch is a binder when printing cotton and cotton/polyester fabrics [19]. Corn starch is commonly used as a sizing agent to improve the handling and weaving properties of the fabric by reducing thread breakage during weaving and enhancing the stiffness and stability of the fabric [20]. Potato starch is utilized as a coating or adhesive for sizing yarn. Coatings based on potato starch can enhance smoothness, increase adhesive potential, improve hydrophilicity and solubility, enhance film-forming ability, and prevent retrogradation phenomena [21].

6.2.2.3 Lignin

Lignin is the highest biopolymer after cellulose and it contains 30% organic carbon. Lignin aids in protecting plants from infections by waterproofing the cell wall, allowing water and other solutes. Consisting of three monomers like p-coumaryl, coniferyl, and sinapyl with different levels of methoxylation, lignins are complex heteropolymers [22]. Various active functional groups, including phenyl, methoxy, aliphatic and aromatic phenyl hydroxy, and carbonyl groups, are found in lignin, enabling simpler chemical alteration and broadening its applications. Several methods, such as sulfonation, oxidation, graft copolymerization, hydroxymethylation, etc., can be employed for its adjustment [23].

Lignin has been utilized in multiple applications, including carbon fiber, fire retardants (FR), stabilizing agents, lubricants, varnishes, plasticizers, and surfactants [24]. Due to its high char output upon decomposition, lignin has recently garnered considerable interest as a bio-based flame-retardant component. Studies suggest that lignin can notably reduce the flammability of various polymers such as PET, ABS, PBS and polypropylene. Moreover, it is assumed that 35–40% of the lignin is consumed by heat, resulting in a substantial char yield [11]. Lignin integrated fire-retardants can be applied on textile through padding, layer by layer (LBL) application, and back-coating. Moreover, bulk polymers are processed with lignin-based intumescent via melt blending, and textiles are coated with it [24]. Fabrics can acquire UV blocking, flame retardant, and antibacterial properties effectively from lignin [23].

Because of lignin's intricate and diverse macromolecular structure, lignin nanoparticles have gained importance and are commonly utilized as antioxidants, UV blockers, drug carriers, anticorrosive nanofillers, adsorbents for dye and heavy metal ion removal, and hybrid nanocomposites [23]. Applying nano lignin to the linen finishing process can produce multipurpose textile products with additional features such as UV protection, antimicrobial properties, and antistatic qualities [25].

6.2.2.4 Pectin

The terms "pectin" or "pectic substances" refers polysaccharides present in plant cell. These compounds play a role in physiological activities like cell development, impacting the firmness and strength of plant tissue. Moreover, they are essential for the defense mechanisms against injuries and plant infections. It consists of up to 17 distinct monosaccharides with more than 20 different linkages. Pectin is a hetero-polysaccharide primarily made up of galacturonic acid residues, with varying levels of acid in the form of methoxyl esters and potentially some neutral sugars in side chains [26].

To reduce the impact of microplastics, a new pectin finishing treatment on polyamide fabrics is recommended. This treatment helps protect the fabric from deterioration during washing. Initially, glycidyl methacrylate was employed to modify pectin, which was then grafted onto polyamide to enhance the fabric's functionality. The number of microfibers released by untreated materials decreased by around 90% in washing experiments on treated fabrics [27]. Furthermore, from pectin an eco-friendly natural gum can be utilized as a thickener [28, 29].

6.2.2.5 Gelatin

Gelatin is a naturally occurring polymer with biocompatibility and biodegradability. Fish, pigskins, hides, and cow bones are the primary commercial resources. Gelatin is not derived from plants. The beneficiation of gelatin could lead to the production of valuable biomaterials if an efficient technology for its transformation into usable products is in place. Due to its non-toxic nature, affordability, and wide availability, gelatin is utilized in textile treatment to enhance dyeability and pretreatment processes like bleaching. Its antibacterial, antioxidant, and antifungal properties make it a suitable finishing agent [30]. Gelatin consists of eighteen amino acids, and the main are proline, hydroxyproline, and glycine. Gelatin's molecular structure is similar to a stick, with primary, secondary, and tertiary helical structures [31]. The surface of wool fabric can be smoothened by transglutaminase (TGase) through a coating procedure [32]. Gelatin can be safely used as an antibacterial finishing agent for cotton through the pad-dry-cure process, utilizing dimethyloldihydroxyethelyene urea and ammonium dihydrogen phosphate [33].

6.2.2.6 Sodium Alginate/alginate

Alginate is a biopolymer, is composed of linear chains of heteropolysaccharides and it's present in the cell walls and intercellular spaces of brown algae [34]. Textiles coated with alginate are perfect for non-implantable materials such as skin grafts, wound healing, the food and pharmaceutical industries, and hygienic textiles. For wound healing applications, alginate coated cotton fibers were developed using sodium alginate from *Sargassum wightii* and *Padina tetrastromatica.* These textiles, coated with alginate film, exhibit enhanced wound healing, antibacterial, and antioxidant properties [35]. Layer-by-layer coating, ionic cross-linking coating, and nanocomposite coating are methods used to apply alginate to textiles for finishing [1]. Cotton fibers coated with (AgNPs)-alginate composite nanocomposite exhibited remarkable antibacterial activity [36]. To enhance the binding efficiency between polyester (PES) fabric and colloidal TiO_2NPs, alginate is employed for fiber surface modification. Even after five washing cycles, the modified polyester materials displayed exceptional antibacterial activity and UV protection efficiency, showcasing their remarkable durability [37]. Through the pad-dry-cure process, SA–TSA nanoparticles were generated and applied to the cotton fabrics. The treated cotton materials showed antibacterial effects against both Gram-positive and Gram-negative bacteria. Moreover, the antibacterial properties could withstand up to 99% exposure over 30 consecutive laundry cycles [38]. Additionally. alginates can also be utilized as fibers which are commonly applied in protective textiles, premium apparel, and undergarments due to their flame-retardant and biocompatible properties [20].

6.3 Techniques for Applying Polymeric Finishes

Functionalizing textiles is a burgeoning method aimed at enhancing textile performance and elevating its value. As shown in Fig. 6.2 various techniques, such as exhaust, pad-dry-cure, coating, spraying, etc. are employed to imbue textiles with novel functionalities and properties [39].

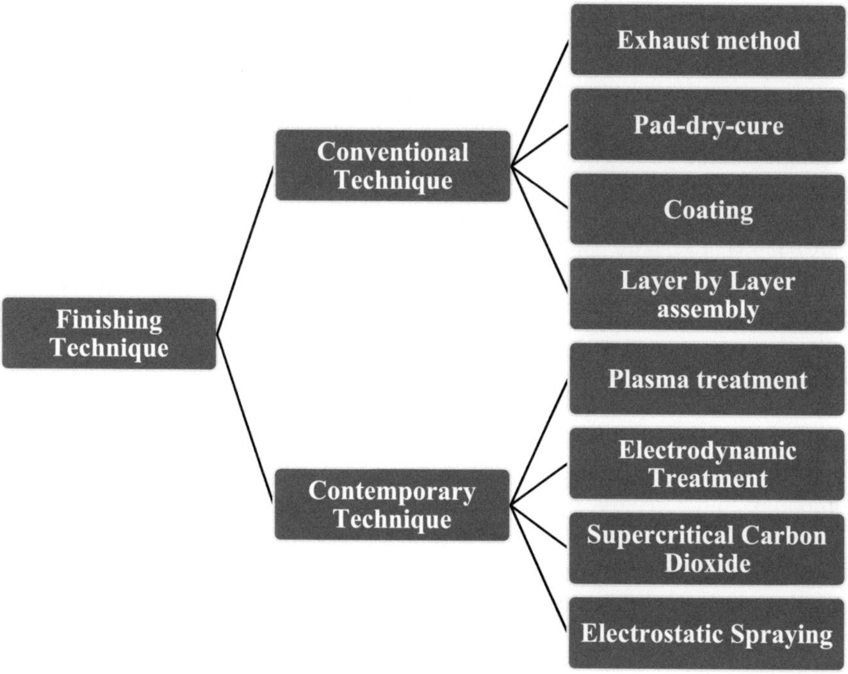

Fig. 6.2 Different application techniques in textile polymer finishing

6.3.1 *Exhaust Method*

In the exhaust method, following the dyeing process using a winch, jigger, or jet dyeing machine, the excess liquid is removed, and the textile undergoes a thorough washing. Subsequently, fresh water is incorporated into the finishing solution, and the material undergoes a predetermined treatment period. Following this, without any additional washing, the material proceeds directly to the drying stage or sometimes drying preceded by squeezing, which is referred to as exhaustion cum padding [1]. In an attempt to develop self-cleaning textile, cotton/polyester blend fabric has been coated by immersing it in a mixture of 1 g/l carboxymethylcellulose (CMC), 1 g/l polyacrylic acid (PA) and 10 g/l to fix EZF solutions at 80 °C for 20 min followed by drying at 120 °C. Which later on exhausted in a dispersion of 0.7 g/l TiO_2 (1%) Fe–N-graphene (2%) at 40 °C for 30 min with constant stirring followed by oven drying at 125 °C [40]. In another study, microgels were created using porous $CaCO_3$ infiltrated with a thermoresponsive polymer [41]. These microgels were applied to cotton using the pad-dry-cure method to fabricate a medical textile for drug delivery.

6.3.2 *Pad-Dry-Cure*

This is the predominant technique utilized in applying polymeric finish in textile substrates. Within a padding apparatus, the fabric undergoes a continuous immersion in a solution. It is subsequently pressed to a specific degree of moisture retention by traversing between a pair of rollers, followed by drying to remove moisture and curing to induce chemical reactions, both of which are accomplished in separate machines. This method constantly conveys the padded fabric through either a cylinder drier or a stenter. This sequence involves (i) Padding, (ii) Drying, and (iii) Curing [1].

6.3.2.1 Padding

The padding is carried out using padding mangle which involves two distinct stages. Initially, the fabric, typically woven, is submerged in the liquor to ensure thorough saturation. Subsequently, the fabric is maneuvered between two rollers to expel air and facilitate the penetration of dye-liquor into the material, with any excess liquor being redirected along the fabric. The former stage is termed "dip," while the compression between the squeeze rollers, as well as the passage through them, is referred to as "nip". In a single dip and nip procedure, the fabric undergoes immersion in the trough once and passes via the nip once. In most finishing cases, the operating speed is up to 50 m/min. However, for higher speeds (e.g., 120 m/min) or heavier fabrics, the preference shifts towards the double-dip and double-squeeze method utilizing a three-bowl padding mangle. The amount of liquor kept in the fabric is quantified by weight [1].

6.3.2.2 Drying and Curing

In the finishing process, water acts as a medium for evenly dispersing chemicals across fabric surfaces and within fibers. Once applied, the water content must evaporate for fabric to dry. Evaporation involves converting water into vapor and removing it from the fabric's surface. Air with lower relative humidity can hold more water vapor, facilitating faster evaporation. However, as humidity nears saturation, evaporation slows. The rate of evaporation depends on fabric temperature and air volume passing through it. Curing involves subjecting material to thermal treatment at temperatures higher than drying. Drying machinery can double for curing if it reaches required temperatures. Often, drying and curing are separate steps, each with specific conditions. Yet, sometimes the line blurs; wet fabric goes in, and cured fabric emerges without clear distinction between processes [1].

A sustainable flame retardant fabric has been developed through the polymerization of trimethyl phosphate with succinic acid, urea, and glyoxal [42]. After polymerization, they applied flame retardant to cotton using the padding process with an 80% pick-up. Then, the fabric underwent drying at 100 °C for 2 min and curing at

180 °C for 2 min. Subsequently, the fabric was washed with a 0.5% detergent and dried at 100 °C for 3 min. In a separate study to develop waterproof textile material, pretreated polycotton were coated with 3% epichlorohydrin and 2% sodium hydroxide. This process achieved 100% pick-up and then dry at 80 °C for 5 min and thermo-fixation at 130 °C for 3 min. Subsequently, the fabric samples underwent several rinses with normal water and were air-dried [43].

6.3.3 Polymer Coating

Polymer coating stands out as a widely employed method for altering textile surfaces to enhance comfort, appearance, and physical attributes. This process involves applying emulsified polymer resins onto textile substrates, followed by fixation through polymerization. Various polymer coating methods have been discussed in Table 6.4.

Table 6.4 Various polymer coating methods

Methods	Description	Refs
Dip coating	The simplest technique involves soaking the cloth in a solution of the low viscosity coating polymer, called dip coating. Using the forces balance applied to the substrate at the bath's exit, the layer's thickness is regulated followed by drying at proper temperature	[44]
Transfer coating	In this instance, the coating is applied to non-stick silicone paper and allowed to dry. Subsequently, the substrate and coated paper are fed through a set of rollers, with the coated layer facing the substrate's surface. The pressure from the rollers and furnaces creates a bond between the substrate and the coated surface, which is then removed, leaving a smooth surface	[45]
Kiss roll coating	The coating components are applied to the substrate's surface using a roll. As the roll rotates, the lower half of the roll is submerged in a bath containing the coating polymer, which transfers the polymer to the substrate's surface where it comes into contact with the upper half of the roll	[46]
Gravure roll coating	This method involves submerging an engraved roll in a coating polymer solution. As the roll rotates, the coating material fills the etched pattern and any excess creates a thin layer on the roll's surface. A squeegee is used to remove the film, and it is then pressed against the substrate to transfer the coated polymer	[47]
Knife coating	Among these methods, knife coating has emerged as a widely adopted approach for textile functionalization due to its affordability, diverse polymer options, and ability to confer lasting properties onto the substrate. This method swiftly covers extensive surfaces by applying polymer coating onto the substrate ahead of a moving knife, ensuring even deposition. The thickness of the polymer layer correlates with the distance between the knife and the fabric	[48]

Diverse techniques are used in textile finishing for polymer coating, spanning from conventional methods like knife coating, back-licking roller coating, and dip coating to innovative approaches such as plasma polymer coating, electrofluidodynamic processes, and supercritical carbon dioxide methods. These techniques are focused on achieving a well-coated substrate with desired properties, typically executed through three sequential steps [39]:

- *Dose:* Dosing controls the number of polymers.
- *Transfer:* Transfer involves depositing the coating material.
- *Fix:* Fixation ensures the adherence of the transferred materials.

6.3.4 Plasma Treatment

The innovative plasma technology involves applying a coating using partially ionized gas containing electrons, ions, photons, atoms, and molecules. This highly reactive plasma process can effectively alter the surface of certain textile substrates, imparting new functionalities. Plasma treatment for textile functionalization has seen significant advancements for last few years [49].

Plasma can be generated either at low pressure or at atmospheric pressure. Low-pressure plasma, typically ranging from 1 to 100 Pa, is often produced using radio frequency or microwave energy. While offering uniform effects and good reproducibility with minimal gas consumption, it's challenging for large-scale textile production due to chamber size limitations [50]. It was found in study that atmospheric pressure plasma for hydrophobic finishing of textile substrates was used [51]. Plasma reaction lasted 12 min (6 min per side), with the flow at 2.5 l/min and 1,3-butadiene at 500 ml/min. post-plasma modification, samples were cured at 140 °C for 150 s.

6.3.5 Layer by Layer Coating

Layer by layer coating contains the sequential deposition of charged electrolytes to construct multilayered films on a substrate. This technique utilized to immobilize phytic acid and chitosan as an environmentally friendly flame-retardant finish for cotton [52]. The coating section was adjusted by varying the pH of the deposition liquids. Fabrics coated with 30 bilayers displayed self-extinguishing properties in vertical flame tests [53].

A flame-retardant phytic acid-based layer coating for cotton have developed using demineralized water to prepare deposition solutions. The deposition process began with activating the fiber surface using a BPEI solution. Subsequently, substrates were consecutively immersed in PA and L9095 solutions, with initial immersions lasting 5 min each and subsequent immersions lasting 1 min. After each immersion, rinsing with DW was conducted to remove unbonded residues, followed by textile wringing. Once the desired number of BL was achieved, fabrics were dried at 80 °C [53].

6.3.6 Electrofluidodynamic (EFD) Treatment

The conventional approaches of biomaterial functionalization are being revolutionized by EFD, which employs electrostatic forces to fabricate 3D models capable of interacting with cells and/or molecules. This novel technology, EFD, is being utilized to modify and coat textile surfaces with bioderived polymers such as polylactid acid (PLA) [54].

6.3.7 Supercritical CO_2-based Method

Supercritical carbon dioxide technology involves depositing copolymers synthesized via radical polymerization simultaneously in supercritical CO_2 with azobis–isobutyronitrile as an initiator, along with crosslinking agents derived from diisocyanate in supercritical carbon dioxide solutions. This process results in achieving a uniform coating, ensuring high hydrophobicity with exceedingly high water contact angles [55, 56].

6.3.8 Electrostatic Spraying

In the electrostatic spray deposition (ESD) process, charged powder coatings are applied onto substrates by spraying. These powders are charged under 30 to 90 kV. Compressed air transports the powder from the feed hopper to the spray-gun tip. The charged powder particles, repelling each other, form a cloud directed towards the workpiece. When the grounded workpiece is introduced, the charged particles adhere temporarily. After application, the substrate is heated in a convection oven, causing the powder to melt and form a film. Subsequent curing solidifies the film.

In their study, electrostatic spraying to deposit polymeric powders on polyamide (PA66) have used where plastic workpieces were grounded and positioned 400 mm away from the ESD gun in a controlled environment (40% relative humidity, 20 °C ± 0.2 °C) [57]. After coating, substrates were removed and placed in a preheated oven at 170 °C followed by baking for 15 min to achieve desired film properties. An electrostatic spraying set up where the set-up contained a syringe pump have used [58].

6.4 Applications of Polymeric Materials in Textile Finishing

6.4.1 Flame Retardant (FR) Finish

In order to prepare FR fabric, several studies on the application of flame retardant treatment using various polymer coating techniques have been conducted [39]. In a study, the phosphorus-based flame retardants for textiles, focusing on thermal degradation, toxicity, bonding, and application on various materials have explored. They developed PVP (PR)-P-DCDA, a novel phosphorus-based flame retardant, and applied it to cotton, polyester, and cotton/polyester blends, significantly improving flame retardancy, thermal stability, and resistance. The commercialized product, Fire-off EBR, shows promising results in reducing flammability and enhancing fire resistance in different fabric blends, indicating industrial potential. Fire-off EBR functions by decreasing CO/PET fabric flammability through esterification and polymerization, yielding P-N polymer. It improves flame resistance by crosslinking the flame retardant with cotton fibers, reducing fabric flammability. FTIR analysis confirms the efficacy of the treatment, while vertical burning tests per ISO standards assess the flame retardant properties [59].

In another study, an eco-friendly flame retardant recipe has been introduced for cotton by polymerizing trimethyl phosphate with crosslinkers like succinic acid, urea, and glyoxal. These are optimized based on temperature, initiator, catalyst, and atmosphere, then applied to cotton fabric samples via padding. Treated samples undergo tests to evaluate flame retardancy, including char-length, char-width, limiting oxygen index, crease recovery, air permeability, shrinkage, smoldering cigarette, phosphorus concentration, formaldehyde content, antimicrobial properties, EDX, and SEM. The optimized TMP-urea-glyoxal blend outperformed commercial Pyrovatex-based flame retardants. The process aims to develop efficient, sustainable flame retardant solutions for cotton while avoiding hazardous chemicals to prioritize environmental and health considerations [42].

Cotton fabrics coated with 15 bilayers of polyvinylamine and phytic acid meet the ISO 15025:2016 standard for protective clothing, demonstrating their suitability for such applications. This coating process enhances the fabric's phosphorus and nitrogen content, providing flame-retardant properties without after flame, afterglow, or ignition. The treated fabric has a mass add-on of approximately 19%, with 2% phosphorus and 1.4% nitrogen content. Analysis confirms the formation of char layer, improving its flame-retardant qualities and reducing pyrolysis gas toxicity. While this method requires fewer steps than a 30-bilayer chitosan/phytic acid coating, challenges remain in the industrial use of layer coatings due to durability and applicability concerns [53].

6.4.2 Drug Delivery System

The development of drug delivery systems using textiles includes choosing appropriate materials like synthetic and natural fibers and integrating drug carriers that react to external triggers to regulate release. Through the adaptation of textile materials with responsive drug carriers, scientists have successfully extended drug release durations and enhanced the management of drug delivery processes. These innovative systems hold promise for transforming skin topical and transdermal drug administration, providing tailored and uninterrupted treatment for patients [60].

A thermosensitive textile drug delivery approach has been introduced to address UVB-induced damage. They created thermosensitive microgels by infusing porous $CaCO_3$ with a thermoresponsive polymer made from carboxyl-terminated poly (N-vinylcaprolactam) grafted onto hydrolyzed chitosan. These microgels were utilized to cotton fabrics using the pad-dry-cure approach to develop a medical textile. The study revealed that the microgels adhered to fiber surfaces and inter-fiber spaces, with the drug showing a release profile dependent on temperature. The medical textile shielded human keratinocyte HaCaT cells from UVB-induced harm by reducing lactate dehydrogenase leakage and boosting superoxide dismutase and glutathione peroxidase levels. This indicates the potential of the thermosensitive microgel-treated cotton fabric as an effective remedy for UVB-induced damage. The drug release mechanism in the thermosensitive microgel-treated cotton fabric is temperature-sensitive. Below the lower critical solution temperature (LCST), non-Fickian diffusion, influenced by microgel shrinkage, governs the release, diverging from Fick's diffusion. Above the LCST, the release pattern becomes pulsatile, toggling drug release on and off with temperature changes, a departure from traditional non-Fickian diffusion processes like erosion [61].

The biological and antibiotic delivery capabilities has been explores of a polypropylene (PP) artificial abdominal wall prosthesis cloth. This enhancement involved treating the PP with chitosan, synthesized per-amino-cyclodextrin and polyacrylic acid through a polymerization process, forming a cross-linked polymer layer on the PP surface. By immersing fabric samples in an antibiotic solution, the researchers observed that pellets treated with a specialized polymer (CS/CD-NH2) retained more antibiotic, enabling slow release over time. This controlled drug delivery system demonstrated improved antibacterial effects compared to untreated PP, showcasing enhanced biocompatibility and cell adhesion, making it a promising solution for efficient drug delivery in textile biomaterial applications [62].

6.4.3 Wound Healing

Wound healing textiles utilize the distinctive characteristics of textiles to enhance and expedite the process of wound recovery. These textiles integrate a range of active substances, including antimicrobials, growth factors, and medications, to establish an optimal healing environment for wounds.

An advanced cellulosic wound dressing for diabetic foot ulcers has been developed. They extracted a bioactive components from the plant and created a alginate/ C. tinctorius extract with antimicrobial and tissue regeneration properties. The study optimized the textile treatment process by adjusting parameters like curing conditions and solution concentration, enhancing the hydrophilicity of the samples. Various tests confirmed the effectiveness of the grafting process without compromising the original mechanical properties. The grafted dressings showed significant antibacterial activity against pathogenic bacteria, especially Gram-positive strains, attributed to the combined action of alginate and CTP polysaccharide. Cell viability tests on the dressings demonstrated improved cell health, indicating enhanced biological activity and a conducive environment for healing and tissue regeneration [62].

6.4.4 Antimicrobial Finish

To prevent bacteria from sticking to textile materials, it is essential to incorporate an antimicrobial agent. These active ingredients aim to inhibit bacterial growth and minimize their negative impacts. The antibacterial treatment involves dispersing the active agents, dissolving them, and then applying the solution either by impregnation or direct application methods [2].

A polypropylene fabric finished has been developed with citrate-hydroxypropyl-βCD polymer (PP-CD) that entraps the photosensitizer 5,10,15,20-tetrakis(4-sulfonatophenyl)-21H,23H-porphine (TPPS). This fabric is designed for controlled release of TPPS, which interacts with microorganisms and generates reactive oxygen species upon light exposure to inactivate them, using antimicrobial photodynamic therapy (aPDT) to combat multidrug-resistant bacteria. The fabric's morphology, optical properties, surface composition, and photodynamic activity were characterized. TPPS release kinetics showed the CD cavity controls elution. The porphyrin-loaded textile was highly effective against Gram-positive Staphylococcus aureus, killing 99.98% of cells with low adhesion, but showed no photodamage to Gram-negative Pseudomonas aeruginosa which adhered well. This approach offers a promising alternative for treating local infections, especially from multidrug-resistant bacteria. The structured release of photosensitizer from the fabric using aPDT provides a targeted, efficient way to combat bacterial infections while reducing resistance development and side effects [63].

A durable and antibacterial textiles finished fabric has been developed with reactive sulfobetaine. The sulfobetaine contains a triazine group that covalently bonds to textiles, providing durable antibacterial activity even after 30 washes. Treated textiles showed 98.0% and 95.2% effectiveness against E. coli and S. aureus, respectively, after 30 washes. The sulfobetaine also exhibited no skin irritation or cytotoxicity, even at high concentrations. This safe and durable antibacterial finishing maintains strong antibacterial properties through repeated laundering [64].

A antimicrobial fabric has been developed using a zwitterionic polyamine finish. These finishes, which do not release biocides or promote microbial adhesion, offer eco-friendly alternatives to existing products. The treated textiles effectively prevented protein and bacterial adhesion, even with a low DS of 20%. Fabrics treated with low DS polymers also showed substantial antibacterial activity, especially against Staphylococcus aureus. This method enables the creation of textile finishes with either purely antiadhesive properties or a combination of antiadhesive and antibacterial effects, offering a sustainable and environmentally friendly option compared to current antimicrobial finishes [65].

6.4.5 Water Repellant Finish

The ability of a fabric to prevent water from entering from the outside is known as impermeability. This can be achieved through laminated membranes or coatings that form a waterproof obstacle against rain or snow. Additionally, the seams and zippers must also be sealed to ensure total impermeability. However, it is crucial to consider the breathability of the fabric, especially during sports or work activities, as this allows for the release of moisture and prevents buildup. The waterproof layer of fabrics plays a vital role in clothing, but its effectiveness can be reduced by washing with detergents, which can lead to abrasion, leaching, or contamination. To restore the water repellent effect, silicone or fluorocarbon coatings can be applied to the fabric surface, allowing water to flow without sticking or wetting [2].

A multifunctional polyacrylate latex, FHPA, has been developed through emulsion polymerization, containing both fluorinated and hydrophilic groups. The synthesized latex was characterized for stability and morphology. Coating polyester fabric with FHPA enhanced its thermal and surface properties, creating a hydrophobic and oleophobic surface with improved water, oil repellency, and soil release. The preface of hydrophilic chains in the fluorinated acrylate polymer enhanced soil release performance. FHPA showed potential as a practical finishing agent for textiles, although it slightly stiffened the polyester fabric [66].

An environmentally friendly oil and water repellent cotton fabrics habe been developed by polymerizing stearic acid with carboxylic acids, eliminating the need for fluorine and formaldehyde. This study aimed to provide a sustainable and safe alternative to traditional water and oil repellents. By combining stearic acid with maleic acid and acrylic acid as crosslinkers, and using sodium hypophosphite as a catalyst and triethanolamine as an enhancer, the researchers optimized the polymerization process. The resulting cotton fabrics exhibited superior oil and water repellency, resistance to creasing and soiling, and antimicrobial properties. Surface analysis techniques such as SEM, EDX, FTIR, XRD, and TGA were employed to characterize the treated fabrics and their properties [67].

6.4.5.1 Hydrophobic Finish

In 2018, researchers led by Jin created a novel fabric coating agent by modifying polysiloxane with fluoroacrylate. This innovative material aimed to enhance the hydrophobic properties of treated textiles while maintaining their softness. The team employed advanced analytical techniques like FT-IR, NMR, and TGA to examine the chemical structure and thermal stability of the modified polysiloxane. The results indicated that the material exhibited excellent thermal stability and significantly improved hydrophobicity, with a water contact angle of 144.7°. Moreover, the treated fabric demonstrated enhanced softness without compromising its color [68].

6.4.5.2 Super Hydrophobic Finish

In 2019, Boticas et al. developed superhydrophobic cotton fabrics by treating them with ZnO nanoparticles. Different incorporation methods were tested on various cotton samples to enhance hydrophobicity. The best results were achieved with a 0.2 M precursor concentration on polymer-coated fabrics, exhibiting superhydrophobic behavior. The study confirmed the successful incorporation of nanoparticles, leading to improved water contact angles and potential for sustainable, cost-effective fabric finishes with enhanced hydrophobic properties [69].

6.4.6 7 Antistatic Finish

Antistatic properties in textiles prevent the buildup of static electricity and quickly dissipate accumulated charges. This is crucial for synthetic fibers like polyester, polyamide, and polyacrylic, which naturally develop static charges. Antistatic finishes can be applied to increase conductivity and reduce friction, minimizing static effects. These finishes can be temporary or permanent and often involve hygroscopic substances that form a conductive layer. Their effectiveness varies with humidity, with lower humidity leading to reduced conductivity and increased static issues [70].

A polymeric stabilizer to polyaniline/polyacrylate composites has been added to improv electrical conductivity and mechanical properties, making them suitable for antistatic finishing. Conductive PANI/PA composite emulsions were prepared using MAS, resulting in stable emulsions with evenly dispersed PANI. The films showed excellent electrical conductivity and mechanical properties compared to pure PA. The PANI/PA composite emulsion was used for antistatic finishing of PET fabrics, resulting in significantly reduced surface resistance, indicating good antistatic performance [71].

6.5 Conclusion

With all the information presented in the book chapter, the idea is to provide a proper demonstration of the variety of polymers that are being used in the textile industry, as well as their intriguing application techniques. The versatility of polymers, such as natural and synthetic kinds, allows the industry to incorporate a broad spectrum of crucial and aesthetic properties that serve some specific purpose in the specific cases where the textile material is used. The addition of these in the textile finishing section not only enhances the quality of the material but also helps the industry to address environmental issues with the help of eco-friendly and sustainable practices. More research on this section can evolve the procedure as more advanced technologies like nanotechnology and bio-based polymers can add new levels to the whole process. More study in this field can pave the way for a new form of textiles that will not only feature high functionalities but also offer sustainability that will keep consumers at ease while using them. A strong focus on the textile R&D section, both financially and scientifically, is needed to ensure the innovation keeps moving forward, and that is the only way for the textile industry to adapt to the ever-changing demands of the world.

Declaration The authors declare that they have no known competing financial interests or personal relationships that could have appeared to influence the work reported in this chapter. Generative AI and AI-assisted technologies have been used in the writing process to improve the readability and language of the work only. After using this tool, the authors carefully reviewed and revised the content to ensure accuracy and quality. The authors take full responsibility for the final version of this publication.

References

1. Roy Choudhury, A. K. (2017). Introduction to finishing. In A. K. Roy Choudhury (Eds.), *Principles of Textile Finishing* (pp. 1–19). Woodhead Publishing.
2. Bouasria, A., Nadi, A., Boukhriss, A., Hannache, H., Cherkaoui, O., & Gmouh, S. (2020a). Advances in polymer coating for functional finishing of textiles. *Polymer Coating for Finishing of Textiles,* 81–82.
3. Hamid Akash, M. S., Rehman, K., & Chen, S. (2015). Natural and synthetic polymers as drug carriers for delivery of therapeutic proteins. *Polymer Reviews, 55*(3), 371–406.
4. Bouasria, A., Nadi, A., Boukhriss, A., Hannache, H., Cherkaoui, O., Gmouh, S. (2020b). Advances in polymer coating for functional finishing of textiles. *Nanomaterials, Enzymes* (pp. 61–86).
5. Das, A., & Mahanwar, P. (2020). A brief discussion on advances in polyurethane applications, *3*(3), 93–101
6. Zafar, F., & Sharmin, E. (2012). Polyurethane. InTechOpen. https://www.intechopen.com/chapters/38589
7. Brenner, W., & Adler, A. (1972). Plasticized polyvinyl chloride finishes for durable flame retardancy for cotton fabrics, *2*(1), 29–37.

8. Islam, M. T., Rahman, M. M., & Mazumder, N. U. S. (2020). Polymers for textile production. *Frontiers of textile materials: Polymers, nanomaterials, enzymes, and advanced modification techniques,* 13–59.
9. Patrick, S. (2005). Practical guide to polyvinyl chloride. iSmithers Rapra Publishing.
10. Nuessle, A., Kine, B. J. I., & Chemistry, E. (1953). *Acrylic resins in textile processing., 45*(6), 1287–1293.
11. Bajaj, P. (2002). Finishing of textile materials, *83*(3), 631–659
12. Özek, H. Z. (2018). Silicone-based water repellents. *Waterproof and water repellent textiles and clothing* (pp. 153–189). Elsevier.
13. Ferrero, F., & Periolatto, M. (2012). Antimicrobial finish of textiles by chitosan UV-curing. *Journal of Nanoscience and Nanotechnology, 12*(6), 4803–4810.
14. Rajeswari, S., Prasanthi, T., Sudha, N., Swain, R. P., Panda, S., & Goka, V. J. W. J. P. P. S. (2017). *Natural polymers: A recent review., 6*, 472–494.
15. Kaushik, K., Sharma, R. B., & Agarwal, S. J. (2016). Natural polymers and their applications, *37*(2), 30–36.
16. Hahn, T., Bossog, L., Hager, T., Wunderlich, W., Breier, R., & Stegmaier, T., et al. (2019). Chitosan application in textile processing and fabric coating, 395–428.
17. Shirvan, A. R., Shakeri, M., Bashari, A. J. (2019). Recent advances in application of chitosan and its derivatives in functional finishing of textiles, 107–133.
18. Jiang, T., Duan, Q., Zhu, J., Liu, H., & Yu, L. (2020). Starch-based biodegradable materials: Challenges and opportunities, *3*(1), 8–18
19. Rani, S., Kadam, V., Rose, N. M., Jose, S., Yadav, S., & Shakyawar, D. B. (2020). Wheat starch, gum arabic and chitosan biopolymer treatment of wool fabric for improved shrink resistance finishing. *International Journal of Biological Macromolecules, 163*, 1044–1052.
20. Omerogullari Basyigit, Z. (2021). Application technologies for functional finishing of textile materials. *Textiles for Functional Applications.* InTechOpen. https://www.intechopen.com/chapters/75103
21. Kovačević, S., Schwarz, I., Đorđević, S., & Đorđević, D. J. P. (2019). Synthetized potato starch—a new eco sizing agent for cotton yarns, *11*(5), 908.
22. Boerjan, W., Ralph, J., & Baucher, M. (2003). Lignin biosynthesis. *Annual Review of Plant Biology, 54*(1), 519–546.
23. Raman, A., Sankar, A., SD, A., Anilkumar, A., Saritha, A. (2022). Insights into the sustainable development of lignin-based textiles for functional applications, *307*(8), 2200114.
24. Mandlekar, N., Cayla, A., Rault, F., Giraud, S., Salaün, F., Malucelli, G., et al. (2018). *An overview on the use of lignin and its derivatives in fire retardant polymer systems., 9*, 207–231.
25. Zimniewska, M., Kozłowski, R., & Batog, J. J. (2008). Nanolignin modified linen fabric as a multifunctional product, *484*(1), 43/[409]-50/[416].
26. Voragen, A. G., Coenen, G.-J., Verhoef, R. P., & Schols, H. A. (2009). Pectin, a versatile polysaccharide present in plant cell walls, *20*, 263–275.
27. De Falco, F., Gentile, G., Avolio, R., Errico, M. E., Di Pace, E., Ambrogi, V., et al. (2018). Pectin based finishing to mitigate the impact of microplastics released by polyamide fabrics. *Carbohydrate Polymers, 198*, 175–180.
28. Ebrahim, S. A., Othman, H., & Hassabo, A. J. (2022). A valuable observation on pectin as an eco-friendly material for valuable utilisation in textile industry, *65*(4), 555–568.
29. Ebrahim, S. A., Othman, H. A., Mosaad, M. M., & Hassabo, A. G. (2023). Eco-friendly natural thickener (pectin) extracted from fruit peels for valuable utilization in textile printing as a thickening agent, *3*(1), 26–49.
30. Hassabo, A. G., Zayed, M. T., Bakr, M., & Othman, H. A. (2022). The utilisation of gelatin biopolymer in textile wet processing, *19*(2), 125–136.
31. Deshmukh, K., Ahamed, M. B., Deshmukh, R., Pasha, S. K., Bhagat, P., & Chidambaram, K. (2017). Biopolymer composites with high dielectric performance: Interface engineering. *Biopolymer composites in electronics* (pp. 27–128). Elsevier.
32. Cui, L., Wang, Q., Wang, P., Huan, Q., & Fan, X. J. (2009). Transglutaminase-mediated crosslinking of gelatin onto wool surfaces to improve the fabric properties, *113*(4), 2598–2604.

33. Ibrahim, H. M., Aly, A. A., Taha, G. M., & Ibrahim, H. I. (2019). Production of antibacterial cotton fabrics via green treatment with nontoxic natural biopolymer gelatin. 62(Special Issue (Part 2) Innovation in Chemistry), 655–669.
34. Homayouni, A., Ehsani, M. R., Azizi, A., Yarmand, M. S., & Razavi, H. (2007). Effect of lecithin and calcium chloride solution on the microencapsulation process yield of calcium alginate beads.
35. Janarthanan, M., & Senthil Kumar, M. J. (2019). Extraction of alginate from brown seaweeds and evolution of bioactive alginate film coated textile fabrics for wound healing application, *49*(3), 328–351.
36. El-Rafie, M. H., Ahmed, H. B., & Zahran, M. K. (2014). Characterization of nanosilver coated cotton fabrics and evaluation of its antibacterial efficacy. *Carbohydrate Polymers, 107*, 174–181.
37. Mihailović, D., Šaponjić, Z., Račić, M., Radetić, T., Jovančić, P., & Nedeljković, J., et al. (2010). Functionalization of polyester fabrics with alginates and TiO_2 nanoparticles, *79*(3), 526–532.
38. Kim, H. W., Kim, B. R., & Rhee, Y. H. (2010). Imparting durable antimicrobial properties to cotton fabrics using alginate–quaternary ammonium complex nanoparticles, *79*(4), 1057–1062.
39. Bouasria, A., Nadi, A., Boukhriss, A., Cherkaoui, O., Hannache, H., & Gmouh, S. (2020). Advances in polymer coating for functional finishing of textiles. In Shabbir, M., Ahmed, S., Sheikh, J.N. (eds) Frontiers of Textile Materials: Polymers, Nanomaterials, Enzymes, and Advanced Modification Techniques, Scrivener Publishing LLC. https://doi.org/10.1002/9781119620396.ch3
40. Dumitrescu, I., Iordache, O.-G., Mitran, C.-E., Perdum, E., SĂNdulache, I.-M., SecĂReanu, L. O., et al. (2020). Attempts to improve the self-cleaning effect of the textile materials. *Industria Textila, 71*(03), 252–258.
41. Mathur, P., Sheikh, J. N., & Sen, K. (2020). Durable flame-retardant wool using sulphamic acid. *Polymer Degradation and Stability*, 174, 109101.
42. Qutab, H. G., Mohsin, M., Ramzan, N., Ahmad, S. W., & Sardar, S. (2020). Development of sustainable flame retardants for cotton fabrics by polymerization of trimethyl phosphate with succinic acid, urea and glyoxal. *Cellulose Chemistry and Technology, 54*(1–2), 135–148.
43. Zaghloul, S., Sharaf, S., Ameduri, B., & Hebeish, A. (2019). Novel fluorinated compound for imparting sustainable functionalities to cellulose-containing substrates. *Cellulose, 27*(2), 629–641.
44. Åkerfeldt, M., Strååt, M., & Walkenström, P. (2012). Electrically conductive textile coating with a PEDOT-PSS dispersion and a polyurethane binder. *Textile Research Journal, 83*(6), 618–627.
45. Shim, E. (2013). Bonding requirements in coating and laminating of textiles. In I. Jones, & G. K. Stylios (Eds.), *Joining Textiles* (pp. 309–351). Woodhead Publishing.
46. Patra, A. K., & Pattanayak, A. K. (2015). Novel varieties of denim fabrics. In R. Paul (Eds.), *Denim* (pp. 483–506). Woodhead Publishing.
47. Subramanian, V., & Sung, D. (2008). Gravure as an industrially viable process for printed electronics.
48. Yilmaz, O., Kucuk, M., Darie-Nita, R. N., & Cheaburu-Yilmaz, C. N. (2023). Halogen-Free waterborne polymeric hybrid coatings for improved fire retardancy of textiles. *Polymers, 15*(23), 4496.
49. Joshi, A. S., Bais, S., Prajapati, R., Kushwah, G., & Prajapat, C. (2015). Application of plasma finishing on cotton fabric, *5*(4), 1–10.
50. Naebe, M., Haque, A. N. M. A., & Haji, A. (2022). Plasma-Assisted antimicrobial finishing of textiles: A review. *Engineering, 12*, 145–163.
51. Samanta, K. K., Jassal, M., & Agrawal, A. K. (2010). Atmospheric pressure plasma polymerization of 1,3-butadiene for hydrophobic finishing of textile substrates. *Journal of Physics: Conference Series*, 208. https://doi.org/10.1088/1742-6596/208/1/012098
52. Laufer, G., Kirkland, C., Morgan, A. B., & Grunlan, J. C. (2012). Intumescent multilayer nanocoating, made with renewable polyelectrolytes, for flame-retardant cotton. *Biomacromolecules, 13*(9), 2843–2848.

53. Zilke, O., Plohl, D., Opwis, K., Mayer-Gall, T., & Gutmann, J. S. (2020). A Flame-Retardant Phytic-Acid-Based LbL-Coating for cotton using Polyvinylamine. *Polymers, 12*(5), 1202
54. De Falco, F., Guarino, V., Gentile, G., Cocca, M., Ambrogi, V., Ambrosio, L., et al. (2019). Design of functional textile coatings via non-conventional electrofluidodynamic processes. *Journal of Colloid and Interface Science, 541*, 367–375.
55. Kazaryan, P. S., Tyutyunov, A. A., Kondratenko, M. S., Elmanovich, I. V., Stakhanov, A. I., Zefirov, V. V., et al. (2019). Superhydrophobic coatings on textiles based on novel poly(perfluoro-tert-hexylbutyl methacrylate-co-hydroxyethyl methacrylate) copolymer deposited from solutions in supercritical carbon dioxide. *The Journal of Supercritical Fluids, 149*, 34–41.
56. Zefirov, V. V., Lubimtsev, N. A., Stakhanov, A. I., Elmanovich, I. V., Kondratenko, M. S., Lokshin, B. V., et al. (2018). Durable crosslinked omniphobic coatings on textiles via supercritical carbon dioxide deposition. *The Journal of Supercritical Fluids, 133*, 30–37.
57. Barletta, M., Gisario, A., & Tagliaferri, V. (2006). Electrostatic spray deposition (ESD) of polymeric powders on thermoplastic (PA66) substrate. *Surface and Coatings Technology, 201*(1–2), 296–308.
58. Dasdemir, M., & Ibili, H. (2016). Formation and characterization of superhydrophobic and alcohol-repellent nonwovens via electrohydrodynamic atomization (electrospraying). *Journal of Industrial Textiles, 47*(1), 125–146.
59. Atakan, R., Bical, A., Celebi, E., Ozcan, G., Soydan, N., & Sarac, A. S. (2018). Development of a flame retardant chemical for finishing of cotton, polyester, and CO/PET blends. *Journal of Industrial Textiles, 49*(2), 141–161.
60. Mehata, A. K., Dehari, D., Priya, V., & Muthu, M. S. (2023). Drug-releasing textile materials: Current developments and future perspectives. *Fiber and Textile Engineering in Drug Delivery Systems* (pp. 1–38). Woodhead Publishing.
61. Sun, X.-Z., & Wang, N. (2020). A thermosensitive textile-based drug delivery system for treating UVB-induced damage. *Cellulose, 27*(14), 8329–8339.
62. El-Ghoul, Y., & Alminderej, F. M. (2020). Bioactive and superabsorbent cellulosic dressing grafted alginate and Carthamus tinctorius polysaccharide extract for the treatment of chronic wounds. *Textile Research Journal, 91*(3–4), 235–248.
63. Castriciano, M. A., Zagami, R., Casaletto, M. P., Martel, B., Trapani, M., Romeo, A., et al. (2017). Poly(carboxylic acid)-Cyclodextrin/Anionic porphyrin finished fabrics as photosensitizer releasers for antimicrobial photodynamic therapy. *Biomacromolecules, 18*(4), 1134–1144.
64. He, L., Gao, C., Li, S., Chung, C. T. W., & Xin, J. H. (2017). Non-leaching and durable antibacterial textiles finished with reactive zwitterionic sulfobetaine. *Journal of Industrial and Engineering Chemistry, 46*, 373–378.
65. Timma, L. M., Lewald, L., Gier, F., Homey, L., Neyer, C., Nickisch-Hartfiel, A., et al. (2019). Nonfouling textiles with tunable antimicrobial activity based on a zwitterionic polyamine finish. *RSC Advances, 9*(17), 9783–9791.
66. Sun, Y., Zhao, X., Liu, R., Chen, G., & Zhou, X. (2018). Synthesis and characterization of fluorinated polyacrylate as water and oil repellent and soil release finishing agent for polyester fabric. *Progress in Organic Coatings, 123*, 306–313.
67. Rabia, S., Muhammad, M., Naveed, R., Waqas, A. S., & Qutab, H. G. (2020). Development of free fluorine and formaldehyde oil and water repellent finishes for cotton fabrics through polymerization of bio-based stearic acid with carboxylic acids. *INDUSTRIA TEXTILA, 71*(2), 145–155.
68. Jin, H., & Xu, W. (2018). The preparation and properties of fluoroacrylate-modified polysiloxane as a fabric coating agent. *Coatings, 8*(1), 31.
69. Boticas, I., Dias, D., Ferreira, D., Magalhães, P., Silva, R., & Fangueiro, R. (2019). Superhydrophobic cotton fabrics based on ZnO nanoparticles functionalization. *SN Applied Sciences, 1*, 1376.
70. Zhang, X. (2011). Antistatic and conductive textiles. In N. Pan, & G. Sun (Eds.), *Functional Textiles for Improved Performance, Protection and Health* (pp. 27–44). Woodhead Publishing.

71. Wu, L., Ge, Y., Zhang, L., Yu, D., Wu, M., & Ni, H. (2018). Enhanced electrical conductivity and competent mechanical properties of polyaniline/polyacrylate (PANI/PA) composites for antistatic finishing prepared at the aid of polymeric stabilizer. *Progress in Organic Coatings, 125*, 99–108.

Chapter 7
Emerging Trends and Technologies in Polymers for Textile Finishing

Md Imran Howlader, A Z M Mofasser, Md Mahbubul Alam, Nadim Ibn Sayed, Niaz Morshed Rifat, and Rony Mia

Abstract This chapter elucidates various types of natural and synthetic emerging polymers and their application techniques in textile finishing processes. There are numerous polymers, such as polyaniline, polypyrrole, and so on, that have the potential to be applied in the multifunctional finishing of textiles. Along with the traditional techniques, emerging technologies like enzyme immobilization, plasma coating, nano-coating, layer-by-layer technologies, supercritical fluid coating, electroless metallization, and so forth for applying polymers as textile finishes are represented as well. Approaches of metal-based nanoparticle coatings have emerged to enrich the antimicrobial properties. Several studies have been conducted to endow textile finishes with eco-friendliness and health safety. For instance, efforts are being made to develop flexible alternatives for cross-linking agents containing formaldehyde contents. The chapter concludes with the functional finishes and their applications for enhancing functionality, safety, and aesthetics of textiles, together with future trends and technologies, practical instances, and the challenges regarding their uses.

Keywords Textile finishing · Functional polymers · Finishing techniques · Sustainable finishing

Md Imran Howlader · N. M. Rifat
Department of Textile and Clothing Technology, Niederrhein University of Applied Sciences, Mönchengladbach, Germany

A Z M Mofasser · Md Mahbubul Alam · N. Ibn Sayed
Department of Textile Engineering, European University of Bangladesh, Mirpur, Dhaka, Bangladesh

R. Mia (✉)
Wilson College of Textiles, North Carolina State University, Raleigh, NC, USA
e-mail: mroni_mia@yahoo.com

M. Shahid et al. (eds.), *Advancements in Textile Finishing*,
https://doi.org/10.1007/978-981-96-6385-9_7

7.1 Introduction

The textile industry relies heavily on traditional polymer-based finishing techniques to enhance textile properties like softness, strength, water repellency, and antimicrobial attributes. However, recent developments indicate a shift towards more sustainable and functional solutions [1]. Bio-based polymers, including cellulose, chitosan, alginate, starches, sericin, lignin, and melanin, are eco-friendly substitutes for synthetic textile finishing agents [2]. Although facing challenges, smart polymers, such as shape memory polymers, phase change materials, and stimuli-responsive hydrogels are gaining popularity for their dynamic adaptability to environmental stimuli, offering opportunities for advanced functionalities like wrinkle resistance, thermo-regulation, and responsive drug delivery in textiles [3, 4]. These emerging trends signal a promising future for sustainable and high-performance textile finishing technologies.

The need for protective clothing has amplified in recent years because of growing concerns about the harm that might result from interacting with chemicals, microorganisms, pesticides, pollutants and UV rays. Today's clothing needs to be lighter and less bulky than existing options. It also needs to be warmer in the winter and colder in the summer [5]. Traditional techniques for applying finishes are employed to introduce functionalities like anti-UV, antimicrobial, self-cleaning, and flame retardant properties [6]. Nevertheless, in most cases, the materials finished by these techniques gain unnecessary weight, less comfort, poor hand-feel, and drape properties, and possess poor durability and mechanical properties. Moreover, the application and disposal of chemicals found in conventional finishes pose significant safety challenges [7]. The limitations of the contemporary finishing methods are recently inspiring researchers to develop more effective replacements for currently established technologies and finishing materials. Furthermore, innovative studies are being conducted to apply finishes that are durable, inexpensive, environment-friendly, and so forth [8].

A variety of novel methodologies have surfaced, purporting to offer flexible alternatives to contemporary techniques. For instance, the layer-by-layer deposition technique is suitable for forming films with precisely controlled thickness [9]. Various nano-coating techniques can be applied to enhance coating qualities like UV shielding, water repellency, self-cleaning capabilities, visual appeal, antimicrobial effectiveness, and durability [10]. For providing an antimicrobial finish and producing conductive clothing, textile materials can be treated in conductive coating methods [11]. Technique of printing nanowires on textile substrates according to a pre-decided patterns has also been discovered [12]. There is a possibility to develop a suitable method for a highly controllable coating using supercritical CO_2. In addition, electroplating and electroless metallization are also promising methods for textile finishing.

An overview of the polymer classes that are recently being implemented or have the potential to be applied to textile materials with a view to conferring desirable functionalities is presented in this chapter. Furthermore, it offers a portrayal of the emerging technologies for applying polymers in textile finishing. In addition, there is an inclusive discussion about the challenges and future prospects regarding the latest developments and technological innovations in polymer-based textile finishing.

7.2 Traditional Polymer-Based Textile Finishing Techniques

Conventional polymer-based textile finishing techniques have long been essential to the textile industry, fulfilling a crucial function in improving the functionalities and properties of textiles [13]. These methods involve the application of polymers onto textile surfaces to acquire specific desired effects. Sizing, for instance, utilizes polymers to strengthen yarns and enhance weaving properties, while thickening agents such as starch derivatives improve the viscosity of textile printing pastes. Softening treatments employ silicone-based polymers to impart a smooth hand feel, while water repellency is achieved through fluoropolymer-based finishes [14]. Additionally, polymers are utilized for imparting antimicrobial properties [15], stiffening fabrics [16], improving abrasion resistance [17], enhancing flame retardancy [18], and more. These traditional polymer-based finishing methods are widely used for augmenting the performance, durability, and aesthetics of textiles. However, current functional finishing technologies, which directly incorporate functional agents into textiles, have limitations in meeting the specific needs of futuristic textile products. The demand for advanced functionality coined with environmental responsiveness calls for the investigation of novel chemistry and new functional finishing polymers. These polymers should enable the surface enhancement of textile materials which can recognize and respond to human and environmental cues with effectiveness, while also providing effective protection against changes in the external environment.

7.3 Emerging Trends in Polymer Classes for Textile Finishing

The evolving landscape of textile finishing is increasingly shaped by the use of various polymer classes, marking a remarkable trend in the industry [5]. Polymers play a crucial role in different processing stages, ranging from thickening and sizing to imparting functional properties for example, stiffness, softness, antimicrobial attributes, flame retardancy, water repellency, and abrasion resistance [19, 20]. Moreover, advancements in textile coating and lamination have extended the application scope of polymers, with materials like Polyvinylidene Chloride (PVDC), Polyvinyl Chloride (PVC), acrylics, fluoro-polymers, silicones, and rubbers finding increased usage, offering enhanced functionalities compared to conventional methods [1]. Concurrently, the rising focus on health, hygiene, and microbial protection has led to the development of antimicrobial polymers such as chitosan and synthetic variants. Additionally, the emergence of smart textiles have driven research into application of biobased polymers, smart polymers [21], conductive polymers [22], and nanofiber and nanocomposites [23] in the finishing of textiles. Despite these advancements, challenges remain in integration of smart materials into textiles, emphasizing the demand for ongoing research to overcome technical barriers and utilize polymers to their fullest extent in shaping the future of textile finishing.

7.3.1 Bio-based Polymers

Bio-based polymers, whether biodegradable or nondegradable, are manufactured from renewable resources like fatty acids, organic wastes and lignocellulosic biomass (including starch and cellulose) [24]. The initial emphasis was on deriving polymers obtained from agricultural sources like corn, potatoes, and other carbohydrates, marking the first generation of bio-based polymers. However, with significant advancements in biotechnology, the concentration has moved away from resources rooted in food [25]. In addition to these synthetic sources, natural bio-based polymers, including polysaccharides (collagen and chitosan), nucleic acids and proteins, are found in nature [26]. In modern classifications (Fig. 7.1), biopolymers are usually sorted into three particular classes. The first one consists of biomass derived polymers, encompassing chemically modified variants such as chitin, chitosan, cellulose, cellulose acetate, starches, modified starches, and natural rubber. The latter class comprises biologically derived polymers synthesized by plants and microbes, exemplified by polyhydroxyalkanoates. The subsequent category includes polymers produced from monomers derived from renewable resources, exemplified by polylactic acid, bio-polyolefins, bio-polyethylene terephthalate, and polybutylene succinate [24, 27]. This classification scheme reflects the diversity of biopolymer sources and their applications, highlighting their importance as sustainable alternatives in various industries.

Fig. 7.1 Categorization of bio-based polymers

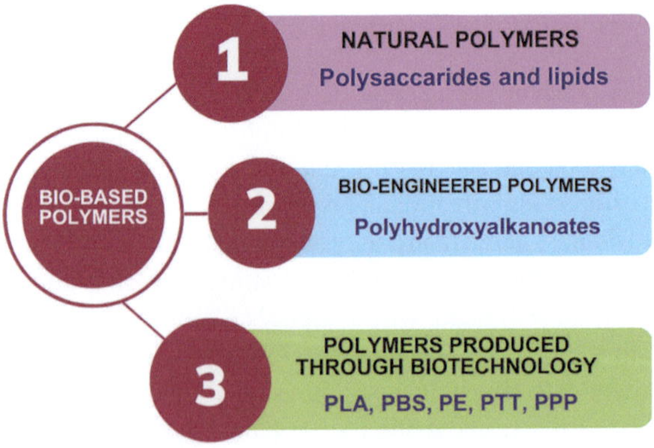

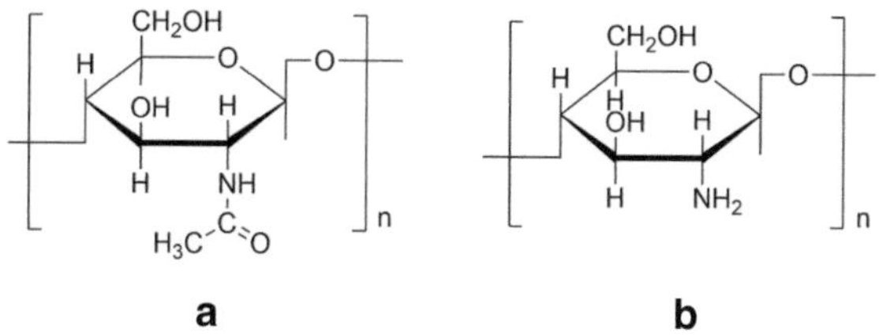

Fig. 7.2 Chemical structures of **a** chitin and **b** chitosan

In the contemporary time, extensive research initiatives have been taken to recognize suitable bio-based polymers for textile finishing due to their eco-friendliness and sustainable characteristics [28–30]. While biobased polymers offer sustainable alternatives, not all are inherently suitable for textile finishing. Some lack essential properties such as durability, washability, and compatibility with common finishing agents. The utilization of biobased polymers in textile finishing is experiencing significant growth, albeit still in its early stages. A wide range of biobased polymers, including chitosan, cellulose, starches, dextrins, sericin, alginates, lignin, melanin, and their derivatives, are being explored for their potential applications [25]. These polymers are derived from agro resources, microorganisms, and biotechnology via conventional synthesis offer inherent advantages such as safety, eco-friendliness, biodegradability, and cost-effectiveness [31].

Chitosan, one of the polysaccharides, is sourced from alkaline deacetylation of chitin. Exoskeleton of crabs, shellfishes, lobsters, prawns, shrimps, and some fungi are the primary sources of chitosan. Among all the biopolymers found in nature, it is the most available biodegradable, renewable, and nontoxic one that is appropriate alternative of some man-made polymers. It is extensively used in finishing of textiles to confer antimicrobial characteristics [32–34] (Fig. 7.2).

Cellulose can be extracted from vegetable biomass and wood. While long cellulose polymer molecules remain insoluble in their natural state, their hydrolyzed derivatives of suitable length exhibit solubility, generating viscous solutions [35]. Cellulose and its modified forms including methyl cellulose, carboxymethyl cellulose, soda cellulose, hydroxyethyl cellulose are applied in the preparation of printing paste, stiffening, coating, back filling, and anti-creasing uses to textile substrates [2].

Starch is one of the most ancient and significant finishing agents that is utilized for the production of cotton based textiles to serve multiple purposes like lustrous appearance, soil and wrinkle resistance, stiffening and weight gain. Additionally, it is used widely in back filling, sizing preparation, and printing paste formulations for its abundant obtainability and affordability [36]. It is derived from wheat, rice, maize and potato for various industrial applications. Starch provides a non-durable stiffening effect on different types of clothing. Since starch imparts a temporary stiffening finish to textiles, the rise of more long-lasting synthetic finishing products has reduced its dominance over the market.

The solubility of starches is limited, demanding particular rheological properties for effective use in industrial contexts. Therefore, these undergo modification via hydrolysis to yield soluble British gum or dextrins through processes like enzymatic reactions, acid hydrolysis, or roasting. The α-1,4-glycosidic linkage creates cyclic structures in cyclodextrins comprised of oligosaccharides and are consist of glucose units [37]. These units possess both lyophilic and hydrophilic parts, with the outer part being hydrophilic and the inner cavity being lyophilic. Depending upon molecular structure, cyclodextrins are divided into three categories: α-, β- and γ-cyclodextrin, as shown in Fig. 7.3. Amidst the textile manufacturing sector, β-cyclodextrin is extensively implemented in microencapsulation [2]. β-cyclodextrin serves as a polymer of imparting novel attributes to textiles. The integration of β-cyclodextrin with textiles has grabbed considerable interests in the recent years for the development of functionalized fabrics. Some of the uses of cyclodextrin in textiles include: elimination of dyes and auxiliaries from dyeing wastewater, slowing down dyeing and finishing processes, regulated dispensation of drugs and flavors, and safeguarding dyes from undesired aggregation and adsorption.

Sericin (Fig. 7.4) is an eco-friendly, proteinous biobased polymer extracted from the silkworm (*B. mori*). Sericin exhibits inherent qualities like, UV resistance, antibacterial activity, biocompatibility, biodegradability, and moisture absorption. The application of sericin has enhanced the functional qualities of man-made fibers [38]. Contemporary researches indicate that application of sericin in the final treatment of natural fibers including wool and cotton results in the acquisition of antimicrobial properties [39].

Alginate is another harmless and eco-friendly biopolymer with remarkable biocompatibility and bioactive characteristics. The alginate available for commercial use is generally extracted from brown algae (*Phaeophyceae*), including *Ascophyllum*

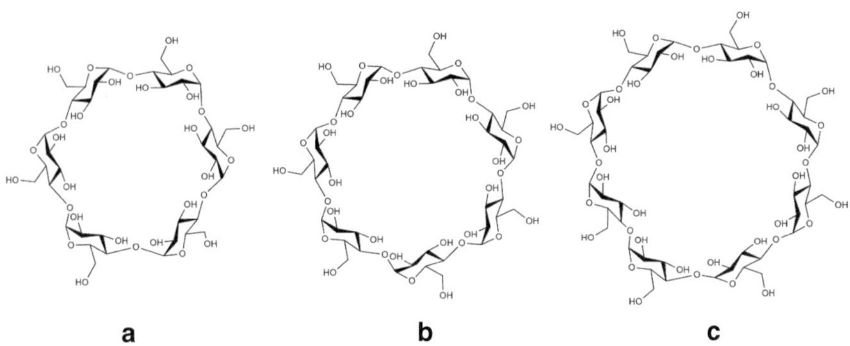

Fig. 7.3 Molecular structures of **a** α-cyclodextrin, **b** β-cyclodextrin and **c** γ-cyclodextrin

Fig. 7.4 Chemical structure of sericin

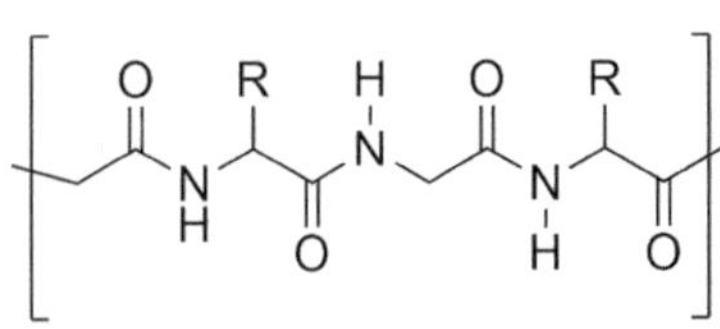

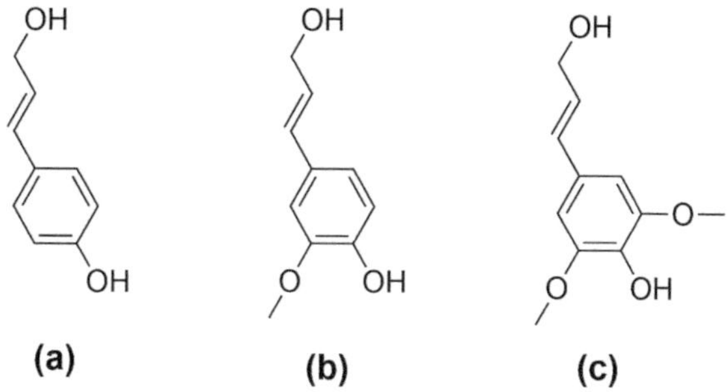

Fig. 7.5 Monomers of lignin: **a** p-coumaryl, **b** coniferyl and **c** sinapyl alcohol

nodosum, Laminaria japonica, L. digitata, L. hyperborea, and *Macrocystis pyrifera* [40]. Alginate is particularly employed in the finishing of textiles to produce wound dressings with antimicrobial properties [41]. The capability of alginate fibers to form gel makes them promising contenders for wound dressings. Alginate, being an insoluble calcium salt, undergoes ion exchange when in contact with sodium ions present in wound exudates, leading to the conversion of alginate from insoluble to soluble polymer form. This process allows alginate to soak up a significant quantity of water, facilitating the moisturization of wound, which promotes healing [10].

Lignin functions as the primary polymeric element found in the cell walls of higher plants, working in conjunction with hemicellulose and cellulose. While researchers have suggested comprehensive blueprints outlining the configuration of lignin [42], its exact structure still remains elusive. It is believed that the structure of lignin can vary according to the method of isolation. This polymer is created through the synthesis of three distinct monomers: coniferyl, p-coumaryl, and sinapyl alcohol, as illustrated in Fig. 7.5 [43]. Textile processing with a nanostructured solution of lignin imparts antibacterial properties, enhances the UV resistance properties of the fabrics, and retains its antistatic attributes without compromising its biophysical and physical properties [44]. Due to its high char yield when burnt, lignin has grabbed attraction as a bio-based flame retardant additive [45].

Melanin, a biobased polymer derived from animal hair and skin, when introduced into cell cultures, exhibits strong antigenic properties. Eumelanin and pheomelanin are the two basic types of melanin, which are illustrated in Fig. 7.6 below. It also exhibits a series of notable properties including free radical scavenging capacities, capability of absorption for UVe-Vise-IRs, and redox reversibility [46]. Melanin, known for its role in shielding skin from UV radiation, has been extracted from the Indian squids' ink sac and applied as a nano-coating on cotton fabric to confer UV resistance. Studies have confirmed the efficacy of squid ink eumelanin as a potent UV absorber, enabling its successful nano-finishing on cotton fabric [47].

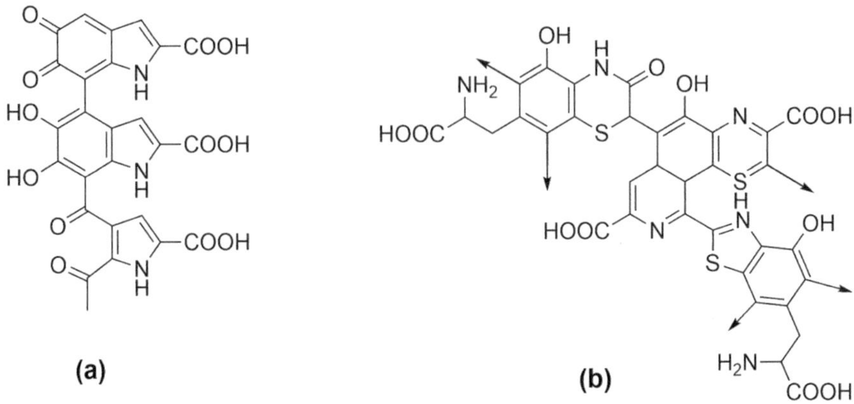

Fig. 7.6 Chemical structures of **a** eumelanin and **b** pheomelanin

7.3.2 Smart Polymers

Smart polymers are substances capable of undergoing significant alterations in their characteristics in response to environmental cues [3]. These polymers undergo reversible and significant chemical or physical changes to counter even minor external fluctuations in environmental circumstances. These stimuli may include changes in temperature, pH levels, exposure to magnetic or electric fields, light, fluctuations in ionic factors, and interactions with biological molecules [3]. This unique responsiveness allows smart polymers to adapt dynamically to their surroundings, making them invaluable for a variety of applications including textile finishing. In the past few decades, smart polymers like SMPs [48], PCMs [49], and stimuli responsive polymer hydrogels [50] have been applied to textile finishing due to their easy processing, lightweight and flexibility.

7.3.2.1 Shape Memory Polymers (SMPs)

SMPs are categorized as intelligent materials that exhibit an impressive capacity to retain a temporarily programmed shape and then return to their original form when subjected to external triggers [51]. Incorporating SMPs within fabric structures has given rise to a new breed of smart textiles characterized by their exceptional properties, including appealing aesthetics, imaginative designs, soft textile displays, enhanced comfort, intelligent drug release systems, advanced wetting characteristics, wound monitoring capabilities, and resilience against extreme environmental variations [48]. Fabrics infused with SMPs exhibit the unique trait of retaining a set structure while also being capable of being programmed to temporarily adopt different shapes. When activated by an external stimulus, they effortlessly revert back to their original, permanent form [3].

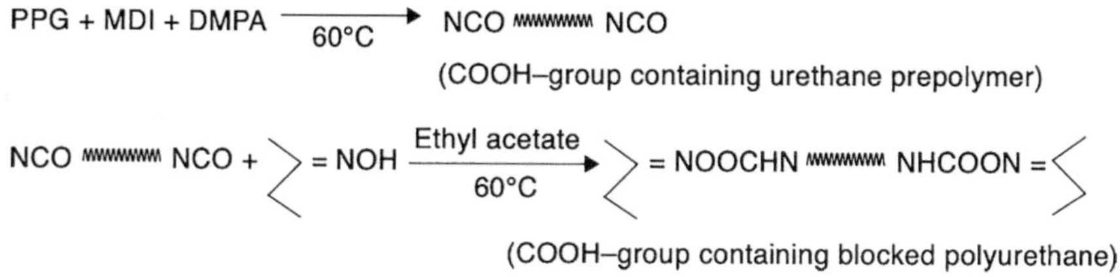

Fig. 7.7 Synthesis route of end-capped SMPU oligomers

SMPs can be produced through the utilization of water-based emulsions, which are formulated via solution polymerization, as illustrated in Fig. 7.7. The synthesis of shape memory polyurethane (SMPU) oligomers involves the use of diisocyanate (MDI), dimethylolpropionic acid (DMPA) poly (propylene glycols) (PPG), and 4,4′-Diphenylmethane. Methyl ethyl ketoxime (MEKO) is subsequently employed to terminate the –NCO groups, followed by neutralization of the free carboxylic groups using Triethylamine (TEA). Ultimately, a mixture of acetone and water is introduced into the reactor to generate shape memory aqueous emulsions. A common technique involves finishing to transition the solution form of SMPs onto textiles.

Cotton textiles treated with shape memory polymers display a wrinkle-free appearance [52]. The fabrics treated with SMPU emulsion maintain a sufficient wrinkle-free effect even after numerous washing cycles, lasting up to hundreds of launderings [51]. Additionally, cotton fabrics treated with SMPU also possess a greater pattern and crease retention abilities, providing the aesthetic appeal of garments. In addition to cotton, SMP-treated wool garments have better dimensional stability compared to untreated garments. This is attributed to SMPU covering wool fibers, reducing directional friction effects in post-finishing. Treated wool garments maintain their size after laundering, unlike untreated garments that shrink. Wool textiles often experience felting as a result of scale entanglement due to friction in specific directions. However, when wool fabric undergoes treatment, it demonstrates a noticeable decrease in felting [51].

7.3.2.2 Phase Change Materials

The organic and inorganic compounds that are capable of absorbing and releasing significant amounts of latent heat while transitions between solid and liquid phases within a low temperature spectrum are referred to as phase change materials (PCMs) [49]. Recently, researchers have shown great interest in incorporating PCMs in textiles by coating or encapsulation to create thermo-regulated smart textiles [49]. The commonly utilized phase change materials predominantly consist of paraffin-derived substances like nonadecane, eicosane, octadecane, and heptadecane, alongside materials based on polyethylene glycol (PEG) Fibers, fabrics, and foams with PCM microcapsule treatment are broadly used in apparels such as gloves, ski suits, socks, and sweaters to enhance their thermophysiological wearing comfort [53]. In a research, the production of polyethylene glycol (PEG)/paraffin composite phase change materials (CPCMs) with stable shapes and improved thermal conductivity was achieved by employing the impregnation method and applying coating technology to the surface of cotton fabric [54].

7.3.2.3 Smart Hydrogels

Hydrogels are natural or synthetic superabsorbent materials, typically composed of hydrophilic polymers, which form intricate three-dimensional structures. Polymer chains of hydrogels are connected together by chemical bonds like hydrogen bonding, ionic and hydrophobic interactions [55]. Though insoluble in water, the three dimensional networks of hydrogels can swell and absorb water. When hydrogels are created using polymers that respond to stimuli, they add attributes and exhibit variations in solvation in reaction to specific stimuli [56]. Described as smart polymeric substances, stimuli-responsive hydrogels (SRHs) adapt their properties in a reversible manner when subjected to variations in environmental conditions [50, 57]. The types of stimuli that trigger responses in smart hydrogels are generally divided into three categories: physical, chemical, and biological [58]. Physical stimuli encompass response to temperature, light, ultrasound, mechanical, magnetic, or electrical fields. Chemical stimuli involve reactions to ionic strength, solvent, electrochemical fields, and pH. Biological cues involve the operational dynamics of molecules, encompassing enzymatic responses and molecular identification at receptor sites [57]. Although SRHs have been used in many other sectors [59], they are still being actively researched the way of using SRHs on textiles [60]. Stimuli-responsive hydrogels have caught the attention of the researchers for the finishing applications of textiles with SRH [61].

Textile finishing commonly employs temperature, pH, or temperature/pH-responsive hydrogels. Temperature-responsive hydrogels are classified as positive and negative systems according to their behavior relative to the lower critical solution temperature (LCST). Examples include cellulose ethers, poly(N-vinyl isobutyramide), poly(N-vinylcaprolactam), poly(N-isopropyl acrylamide), copolymers (PEO-PPO), pyrrolidone, poly(ethylene oxide)-poly(propylene oxide) and poly(N-alkyl acrylamide) [50, 62]. One of the most studied and utilized temperature sensitive hydrogel is poly(N-isopropylacryl- amide) PNIPAAm (Fig. 7.8). In aqueous medium PNIPAAm hydrogel possesses a LCST, generally around 32–34 °C, which matches the human body temperature.

pH-responsive hydrogels go through significant fluctuations in volume and characteristics in reaction to changes in environmental pH levels, with polyelectrolytes being preferred for their preparation [63]. Natural polymers like collagen and chitosan containing ionizable groups like $-NH_2$ and $-COOH$ are commonly used, alongside synthetic pH-responsive hydrogels including weak poly acids like PAA, which receive protons at acidic medium levels and release them at neutral and alkaline condition [64].

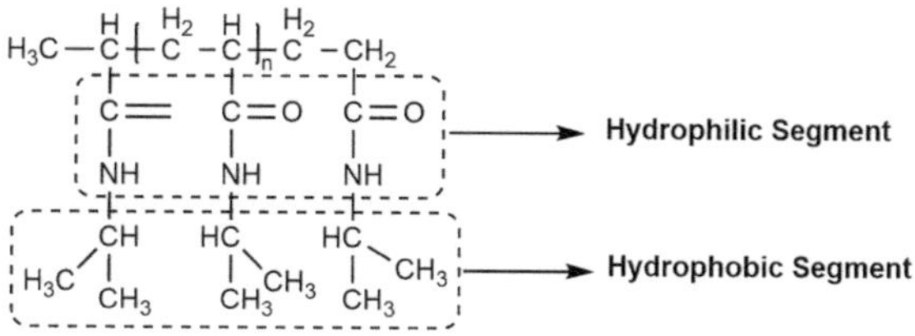

Fig. 7.8 The molecular structure of PNIPPAm

7.3.3 Conductive Polymers

In the area of organic materials, electro-conductive polymers are distinguished by their ability to demonstrate conductive or semi-conductive behavior [22, 65]. Although polymers are typically non-conductive, certain alterations can induce intrinsic or extrinsic conductivity within them [66]. ICP (Intrinsically Conductive Polymers), commonly termed conducting polymers, exhibit electrical conductivity and properties on par with metallic materials. Polymers exhibit conductivity through the interconnected conjugated double bonds, which allow doped charge carriers to move freely within the material [67]. For example, PEDOT (poly-3,4-ethylenedioxythiophene), which can be doped with substances such as PSS (polystyrene sulfonate). This process yields compounds characterized by the presence of liberated electrons within the conduction band, thus enabling charge transfer [65]. Recently, significant interest has been focused on PAc (polyacetylene) (Fig. 7.9a), PPy (polypyrrole) (Fig. 7.9b), PPS (polyphenylene sulfide) (Fig. 7.9e), PAni (polyaniline) (Fig. 7.9f), PFu (polyfuran) (Fig. 7.9j), PPP (poly para-phenylene) (Fig. 7.9h), PSe (polyselenophene) (Fig. 7.9g) and PTh (polythiophenes) (Fig. 7.9c) among the wide array of intrinsically conductive polymers. These materials have emerged as prime candidates for the development of conductive textiles across various domains, owing to their distinctive blend of characteristics [68, 69]. These include their capacity for electrical conductivity, effectiveness in electromagnetic shielding, lightweight nature, flexibility, high resilience, compatibility with biological systems, optical transparency, cost-effectiveness, robust environmental durability, ease of processing in solution, strong adhesion to different substrates, and straightforward preparation and application methods [67, 70, 71].

7.3.3.1 Polypyrrole (PPy)

PPy, a polymer made up of 2,5-crosslinked pyrrole units, is produced through electrochemical or oxidative polymerization, categorizing it among electrically conducting or conjugated polymers within the PHC group [67, 72, 73]. PPy exhibits added strengths when compared to alternative ICPs, including its water solubility, efficient oxidizing properties, and film resistance to various environmental factors [74, 75]. Polypyrrole (PPy) exhibits electrical conductivity facilitated by both ions and electrons, with the relative contribution governed by the size and mobility of ions within

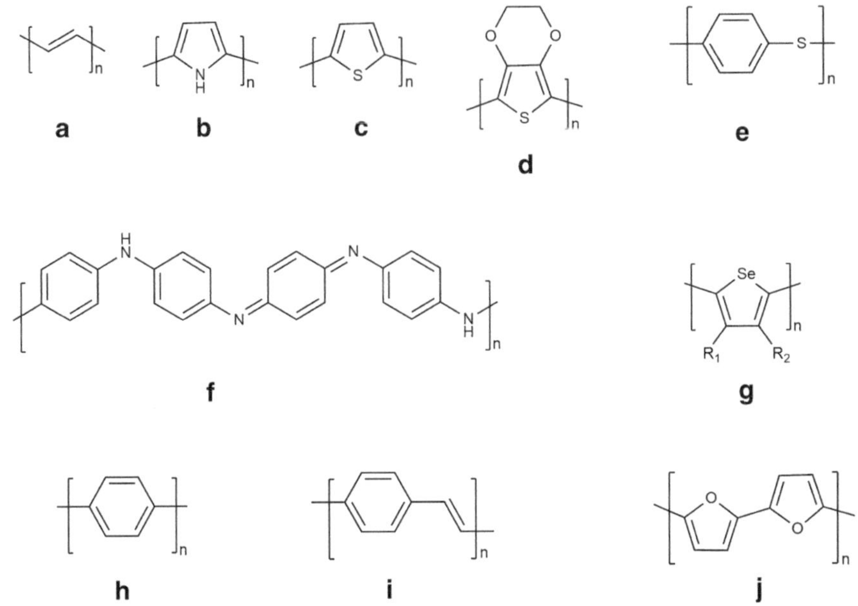

Fig. 7.9 Intrinsically conductive polymers used in textiles. **a** PAc (polyacetylene), **b** PPy (polypyrrole), **c** PTh (polythiophenes), **d** PEDOT (poly-3,4-ethylenedioxythiophene) **e** PPS (polyphenylene sulfide), **f** PAni (polyaniline), **g** PSe (polyselenophene), **h** PPP (poly para-phenylene), **i** Poly (para-phenylene vinylene) and **j** PFu (polyfuran)

the material (Fig. 7.10). Electrons traverse along the polymer chain and transition between chains due to the presence of a conjugated system and doping with negative ions. Consequently, PPy demonstrates robust conductivity, reaching approximately 10^4 S/cm at elevated temperatures. Although doping enhances band gap and conductivity, PPy's electrical conductivity remains inferior to certain other conductive polymers [75, 76]. PPy finds extensive application in the production of conductive textile materials due to its inherent resistance to environmental factors. Its water solubility, ease of oxidation, and sustained conductivity make it a favored choice for long-lasting performance [67].

7.3.3.2 Polyaniline (PAni)

PAni represents a radical-cation compound derived from a conjugated polymer featuring oxidatively linked aniline segments and an acidic component. It demonstrates inherent conductivity in its salt state, requiring no additional substances. PAni exhibits solubility in NMP (N-methyl-2-pyrrolidone). The synthesis of PAni is affected by the factors such as, the monomers chosen, the type of solvent used, the conducting salt, and the oxidation conditions, leading to variations in chemical composition and morphology [67]. PAni exhibits different colors and attributes

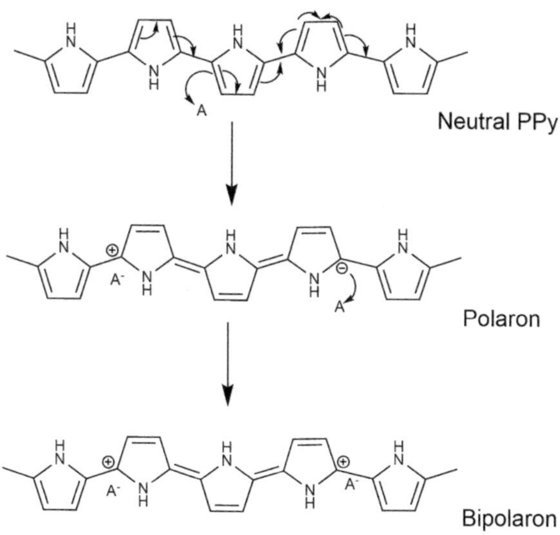

Fig. 7.10 Polaron and bipolaron of PPy

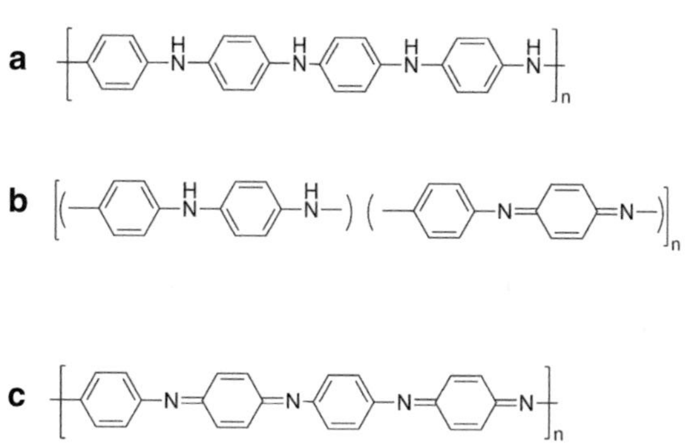

Fig. 7.11 Oxidation states of polyaniline. **a** Colorless or light yellow oxidation state, **b** Blue oxidation state, and **c** Green oxidation state

contingent upon its oxidation level: colorless or light yellow oxidation state (Fig. 7.11a), the stable "green polymer" exists in a reduced form, but it readily oxidizes when exposed to air, leading to a non-conductive state; blue oxidation state (Fig. 7.11b), the polymer exhibits stability, unlike its neutral form, which lacks conductivity. and can be achieved by neutralizing the "green" type of PAni using bases; the most stable state, the polymer assumes the green oxidation state (Fig. 7.11c), a prevalent occurrence in PAni chemistry, often denoted as conducting or organic metal status [77–79].

7.3.3.3 Poly(3,4-Ethylenedioxythiophene)

In recent times, Poly(3,4-Ethylenedioxythiophene)-PEDOT and its derivatives have emerged as notable contenders for their favorable attributes, including ease of solution processing, electrical conductivity, and optical transparency when in their conductive form. However, their limited water solubility impedes their potential applications. To surmount this challenge, researchers often opt to combine PEDOT with water-soluble polymers like PSS, resulting in the formation of PEDOT:PSS, which exhibits superior conductivity, transparency, and stability. The industrial synthesis process entails the mixing of EDOT and PSS with an oxidizing agent, leading to the production of highly conductive PEDOT. Generally, PEDOT:PSS hydrogel particles present outstanding processing features, enabling the manufacture of thin, transparent films with conductivity [80]. Because of their solution-processable nature, consistent transparency and electrical conductivity, when in a conductive state, and relatively low cost compared to other inherently conductive polymers (ICPs), PEDOT is commonly preferred in order to produce electrically conductive textile materials [67, 81].

7.3.4 *Nanofibers and Nanocomposites*

Polymeric nanocomposites (NCs) consist of polymer blends at the nanoscale, either with or without incorporating additional substances like organic or inorganic nanoparticles [23]. When a bio-based polymers are utilized in the production of nanocomposites, the term "bionanocomposites (BNCs)" emerges as a more fitting and articulate descriptor. BNCs, alternatively known as bio-hybrids, nano-biocomposites, green composites have garnered significant attention in scientific literature in recent years [23]. The use of layered compounds in nanocomposites (NCs) has become increasingly popular recently, driven by their remarkable properties [82].

7.3.4.1 Layered Double Hydroxides (LDHs)

LDHs have especially attracted attention as nanofillers, enabling the creation of innovative composites with distinctive characteristics [83, 84]. LDHs, commonly referred to as anionic clays, comprise lamellar hydroxides composed of more than one metallic cations alongside anionic species and water molecules within their interlayer regions [85]. These substances bear a resemblance to hydrotalcite, denoting the mineral's water content (hydro) and its talc-like appearance (talcite), a naturally occurring mineral known for its ease of crushing into a white powder akin to talc [86]. LDHs are often represented by the chemical formula $[M(II)_{(1-x)}M(III)_x(OH)2]^{x+}(A^{(n-)})^{x/n}.mH_2O$, where M(II) denotes divalent metal cations like Ca^{2+}, Mg^{2+}, Cr^{2+}, Mn^{2+}, Fe^{2+}, and Co^{2+}, and M(III) indicates trivalent metal cations such as Al^{3+}, Fe^{3+}, Cr^{3+}, and Mn^{3+} and the symbol A^{n-} denotes the anion component of the compound [23].

7.3.4.2 Carbon Nanotubes (CNTs)

The properties of carbon nanotubes (CNTs) encompass a range of unique characteristics, including mechanical and electronic attributes, among others. For example, a lone carbon nanotube demonstrates a modulus surpassing several thousand gigapascals (GPa), alongside a tensile strength ranging from several tens of gigapascals (GPa) [87]. Achieving alignment of CNTs for reinforcement in composite fabrication, however, faces difficulties, resulting in the inability of the resulting nanocomposite to meet mechanical property expectations. In response, various research groups have explored the incorporation of CNTs into polymer nanofibers via electrospinning to address this limitation [88, 89]. CNTs or their aggregates aligned through the spinning process is projected in the direction of the fiber, facilitated by the interplay of dielectrophoretic forces arising from the contrast in dielectric characteristics between CNTs and solution of the polymer, coupled with the substantial shear forces exerted during spinning [89]. Ko et al. [88] employed electrospinning to fabricate ultrafine fibers from a polyacrylonitrile solution containing dispersed carbon SWNTs. Park et al. [89] incorporated carbon SWNTs, with diameters of 1.2–1.6 nm and lengths of 3 mm, into a polyimide solution for electrospinning, yielding ultrafine fibers with a wide range of diameter (500−2 mm). However, no mechanical characterization or comparison with pure polymer nanofibers was provided. Recent studies have investigated the synthesis of composite nanofibers by incorporating inorganic and organic materials into polymer solutions for electrospinning, resulting in diverse combinations like polycaprolactone/gold or ZnO, PVA/Silica [90], Polyacrylonitrile (PAN)/TiO2, and Nylon6/montmorillonite (Mt) [91] ultrafine fibers, with recent efforts focusing on alumina-borate/PVA composite nanofibers using sol–gel processing and electrospinning [92].

7.4 Emerging Technologies for Applications of polymers in Textile Finishing

In the preceding years, the need for protective clothing has surged because of concerns over microbial exposure, chemicals, UV radiation, and pollutants. However, traditional finishing methods often result in heavy, uncomfortable garments with reduced durability. To address these challenges, innovative coating technologies have emerged. These methods, such as enzyme immobilization, nanocoatings, and plasma deposition, enhance textile surfaces without compromising comfort or performance [5]. These technologies diverge from traditional finishes by introducing specialized functionalities to textile surfaces through micro or nano-level alterations, preserving bulk properties.

7.4.1 *Enzyme Immobilization*

Enzymatic methods offer an eco-friendly and efficient alternative to traditional chemical finishing techniques, functioning under mild conditions, exhibiting substrate specificity, being environmentally safe, biodegradable, and devoid of hazardous byproducts, simple biotechnological approaches enable the large-scale manufacturing of enzymes for industrial purposes as well [5]. Enzymatic processes like desizing, bio-washing, bio-polishing, bio-scouring, and bleaching have become established commercial practices for treating cotton (Fig. 7.11). Proteinase subtilisin enzymes have emerged as a contemporary approach for modifying cotton via transesterification [93]. The production of versatile wool textiles, endowed with antioxidant, antibacterial, and hydrophobic attributes, was achieved through the incorporation of alkyl gallates via a laccase-catalyzed process [94]. Degumming of silk with proteases [95], treatment of polyester with cutinases and esterases [5], nylon with amidase, and polyacrylonitrile with nitrilase [96] have been reported in recent studies.

7.4.1.1 Methods of Enzyme Immobilization

Traditional techniques frequently used for covalent bonding, encapsulation, entrapment, crosslinking and enzyme immobilization include adsorption (Fig. 7.12) [97].

Adsorption

This approach encompasses the treatment of the substrate with enzymes, ensuring the pH and ionic strength are conducive, and then allowing for incubation. It was demonstrated that carbon nanotubes can adsorb alpha-chymotrypsin and soybean peroxidase. This discovery opens up possibilities for various applications, including biomedical devices, biosensors, and the advancement of self-healing and self-cleaning materials [5]. A pioneering immobilization technique has emerged, featuring the application of polyketone polymer as an innovative support medium [98]. The polyketone polymer ($–[–CO\text{-}CH2CH2–]_n–$), synthesized through the copolymerization of ethane and carbon monoxide, has found application in immobilizing various enzymes, including horseradish peroxidase, as well as amine oxidases sourced from lentil seedlings and bovine serum [99]. The simple immobilization process was conducted in a low concentrated aqueous buffer, mildly blending the proteins with the polymer, eliminating the need for bifunctional agents or spacer arms, and relying solely on hydrogen bonds between the polypeptidic chain's amine groups and the polymer's carbonyl groups (Fig. 7.13).

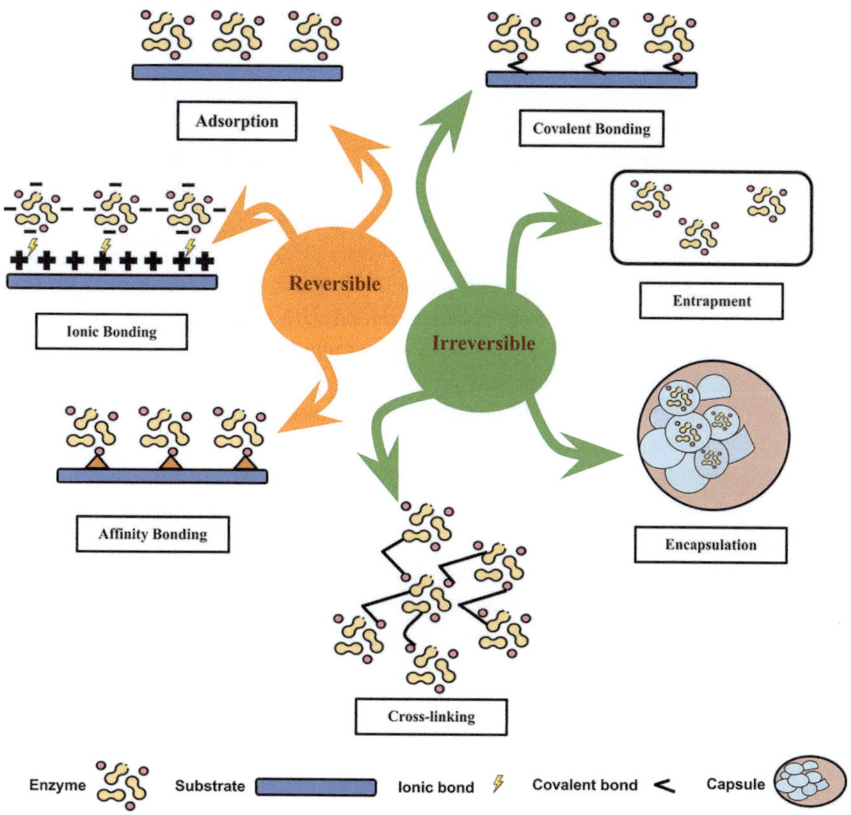

Fig. 7.12 Different methods of enzyme immobilization

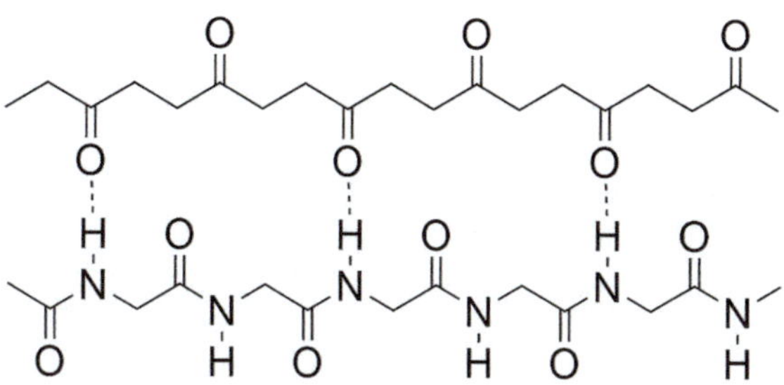

Fig. 7.13 Bonds (Hydrogen) between the –NH groups of the polypeptidic chain and carbonyl groups

Covalent Bonding Methods

Covalent bonds are necessary to achieve more stable attachment, typically formed by reacting with reactive sites on the protein surface, which can be applied to unmodified proteins due to their reliance on naturally occurring functional groups. For instance, lysine residues' exposed amine groups immediately react with supports consisting of active esters like N-hydroxysuccinimide (NHS) esters, forming stable amide bonds [99]. However, NHS esters are susceptible to instability in aqueous conditions, potentially leading to modest immobilization yields due to ester hydrolysis. Alternatively, aldehyde groups can be bonded with exposed amino groups and subsequently reduced to form stable secondary amine linkages, offering an alternative approach. The amine group nucleophilicity enables its reaction with epoxide-functionalized materials, such as diglycidyl ethers. These epoxides offer stability to hydrolysis at neutral pH, facilitating material handling but potentially leading to slow or incomplete coupling. Optimal supports should possess a high density of epoxy groups and stable conditions to allow prolonged enzyme-support reactions, with minimal spacer arms to preserve enzyme structure. Innovative heterofunctional supports or novel applications of existing supports could accelerate the adoption of various industrial bioprocesses, currently hindered by insufficient biocatalysts [100, 101]. Cysteine residues, featuring thiol functionality, are commonly employed in protein immobilization and readily participate in conjugate addition reactions with unsaturated carbonyls, such as maleimides, forming stable thioether bonds. It has been revealed that maleimide groups exhibit a strong preference for thiol conjugate addition at physiological pH (6.5–7.5) due to the predominant protonation of amines, rendering them unreactive (Fig. 7.14) [99].

Physical Entrapment

Immobilizing enzymes within gel matrices involves managing gel porosity to hinder enzyme leakage while ensuring unimpeded enzyme action. Li et al. [102] documented the fixation of papain enzyme onto cotton using the sol–gel technique. After six successive applications, the enzyme maintained 30% of its initial activity, while polyelectrolyte multilayer films demonstrated efficacy in enzyme immobilization onto textile surfaces. Singh et al. [103] introduced a method for fabricating cotton-based polyethylenimine films incorporating organophosphorus hydrolases. Entrapment for enzyme immobilization garners significant research attention due to

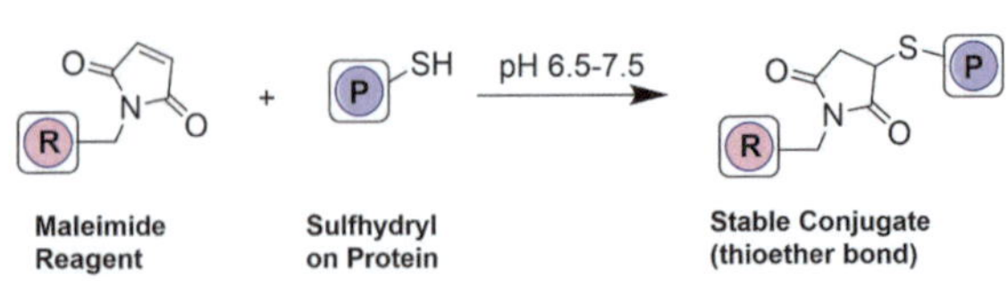

Fig. 7.14 Maleimide forming chemical conjugation to a sulfhydryl reaction

its ability to maintain gentle pH and temperature conditions, leading to prolonged enzyme stability.

Enzyme Crosslinking

Crosslinking of enzymes, facilitated by agents such as glutaraldehyde or polyamines, enhances bonding in immobilization processes, albeit potentially causing fabric harshness. In addition to traditional biochemical methods, recent advancements in enzyme immobilization technologies involve the utilization of single enzyme nanoparticles, encapsulating each enzyme molecule within a nanocapsule to enhance both stability and activity [104]. Elnashar et al. [97] have extensively explored alternative methods, including enzyme utilization, microwave exposure, and ionic liquid application for immobilization.

7.4.2 *Plasma Treatment*

The application of plasma to modify textile surfaces has a 40 years' history, yet its integration into large-scale textile production has progressed at a considerably slow pace. This can be elucidated by the historical reliance on low-pressure plasma techniques, limiting fabric treatment to batch processes, while recent advancements in atmospheric-pressure systems enable continuous processing [105]. Plasma comprises an ensemble of particles, including balanced quantities of free radicals, positive ions, electrons, uncharged species and ultraviolet (UV) radiation, generated through gas or vapor excitation within electromagnetic fields. Utilizing diverse plasma methods, researchers enhance fiber surfaces, offering flexibility in textile processing applications, either independently or as preparatory steps for subsequent treatments [106–109]. The formation of ultra-thin layers on textiles can be obtained through plasma techniques, including physical vapour deposition (PVD) and plasma enhanced chemical vapour deposition(PECVD) modes. The initial method employs plasma to vaporize coating material, which subsequently deposits onto the substrate, resulting in a thin, resilient film. In PECVD, plasma treatment of the substrate primes its surface for accelerated coating via CVD. Thermal stress on substrates is alleviated with the PECVD approach in contrast to CVD, while still obtaining comparable film quality; exemplified by silicon film deposition at 300–350 °C versus 650–850 °C in CVD. Horrocks et al. [110, 111] applied a clay and hexamethylene disiloxane (HMDSO) silicon coating on cotton treated with atmospheric plasma to enhance flash fire resistance, whereas Alongi et al. [112] found that cold oxygen plasma treatment followed by talcite and silica nanoparticles application on cotton fabric provides a superior flame retardant finish compared to untreated fabric. Malkov et al. [113] employed pulsed plasma-enhanced chemical vapor deposition (CVD) to deposit thin and even coatings of poly(allyl alcohol) onto silk, wool, and cotton fibers. Improvement in tensile strength was noted in both wool and silk subsequent to the treatment. The application of PECVD to apply TiO_2 films onto cotton knits has been explored in the context of photocatalytic functions [114]. Advancements in plasma-enhanced coatings extend to nonwoven materials for industrial settings as well. Particularly, polypropylene nonwoven fabrics have been subjected to carbon coatings through glow discharge plasma, utilizing propane-butane in the gaseous state to fulfill electromagnetic shielding needs [115]. Textiles coated with a $SiO_xC_yH_z$ thin film via PECVD utilizing hexamethyldisiloxane as a precursor compound demonstrated remarkable enhancement in wear resistance [116].

7.4.2.1 Improvement of Antibacterial Activity

Antibacterial properties of cellulose fabrics were efficiently obtained utilizing a DC magnetron sputtering apparatus, which applied a copper coating onto the fabric's surface devoid of any chemical or wet treatment. Furthermore, this process exhibited significantly shorter duration compared to conventional methods, which typically entail over 100 min involving detergents, metallic salts, and multiple washing cycles. Remarkably, the antibacterial efficacy persisted even after subjecting the fabrics to at least 30 laundering cycles [117]. Modification of cellulose fibers through plasma-induced cationic groups presents a viable approach for enhancing silver particle absorption and consequently improving antibacterial activity [118]. Plasma treatment of PP fibers enhances adhesion of copper particles, leading to heightened antibacterial effectiveness [119].

7.4.2.2 Decolorization of Denim

The efficacy of argon low temperature plasma in decolorizing denim was evidenced by its ability to produce a smooth surface finish and maintain fiber integrity, offering a shorter processing time of 15 min compared to over 90 min required for stonewashing, without the reliance on expensive materials or specialized enzymes [120]. Decolorization of denim through DC magnetron sputtering with argon or oxygen gasses indicates argon's effectiveness over oxygen initially, but post-treatment washing enhances appearance in oxygen-treated denim [121].

7.4.2.3 Improved Dyeability

Low-temperature plasma treatment of polypropylene fabrics enhances dye-ability: N_2 treatment enhances affinity for anionic dyes, whereas O_2 treatment increases cationic dye exhaustion, enabling straightforward dyeing with O_2-treated cationic dyes and offering new opportunities for polypropylene fabrics in textile manufacturing, underscoring the potential of plasma technology for textile modification [122–124].

Numerous studies have shown that the plasma treatment in textile industries has various applications such as improvement of hydrophilic properties [125], water repellency [126], anti-felting properties of wool [127], and flame retardant properties [128].

7.4.3 Layer by Layer Assembly Technique

Active materials can be deposited in order to produce thin composite films on the surface of textiles in an effective way by layer by layer (LbL) assembly techniques [5, 9]. The concept of LbL assembly was revealed in the 1960s with the development of films comprising boehmite fibrils (cations) and silica particles (anions) [9]. The technique utilizes solutions of poly-anions and poly-cations that are able to form films of multiple layers on the surface of the substrate by creating weak electrostatic attractions among the layers. The surface of the substrate is charged opposite to the ionic nature of the polyelectrolytes that construct the first layer of the film, as schematized in Fig. 7.15 [9]. The polyelectrolytes can be deposited onto the substrate surface either by immersing the substrate into the solutions or by spraying the solutions on the substrate surface [129]. Currently, some advanced deposition methods have been discovered that involve layer-producing polymers with bond-creating sites [130].

Complexation of polyelectrolytes is a LbL deposition method of developing films with tunable configuration by alternately immersing the substrate into the solutions containing polymers that can interact with each other and create polymer complexes while combined together. The thickness of the film produced in this way depends on the number of monolayers adsorbed on the substrate surface in each deposition cycle. Various functionality of the film can be endowed by incorporating several compounds

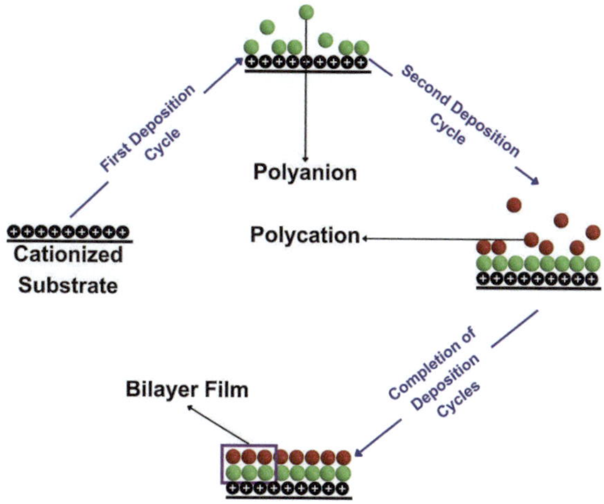

Fig. 7.15 Scheme of typical LbL deposition process on cationized textile substrates

like organic dyes, inorganic nanomaterials, biologically active molecules, electroactive polymers, electrochemically active species and semiconductor quantum dots [131]. LbL deposition films can also be produced by *stereocomplex formation* that refers to the generation of precisely arranged polymeric structures by creating weak Van der Waals forces among the synthetic polymers that have well-defined structures [132]. For example, Li et al. [133] presented polymeric helices formed by stereocomplexation between polylactic acids having opposite chirality that exhibited enhanced thermal and mechanical features.

It is possible to make multilayer films by means of *hydrogen bond formation* among the polymeric layers. For instance, Stockton et al. [134] developed a hydrogen bonding-based LbL deposition film by producing layers of nonionic polymers on polyaniline. The exploration of Benjamin et al. [135] illustrated an approach of producing self-assembled multilayer-films from copolymers of polyphenylenevinylene analogs having hydroxy moieties that can make hydrogen bonds with amine groups present in polyethyleneimine.

LbL disposition film can also be prepared by *host–guest interaction* between the polymeric layers [130]. This type of film is produced by either alternate deposition of anionic guest (polyelectrolytes) and cationic host (macrocyclic molecules) or cationic guest and anionic host. Tieke et al. [136] demonstrated that some building blocks of supramolecular chemistry like cyclodextrins, polyazacrown ethers and calixarenes possesses the ability to make host–guest complexes with substantially lower molecular weight compounds and ions.

Charge transfer among the electron acceptor and donor groups of layer forming polymers is another approach to develop LbL films. For example, Shimazaki et al. [137] presented that nanolayered films can be produced by self-assembly of polymethacrylates that consist of 3,5-dinitrobenzoyl groups (electron acceptor) and carbazole groups (electron donor). In this case, the intermolecular bindings are formed by the charge transfer interactions among the electron donor and acceptor groups. In this method various functionalities can be endowed by familiarizing numerous hydrophobic functional groups to the films as a consequence of using organic solvent during the film production [138].

7.4.3.1 Application of LbL Assembly in Textile Finishing

A technique of introducing ZnO nanoparticles to the multilayer film produced on cotton fabric surface was unraveled in 2010, where the fabric surface is coated with 10 to 16 ZnO nanoparticle layers by LbL deposition method posterior to cationization of the surface by 2,3-epoxypropyl trimethylammonium chloride. The finished fabric obtained in this way gains mentionable UV protective and antimicrobial properties [139]. Following the same method and using TiO_2 nanoparticles instead of ZnO nanoparticles can provide even better UV blocking. Moreover, the fabric with TiO_2

nanoparticle layers exhibits self-cleaning characteristics [140]. Durable UV protections can be applied to cotton fabrics by producing layers of fluorescent brightening agents and polydiallyldimethylammonium chloride in LbL deposition technique [141]. Flame retardant finish can be given to cotton fabrics by forming LbL films comprising sodium montmorillonite clay and branched polyethylenimine [142]. Recently, research has been conducted for the improvement of flame retardancy properties of cotton fabrics by formation of LbL deposition film of zinc phosphate compounds, where polyacrylic acid has been used as polyelectrolyte layers [143]. LbL assembly is utilized to create cellulose nanocrystal-graphene layers adhered by polyelectrolyte complexes on cotton fabric surface that provide superior electrical conductivity [144]. In the same way, cellulose nanocrystal-chitosan membranes can be formed which exhibit substantially higher oxygen barrier properties [145]. Remarkable fire retardancy can be induced by adsorption of P/N-dual-doped cellulose nanocrystals and Mxene (General formula-(transition metal)$_{n+1}$(carbon/nitrogen)$_n$) modified by tannic acid layers on cotton fabric surface [146].

Superior flame retardancy can be attributed to polyester fabrics (specifically Polyethylene terephthalate fabrics-PET) by LbL deposition technique where the fabric is immersed sequentially into positively charged suspension of silica nanoparticles coated with ammonia and negatively charged suspension of α-zirconium phosphate [147]. Furthermore, colloidal silica (negatively charged) and alumina capped silica (positively charged) can be used as the layers to reduce flammability of PET fabrics. In this process, washing of the substrate after each deposition cycle is important to remove extra ions before starting the consecutive deposition cycle (Fig. 7.16) [148].

An UV and chemical hazard protective finish can be provided by forming a film of TiO_2 nanoparticles to polyvinylidene chloride sheets that can be applicable for military personnels [149]. Alternate immersion of polyamide fabrics like silk and nylon into the solutions of polydiallyldimethylammonium chloride and polymethacrylic acid coated silver nanoparticles to impart antimicrobial characteristics. The layers appear to be more uniform on the fabric surface silk than that of nylon [150].

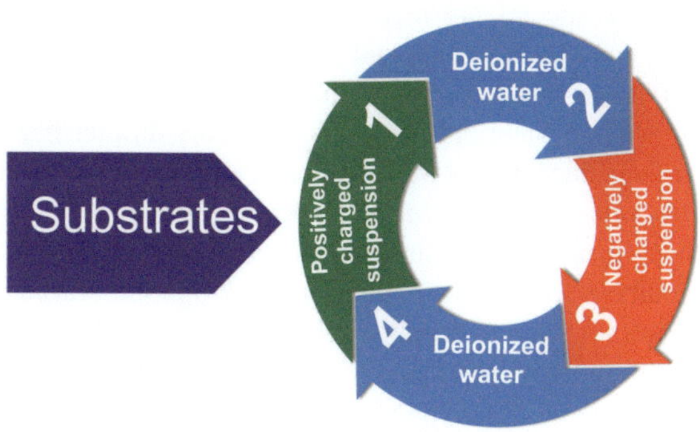

Fig. 7.16 Schematic presentation of process flow of LbL deposition on textile substrates

The weakness of LbL assembly is that it involves frequent washing stages, especially after each deposition cycle that results in generation of high amounts of waste solutions and high time-consumption [151].

7.4.4 *Nano-coatings*

The approach of forming nanoparticle coating on a substrate surface with a view to endowing high-level attributes such as breathability, UV protection, water repellency, antimicrobial properties, self-cleaning, durability, drape, clothing comfort, aesthetics and so forth, is referred to as nano-coating technique [10, 152]. Nano-coating can be applied on textiles in two techniques such as so-gel nano-coating and cross-linking, nanowire composite coating, electroless plating and so on.

7.4.4.1 Sol–Gel Nano-coating Methods

The consequences of the significance of sol–gel technology in textile finishing are its diminished environmental ramifications, lower chemical consumption and cost-effectiveness [153]. Sol–gel coating refers to the process that commences with the formation of sol, a colloidal suspension and transition of sol into gel by establishing networks within a continuous liquid phase [154]. The sols are typically applied on textiles in a pad-dry-cure method and after that, gelation of the applied sols is done by resting the coated substrate for a certain period of time, for example, 14 days [155].

Sol–gel method can be utilized to impart photocatalytic and superhydrophobic self-cleaning properties to textile substrates [154]. One example is that photocatalytic self-cleaning can be induced by the incorporation of TiO_2 nanoparticle sols which behave like a semiconductor as they produce negative (electrons) and positive (electron holes) species under the influence of UV radiation. The negative and positive species are responsible for breaking down water molecules and oxygen into superoxide anions and hydroxyl radicals respectively that are able to clean the pollutants from the textile surfaces [156].

When a water droplet holds a mercury droplet-like spherical shape on a substrate surface due to relatively less adhesion between the substrate and water than the cohesion among the water molecules, the substrate is said to be hydrophobic. When the hydrophobic substrate has a water contact angle ranging from 150° to 180°, the surface is specifically called superhydrophobic surface [157]. When a coated surface possesses superhydrophobicity, it gains self-cleaning properties [158]. In order to instill superhydrophobicity to a surface, one way is to impart microstructured or nanostructured roughness to the surface [159]. Another way is to customize the nanostructured surface with the chemical agents that are able to lower the surface energy [160]. Superhydrophobic self-cleaning properties can be inculcated in textiles by combining hydrophobic agents (such as silane monomers consisting of long alkyl chains) and additives (such as polysiloxanes) with the sols and forming a sol–gel coat on the surface. Not only hydrophobicity but oleophobicity can be introduced to textile substrates using oleophobic agents (like polymers containing perfluorinated alkyl side chains) fluorine-based additives (like monometic fluoroalkylsilane) while preparing the sols and subsequently developing coat in sol–gel coating technique [161]. There are research scopes to find alternatives for fluorine-based compounds in oleophobic coating since these are detrimental to health and environment.

7.4.4.2 Impregnation and Cross-Linking Method of Nano-coating

For improving the endurance of textile materials (particularly cellulosic and protein fibers and their manufactured assemblies) the incorporation of chemical spacers and cross-linking agents in the impregnation process is an impactful approach as it assists to develop bondings with the functional groups situated on the surface of textile materials [162]. In this method, nanoparticles like TiO_2 [163], zirconium dioxide (ZrO_2) [164] and carbon nanotubes [165] can be imparted to a substrate surface. As proteinous materials do not have sufficient reactive sites compared to the cellulosic materials, preparatory processes like oxidation of the surface with $KMnO_4$ in a weak acidic environment; treating with solution of succinic acid and dimethylformamide and so forth are the prerequisites for nano-coating in the cross linking method [163, 166].

Figure 7.17 [167] demonstrates the chemistry of incorporating titanium dioxide nanoparticles on the surface of cellulosic materials with the assistance of a spacer consisting of two terminal carboxylic acid (−COOH) groups. One of the terminal −COOH groups of a spacer molecule esterifies one of the hydroxyl groups at carbon number 6 in the glucose units of the cellulose. The other terminal −COOH group of that spacer helps to bind the TiO_2 nanoparticles by electrostatic force.

7.4.4.3 Nanowire Composite Coating

Enhanced mechanical properties, electrical conductivity and body heat retention are offered by coating textiles with nanowires. To impart heat retention property, the gaps

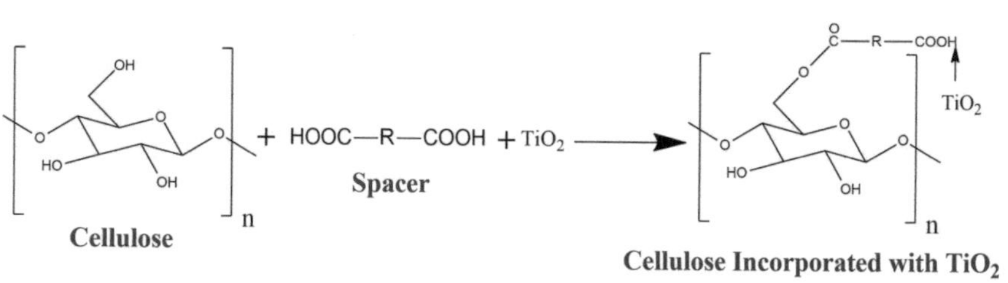

Fig. 7.17 The integration of TiO_2 nanoparticles onto the surface of cellulosic materials

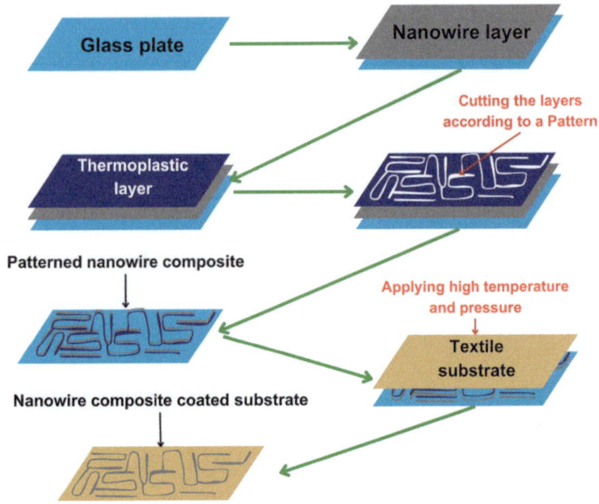

Fig. 7.18 Schematic presentation of nanowire composite coating method

among the nanowires are kept too small to pass the infrared radiations that are emitted from the human body. One of the prime advantages of nanowire coated textiles is that these resemble the wearability, breathability and flexibility of the normal uncoated cloths [168].

As per the method of coating demonstrated by Yao et al. [12], nanowires are suspended in ethanol to prepare nanowire ink that is then applied on a glass surface following Meyer rod coating technique. subsequently, the coat is dried on a hotplate. Coating and drying process is repeated several times to achieve optimum density of the nanowire network. Then, over the nanowire network, another dispersion containing thermoplastic material (such as polyurethanes/dimethylformamide dispersion) is applied by spin coating. The solvent of this dispersion evaporates on drying and the thermoplastic material is solidified and encapsulates the nanowire network. After that, by means of laser cutter, the nanowire composite is cut into a suitable pattern. Finally, a heat press machine is used to press the glass material with patterned nanowire composite onto the substrate with high pressure and temperature and thus the pattern is transferred from the thermoplastic film to the substrate surface. The process is illustrated in Fig. 7.18.

7.4.4.4 Conductive Coating

Apart from dip-coating the substrate with conductive materials, typical approaches to producing conductive textiles also involve spinning synthetic fibers after incorporating conductive additives into polymer solutions, spinning yarns from metal fibers and electroless metalization on the surface of yarns or fabrics.

Kayabaşı et al. [169] prepared a conductive nanoparticle dispersion using ethanol as a solvent and precisely applied it to a strand of textile fiber during drafting operation through a feed pump and a syringe. The fibers are then ring spun to produce conductive yarn consisting of conductive additives in both the interior and exterior parts.

According to the exploration of Hassan et al. [11], the electroless plating comprises three stages such as sensitization, activation, and metallization. Sensitization involves a mixture consisting of $SnCl_2$ and HCl in which the textile substrate (such as cotton fabric) is immersed so that the Sn^{2+} ions can be adsorbed onto the substrate surface. Activation, the subsequent stage, involves the treatment of the sensitized surface of textile substrate with another mixture of $AgNO_3$ and NH_4OH with a view to activate the surface for metallization process. After the activation stage, the substrate is transferred to a plating bath that contains a chelating agent, a catalyst, and a salt (consisting of the metal that is aimed to plate on the surface). The functions of chelating agent and catalyst (reducing agents like CH_2O) are to hinder the formation of $Cu(OH)_2$ in the plating bath and to facilitate the deposition of metals onto the substrate surface respectively. The metallized substrate is eventually washed by deionized water at a temperature near 40°C for around 20 min and dried.

7.4.4.5 Magnetron Sputter Coating

Magnetron sputter coating is a technique in which the disposition of thin films of metals or metal oxides on a substrate is carried out by applying high voltage. It involves a vacuum system where reactive gas molecules are generated from low-pressure gas under the influence of high voltage. The metal target (cathode) is bombarded by the reactive gas molecules, leading to the transfer of metal atoms onto the substrate (anode), as shown in Fig. 7.19 [170]. In case of longer sputtering time, the film gets higher compactness, grain sizes and conductivity [171].

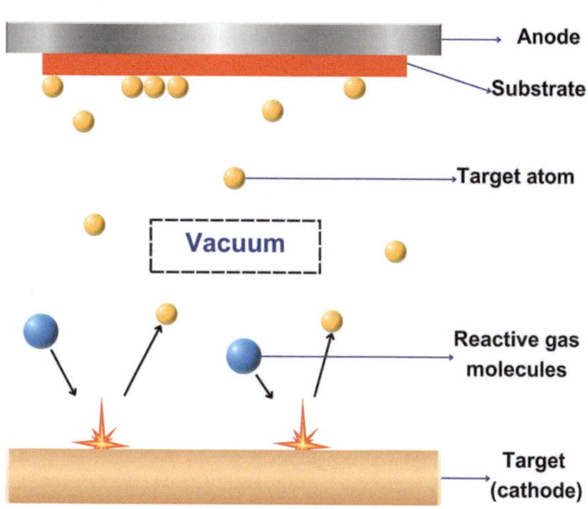

Fig. 7.19 Schematic presentation of magnetron sputter coating

7.4.5 *Supercritical Coating*

Within a supercritical carbon dioxide polymer coating setup, coating polymer may exist as a solute that is dissolved within CO_2, as a solvent that dissolves CO_2 or as a dispersant within CO_2. For instance, Shim et al. [172] illustrated a coating method using a suspension of coating polymers in supercritical CO_2 which can be uniformly sprayed on a substrate. To achieve better uniformity of film, a siloxane based surface active agent is added to the mixture.

Decompressive atomisation is a technique to generate small droplets (20–50 μm) consisting of coating materials that eventually create a coating on the substrate, getting a form of highly uniform atomized spray. The consequence of superior uniformity of the spray is the production of substantial internal forces that surpass both surface tension and cohesive forces due to quick release of pressure and rapid evaporation of CO_2 [173].

Solutions of metal–organic precursors can be prepared using supercritical CO_2 as media that are capable of forming metal and metal oxide-films on the substrate surface. The adsorbed metal films impart antimicrobial properties to the textiles. Bioactive materials can be incorporated in textiles in order to confer antimicrobial properties with the assistance of a supercritical CO_2 medium [174]. A suspension of silver metal in supercritical CO_2 can provide antifungal finish to the substrate. Supercritical CO_2 can also act as a medium for electroless plating on textile materials [175]. Furthermore, supercritical CO_2 can be utilized as a medium to give a silicon-based finish to cotton fabric, employing modified dimethylsiloxane polymers. In this process, films are produced by creation of three dimensional networks between dimethylsiloxane polymers and cross linking agents [176]. To obtain durable antimicrobial properties, chitosan-lactic acid salt can be applied via supercritical CO_2 media on the surface of polyethylene terephthalate fabrics [177] to impart water or oil repellency, the same media can also be used to incorporate organic fluorine to polyester fabrics [178, 179].

7.5 Functional Finishing of Textiles with Emerging Polymers

7.5.1 *Easy-Care and Wrinkle-Resistant Finishes*

In order to recover dry and wet wrinkles in fabrics after washing and wearing, easy-care finishing techniques are applied. Additionally this technique enhances shrinkage and swelling resistance properties of the fabrics [180]. Since the 1920s, easy care finishing of cellulosic substrates has been carried out by cheap organic cross-linking agents like urea–formaldehyde and melamine–formaldehyde. Formaldehyde compounds, being volatile, can pose a direct risk to workers directly engaged in their production as well as individuals wearing the final product through inhalation. Moreover, these toxic substances can penetrate the skin of the wearer causing serious harm to his/her health. Therefore, safety and health concerns regarding the application of formaldehyde in the easy care textiles emerged in the 1960s and 1970s. Considering environment and health safety, a number of researches has been conducted to develop cross-linking agents having less or no formaldehyde content [181].

N, N-dimethylol-4,5-dihydroxyethylene urea (DMDHEU), a product synthesized from glyoxal, Urea and trace amounts (around 0.3%) of formaldehyde, inhibits shrinkage and appearance of wrinkles in cellulosic materials as it cross-links under stress and constraints the mobility of the molecules of the fibers [182]. From gloxal and N, N-dimethyl urea, a non-formaldehyde cross-linking agent called N, N-dimethyl-4,5-dihydroxyethylene urea (DMeDHEU), has been synthesized. Though DMeDHEU is safer for health and environment, it is much more expensive compared to DMDHEU. Furthermore, DMeDHEU is comparatively less reactive than other finishing agents, leading to implementation of powerful catalysts or intensified conditions of reaction. To obtain easy-care finish with better cost effectiveness, and minimal formaldehyde content, an equally proportioned blend of DMDHEU and DMeDHEU can be used [181]. 1,2,3,4-butanetetracarboxylic acid (BTCA), another finishing agent for easy-care of textiles that is free from formaldehyde, provides satisfactory recovery of wrinkles. Nevertheless, it is more expensive than DMeDHEU and less durable to wash [183]. The cross-linking properties of BTCA can be elevated by the application of phosphorus catalyst. It offers minimal strength degradation and long-lasting durability up to 20 washes. In addition, it does not cause discoloration of dyed fabrics during or after the finishing process [182].

Citric acid is an inexpensive compound that has the potential to be utilized in easy-care finishing of cellulosic materials [184]. For controlling shrinkage and enhancing easy care properties of wool fabrics, a mixture of formic acid, and citric acid has been reported to be effective [185]. The drawback of applying polycarboxylic acids is that it needs a significant quantity of hydrose, an expensive reducing agent that strips color from the dyed fabric.

Some other easy-care finishes such as carbamates, dimethylol ethylene, diepoxides, diisocyanates, urea, propylene urea, triazons, urons, and diglyoxal urea are expensive and have some technical shortcomings. Consequently, they have not yet been used in commercial scale finishing processes [181]. Silicone softeners have been reported as a formaldehyde-free and environment-friendly finishes that possess upgraded easy-care properties [186, 187].

7.5.2 *Water and Stain Repellency*

For water repellent finishing, sometimes, silicone or fluorocarbon are recommended to apply on fabrics [188, 189]. Xue et al. [190] have prepared superhydrophobic cotton materials by padding with polyacrylates and polyphosphate ammonium. The self-cleaning properties of the film have endowed water repellency to the coated textiles. Furthermore, the coating is capable of reducing flammability of the materials. Their further study have revealed newer self-cleaning materials having superior chemical stability which have been produced by incorporating polymers based on polyhedral oligomeric silsesquioxane [191]. In 2018, Cai et al. [192] showed a straightforward and environmentally sustainable approach of coating fabrics by dipping them into a dispersion containing polydimethylsiloxane. The materials finished by this method have superhydrophobicity and the surfaces of the materials have produced a 145° angle with the droplets of water. Point to be noted that the coated materials that exhibit self-cleaning properties obviously display water and stain repellency.

7.5.3 Flame Retardancy

Flame retardant finishes are essential for ensuring the safety and compliance of textiles in various applications, like protective clothing, apparel, and upholstery. The development of effective flame-retardant formulations with high performance necessitates specific attributes affordability, safety and eco-friendliness, substantial compatibility, enduring stability in natural or industrial settings, and outstanding flame-retardant capabilities. The initial phase in formulating such compositions involves the selection of appropriate combinations of flame-retardant constituents [193]. Traditional flame retardants often contain halogenated compounds or heavy metals, raising environmental and health concerns [194]. Emerging polymer-based flame retardants provide viable alternatives with superior efficacy and sustainability. Phosphorus-based polymers, such as polyphosphazenes and phosphorus-containing polyesters, demonstrate excellent flame-retardant properties by forming intumescent char layers upon exposure to heat, effectively inhibiting flame spread and heat release. Nanocomposite approaches are also incorporating flame-retardant nanoparticles within polymeric matrices offer synergistic effects, enhancing flame retardancy while minimizing additive loading and environmental impacts [195]. Bio-derived flame retardants possess a deluge of functionalities that offer opportunities to scientists for additional modifications or utilizing non-covalent interactions. Inexpensive and readily available in nature, bio-based flame retardants like phytic acid and cellulose exhibit low toxicity levels, thus validating the consideration as viable components in flame-retardant formulations [193].

7.5.4 Antibacterial and Antimicrobial Properties

In current times, antimicrobial textiles are essential in hospitals, environments, and high-microbial-risk areas, safeguarding against the transfer of harmful microorganisms present on garments worn by patients, healthcare staff, and practitioners, presenting significant commercial prospects in infection control [196]. Textiles with antimicrobial properties can be categorized based on their ability to combat specific types of microbes, like fungi, bacteria, or viruses. Antimicrobial textiles exhibit specificity towards microbes, including antibacterial, antifungal, or antiviral properties, with some capable of targeting multiple types of microorganisms simultaneously [197]. Surface modification strategies encompassing polymerization, plasma treatment, electrospinning, microencapsulation, nanotechnologies, and sol–gel techniques have been employed to confer advanced functionalities onto textiles, such as antibacterial properties, water repellency, and flame retardancy [198]. The adoption of antimicrobial textiles for winter clothing is driven by the prolonged storage and minimal washing frequency, factors that predispose to microbial growth, making them an essential choice for maintaining hygiene and microbial control.

Silver nanoparticles (AgNPs) integrated into textile coatings, whether natural or synthetic, exhibit robust antimicrobial properties against a wide spectrum of microbes, coupled with prolonged durability, heightened dye absorption, and minimal cytotoxicity to human cells [199]. Silver nanoparticles (approximately 10 nm in diameter) have also shown antiviral effects against SARS-CoV-2, the virus responsible for COVID-19 [200]. Apart from silver, a range of metals and metal oxide nanoparticles, including Zn, Ti, Sn, Au, and Cu have been utilized in various natural and man-made textile applications [201]. The synthesis and utilization of nanostructured inorganic materials play a crucial role in the innovation of antimicrobial textiles, exemplified by the transformation of selenium nanoparticles into brooms over time through almond skin extract, with resulting cotton-coated fabric exhibiting activity against *Bacillus subtilis* [202]. The development of a functional nanohybrid material, blending oligochitosan sourced from crab shells with nano-silica extracted from rice husk, demonstrated remarkable efficacy against *Phytophthora infestans* fungal threats, indicating a significant stride in sustainable agricultural innovation [203]. Cotton fabric treated with seaweed-capped zinc oxide nanoparticles (SW-ZnO NPs) using the pad-dry-cure method, exhibiting antibacterial action against both gram-negative (*E. coli* and *K. aerogenes*) and gram-positive (*S. aureus* and *S. pyogenes*) bacteria [204].

Developing sustainable antimicrobial textile finishing methods involves harnessing natural compounds like cyclodextrins, renowned for their hydrophilic outer layer and lipophilic central cavity [205]. The incorporation of sugarcane bagasse lignin onto fabric resulted in antibacterial effects against *Staphylococcus epidermidis* [206]. Based on the minimum inhibitory concentration (MIC), the lignin coating concentration was adjusted to hinder bacterial growth within 6 h of interaction [207]. Applying chitosan hydrogel onto cellulose-based fabrics demonstrated antimicrobial effectiveness against *S. aureus*, *L. monocytogenes,* and *E. coli* [208].

7.5.5 UV Protection

Finishing chemical refers to UV shielding agents that are applied on textile fabric surfaces to secure human body surface from the damages caused by the radiations of

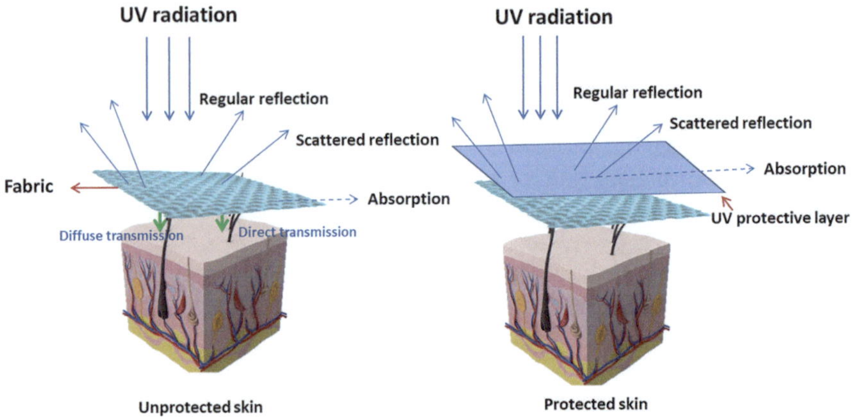

Fig. 7.20 The interactions between UV radiation, fabric and human skin

ultraviolet rays [209]. As UV radiation exhibits higher energy, they can cause skin damage, skin aging, skin cancer and DNA degradation etc. as well as UV radiations are capable of degrading polymers of textile fabrics and impart color fading and other physical and chemical changes [210]. Consequently, UV protective textile finishes have been gaining popularity for producing sportswear, wearable sensors and high altitude apparel and covering clothing [209]. Among three type of UV radiations, UV-A, UV-B and UV-C have the weave length (315 to 400 nm), (280 to 315 nm) and (200 to 280 nm) respectively, the maximum portion of UV-C is absorbed in the stratosphere and UV-B and UV-A enter into the atmosphere where UV-A can deeply penetrate into the skin [211]. Figure 7.20 shows the interactions between UV radiation, fabric and human skin.

The value of Ultraviolet protection factor (UPF) is the measure of UV protection capability of textile materials. EN 13,758 standard suggests UPF value for protective clothing is more than 40 and the UVA a transmission is less than 5% [212]. In the textile industry, for the enhancement of UV protection of fabrics, three techniques are commonly applied such as application of TiO_2, fluorescent whitening agents (FWAs), and application of UV absorbers. When UV radiation enters into any material, the radiation is scattered if the material is finished by TiO_2 so that the UV permeability decreases [213]. FWAs help the textile materials to achieve higher UPF value by whitening and visible blue emission [214]. Furthermore, the UV absorbers are organic and inorganic substances which are capable of absorbing short wavelength rays [215]. UV absorbing compounds consist of aromatic rings of long chain structure and side chains with pi-bonds and $C = O$ groups [216]. The organic UV absorbers like tannin, flavone, luteolin, baicalin, aloin and lutein are used for UV protective finish which are shown in Fig. 7.21. The commercially available FWAs used in textile finishing are shown in Fig. 7.22 [209].

The most popular inorganic UV absorbers are oxides of chromium, titanium and iron, and carbon black [217]. Besides the metal oxides, some pigments and dyes

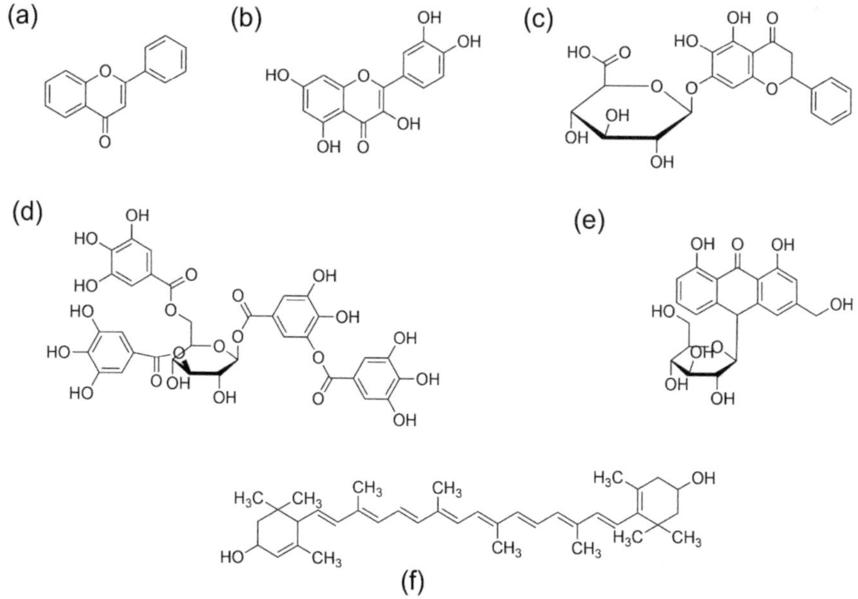

Fig. 7.21 Natural organic UV absorbers: lavone (**a**), luteolin (**b**), baicalin (**c**), tannin (**d**), aloin (**e**) and lutein (**f**)

Fig. 7.22 Fluorescent whitening agents

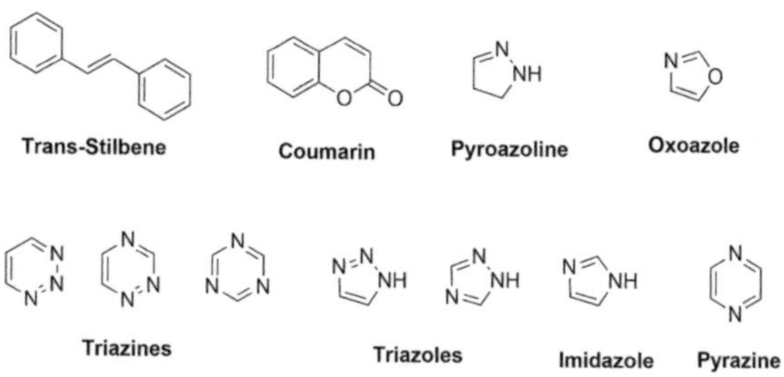

are directly applied on textile substrates to enhance UV protection capacity. Broasca et al. used ZnO for UV absorbing of textile material and found that the final sample became hydropic in nature and capable of absorbing UV radiation [218].

7.5.6 Conductive Textile Fibers/Yarn

Generally, textile materials are non-conductive in nature. To impart conductivity on textile materials, conductive nanoparticles, polymers or compounds can be employed on the surface of textiles. In addition, these conductive materials further can be used in wearable electronics [219]. There are some factors which are closely related with the conductive behavior of textiles, for example fabric density, roughness, amount of pores as well as the uniformity of applied conductive materials. In comparison with rough surfaces, smooth surfaces exhibit better conductivity [220].

There are two types of conductive polymer used in textile. Intrinsic conductive fibers/yarns which are completely conductive material based. They are produced by using electrospinning, wet spinning and melt spinning [65]. After producing the yarn, it turned into fabric by weaving, knitting or braiding. The materials used for intrinsic conductive textile are copper, aluminium titanium, nickel, ferrous alloys and stainless steel [221]. The extrinsics are produced by combining conductive and nonconductive materials. Conductive materials are applied on textile substrates by mixing, blending or coating. Conductive filler like metallic nanowires, metallic powder, carbon black, carbon nanotube, indium tin oxide and nonconductive material polyethylene, polypropylene, and polystyrene are used for producing extrinsic [222].

7.6 Practical Instances and Applications

Water repellent and waterproof fabrics can be used as an exterior layer for waterproof garments. Besides their traditional use like umbrella, suit, coat etc. they also used as functional textile such as automotive engine filter, swimsuit, cardboard, rain coat etc. [223]. Flame retardant textile materials are mainly used to protect a person, device or property from fire. During the burning process of flame retardant textile non-flammable gas like CO_2 and H_2O is produced which hinder the spread of flame. Furthermore, when fabric is treated with flame retardant material the fabric reduces the absorption of heat energy in the burning process [224]. Flame retardant garments are used in different fields, including medical, automotive and military applications [225]. UV protective clothing are mainly used for protecting human skin from cancer causing radiations so that hats, awnings, cover sheets, parachutes and tents are produced from UV protective textiles [213]. Conductive textiles are used in EMI shielding, heating pads, sensors, actuators, antistatic materials, automotive, wearable health monitoring, and flexible electrode [226].

7.7 Conclusion

Textile polymer finishing is mainly shifted to functional finishing rather than traditional finishing which manifest both challenges and opportunities. Besides the properties like comfort, health and hygiene, easy-care, customers also demand protection from thermal, chemical, mechanical and biological damage. After finishing with functional polymers, textile materials become flame retardant, antimicrobial, UV protective, wrinkle resistant, waterproof and many more. So, researches are being conducted for the improvement of overall functionality of textile materials. Smart, interactive and intelligent textiles will tentatively be produced by using textile finish in future.

Declaration of Competing Interest The authors declare that they have no known competing financial interests or personal relationships that could have appeared to influence the work reported in this chapter. Generative AI and AI-assisted technologies have been used in the writing process to improve the readability and language of the work only. After using this tool, the authors reviewed and edited the content as needed and take full responsibility for the content of the publication.

References

1. Bouasria, A., Nadi, A., Boukhriss, A., Hannache, H., Cherkaoui, O., & Gmouh, S. (2020). Advances in polymer coating for functional finishing of textiles. In *Frontiers of Textile Materials: Polymers, Nanomaterials, Enzymes, and Advanced Modification Techniques* (pp. 61–86).
2. Jajpura, L., Rangi, A., & Khandual, A. (2020). Natural finishes, technologies and recent developments. In *Sustainable Technologies for Fashion and Textiles* (pp. 209–229).
3. Hu, J., & Lu, J. (2014). Smart polymers for textile applications. In *Smart Polymers and their Applications* (pp. 437–475). Elsevier.
4. Zannat, A., Uddin, M. N., Mahmud, S. T., Mia, R., & Ahmed, T. (2024). Natural dyes and pigments in functional finishing. In *Renewable Dyes and Pigments* (pp. 271–287). Elsevier
5. Gulrajani, M., & Deepti, G. (2011). *Emerging techniques for functional finishing of textiles.*
6. Mia, R., Mofasser, A., Bhat, M. A., Howlader, M. I., Sayed, N. I., & Bakar, M. A. (2024). Utilizing neem wood waste as a natural dye in the development of bio-based functionalized colored jute fabric. *Sustainable Chemistry and Pharmacy, 38*, Article 101489. https://doi.org/10.1016/j.scp.2024.101489
7. Haque, M. A., et al. (2022). Sustainable dyeing and functionalization of wool fabrics with black rice extract. *Resources, Environment and Sustainability, 7*, Article 100045. https://doi.org/10.1016/j.resenv.2021.100045
8. Saha, P. K., Mia, R., Zhou, Y., & Ahmed, T. (2021). Functionalization of hydrophobic nonwoven cotton fabric for oil and water repellency. *SN Applied Sciences, 3*(5), 586. https://doi.org/10.1007/s42452-021-04582-9
9. Magovac, E., Vončina, B., Jordanov, I., Grunlan, J. C., & Bischof, S. (2022). Layer-by-layer deposition: A promising environmentally benign flame-retardant treatment for cotton, polyester, polyamide and blended textiles. *Materials, 15*(2), 432.
10. Syduzzaman, M., Patwary, S. U., Farhana, K., & Ahmed, S. (2015). Smart textiles and nano-technology: A general overview. *Journal of Textile Science and Engineering, 5*(1), 1–7.
11. Hassan, Z., Atalay, O., & Kalaoglu, F. (2022). Conductive cotton fabric using laser pre-treatment and electroless plating. *The Journal of The Textile Institute, 113*(5), 737–747.
12. Yao, S., Yang, J., Poblete, F. R., Hu, X., & Zhu, Y. (2019). Multifunctional electronic textiles using silver nanowire composites. *ACS Applied Materials & Interfaces, 11*(34), 31028–31037.
13. Bajaj, P. (2002). Finishing of textile materials. *Journal of Applied Polymer Science, 83*(3), 631–659.
14. Paz-Gómez, G., et al. (2019). Water-repellent fluoropolymer-based coatings. *Coatings, 9*(5), 293.
15. Varesano, A., Vineis, C., Aluigi, A., & Rombaldoni, F. (2011). Antimicrobial polymers for textile products. *Science Against Microbial Pathogens: Communicating Current Research and Technological Advances, 3*, 99–110.
16. Schindler, W. D., & Hauser, P. J. (2004). 4—Hand building finishes. In W. D. Schindler, & P. J. Hauser (Eds.), *Chemical finishing of textiles. Woodhead publishing series in textiles*

(pp 43–50). Woodhead Publishing. ISBN 9781855739055. https://doi.org/10.1533/9781845690373.43. https://www.sciencedirect.com/science/article/pii/B9781855739055500082
17. Yang, C. Q., Qian, L., & Lickfield, G. C. (2001). Mechanical strength of durable press finished cotton fabric. Part IV: Abrasion resistance. *Textile Research Journal, 71*(6), 543–548. https://doi.org/10.1177/004051750107100612
18. Bourbigot, S. (2008). Flame retardancy of textiles: New approaches. In *Advances in Fire rRetardant Materials* (pp. 9–40). Elsevier.
19. Mahltig, B., & Fischer, A. (2010). Inorganic/organic polymer coatings for textiles to realize water repellent and antimicrobial properties—a study with respect to textile comfort. *Journal of Polymer Science Part B: Polymer Physics, 48*(14), 1562–1568.
20. Zhara, H., Adeel, S., Özomay, Z., & Mia, R. (2024). Properties and performance relationship of biopolymers in textile industry. In *Biopolymers in the Textile Industry: Opportunities and Limitations* (pp. 87–121). Springer.
21. Lv, H., Leng, J., Liu, Y., & Du, S. (2008). Shape-memory polymer in response to solution. *Advanced Engineering Materials, 10*(6), 592–595.
22. Maity, S., & Chatterjee, A. (2018). Conductive polymer-based electro-conductive textile composites for electromagnetic interference shielding: A review. *Journal of Industrial Textiles, 47*(8), 2228–2252.
23. Mallakpour, S., Radfar, Z., & Hussain, C. M. (2021). Current advances on polymer-layered double hydroxides/metal oxides nanocomposites and bionanocomposites: Fabrications and applications in the textile industry and nanofibers. *Applied Clay Science, 206*, Article 106054.
24. Babu, R. P., O'connor, K., & Seeram, R. (2013). Current progress on bio-based polymers and their future trends. *Progress in biomaterials, 2*, 1–16.
25. Pan, Y., Farmahini-Farahani, M., O'Hearn, P., Xiao, H., & Ocampo, H. (2016). An overview of bio-based polymers for packaging materials. *Journal of Bioresources and Bioproducts, 1*(3), 106–113.
26. Azmana, M., Mahmood, S., Hilles, A. R., Rahman, A., Arifin, M. A. B., & Ahmed, S. (2021). A review on chitosan and chitosan-based bionanocomposites: Promising material for combatting global issues and its applications. *International Journal of Biological Macromolecules, 185*, 832–848.
27. Mia, R., Sk, M. S., Oli, Z. B. S., Ahmed, T., Kabir, S., & Waqar, M. A. (2021). Functionalizing cotton fabrics through herbally synthesized nanosilver. *Cleaner Engineering and Technology, 4*, Article 100227. https://doi.org/10.1016/j.clet.2021.100227
28. Sharif, R., Mohsin, M., Ramzan, N., Sardar, S., & Anam, W. (2022). Synthesis of bio-based non-fluorinated oil and water repellent finishes for cotton fabric by using palmitic acid, succinic acid, and maleic acid. *Journal of Natural Fibers, 19*(16), 14077–14088.
29. Patti, A., & Acierno, D. (2022). Towards the sustainability of the plastic industry through biopolymers: Properties and potential applications to the textiles world. *Polymers, 14*(4), 692.
30. F. H. H. Abdellatif, & Abdellatif, M. M. (2021). Utilization of sustainable biopolymers in textile processing. In *Green Chemistry for Sustainable Textiles* (pp. 453–469). Elsevier.
31. Ali, S. W., Chowdhury, A., Banerjee, S., & Bairagi, S. (2022). Bio resources mediated technological advancements in chemical finishing of textiles. In *Applications of Biotechnology for Sustainable Textile Production* (pp. 187–220). Elsevier.
32. Lim, S.-H., & Hudson, S. M. (2003). Review of chitosan and its derivatives as antimicrobial agents and their uses as textile chemicals. *Journal of macromolecular science, part C: Polymer reviews, 43*(2), 223–269.
33. Valle, J. A. B., Valle, R. D. C. S. C., Bierhalz, A. C. K., Bezerra, F. M., Hernandez, A. L., & Lis Arias, M. J. (2021). Chitosan microcapsules: Methods of the production and use in the textile finishing. *Journal of Applied Polymer Science, 138*(21), 50482.
34. Zhou, B. C.-E., Kan, C.-W., Sun, C., Du, J., & Xu, C. (2019). A review of chitosan textile applications. *AATCC Journal of Research, 6*(1_suppl), 8–14.
35. Mia, R., et al. (2022). Natural dye extracted from Triadica sebifera in aqueous medium for sustainable dyeing and functionalizing of viscose fabric. *Cleaner Engineering and Technology, 8*, Article 100471. https://doi.org/10.1016/j.clet.2022.100471

36. Khandual, A., Pai, R. S., Jajpura, L. R., & Das, S. (2004). Sizing processes and its application. *Colourage, 51*(11), 33–40.
37. Bhaskara-Amrit, U. R., Agrawal, P. B., & Warmoeskerken, M. M. (2011). Applications of b-cyclodextrins in textiles. *AUTEX Research Journal, 11*(4), 94–101.
38. Gulrajani, M., Brahma, K., Kumar, P. S., & Purwar, R. (2008). Application of silk sericin to polyester fabric. *Journal of Applied Polymer Science, 109*(1), 314–321.
39. Belhaj Khalifa, I., Ladhari, N., & Touay, M. (2012). Application of sericin to modify textile supports. *Journal of the Textile Institute, 103*(4), 370–377.
40. Lee, K. Y., & Mooney, D. J. (2012). Alginate: Properties and biomedical applications. *Progress in Polymer Science, 37*(1), 106–126.
41. Qin, Y. (2004). Absorption characteristics of alginate wound dressings. *Journal of Applied Polymer Science, 91*(2), 953–957.
42. Freudenberg, K., & Neish, A. C. (1968). *Constitution and biosynthesis of lignin.*
43. dos Santos Abreu, H., do Nascimento, A. M., & Maria, M. A. (1999). Lignin structure and wood properties. *Wood and Fiber Science*, 426–433.
44. Zimniewska, M., Kozłowski, R., & Batog, J. (2008). Nanolignin modified linen fabric as a multifunctional product. *Molecular Crystals and Liquid Crystals, 484*(1), 43/[409]–50/[416].
45. Li, B., Zhang, X., & Su, R. (2002). An investigation of thermal degradation and charring of larch lignin in the condensed phase: The effects of boric acid, guanyl urea phosphate, ammonium dihydrogen phosphate and ammonium polyphosphate. *Polymer Degradation and Stability, 75*(1), 35–44.
46. Kumar, P., Di Mauro, E., Zhang, S., Pezzella, A., Soavi, F., Santato, C., & Cicoira, F. (2016). Melanin-based flexible supercapacitors. *Journal of Materials Chemistry C, 4*(40), 9516–9525. https://doi.org/10.1039/C6TC03739A
47. Di Mauro, E., Xu, R., Soliveri, G., & Santato, C. (2017). Natural melanin pigments and their interfaces with metal ions and oxides: Emerging concepts and technologies. *MRS Communications, 7*(2), 141–151.
48. Hu, J., Meng, H., Li, G., & Ibekwe, S. I. (2012). A review of stimuli-responsive polymers for smart textile applications. *Smart Materials and Structures, 21*(5), Article 053001.
49. Mondal, S. (2008). Phase change materials for smart textiles–an overview. *Applied Thermal Engineering, 28*(11–12), 1536–1550.
50. Bashari, A., Hemmati Nejad, N., & Pourjavadi, A. (2013). Applications of stimuli responsive hydrogels: a textile engineering approach. *Journal of the Textile Institute, 104*(11), 1145–1155.
51. Thakur, S. (2017). Shape memory polymers for smart textile applications. *Textiles for Advanced Applications, 10.*
52. Hassan, M. S. (2009). Crease recovery properties of cotton fabrics modified by urea resins under the effect of gamma irradiation. *Radiation Physics and Chemistry, 78*(5), 333–337.
53. Pause, B. (2000). Textiles with improved thermal capabilities through the application of phase change material (PCM) microcapsules. *Melliand Textilberichte International Textile Reports, 81*(9), 753–754.
54. Zhang, W., et al. (2021). Thermally regulated cotton fabric coated with expanded graphite stabilized paraffin mixture as composite phase change material. *Pigment & Resin Technology, 50*(4), 293–301.
55. Langer, R. (1998). Drug delivery and targeting. *Nature, 392*(6679 Suppl), 5–10.
56. Jocić, D. (2010). Functional finishing of textiles with responsive polymeric systems. In D. Jocić (Ed.), *Surface modification systems for creating stimuli responsiveness of textiles. Workshop proceedings: 6th framework programe ADVANBIOTEX of the EU* (pp. 37–59). University of Twente.
57. Samal, S. K., Dash, M., Dubruel, P., & Van Vlierberghe, S. (2014). Smart polymer hydrogels: Properties, synthesis and applications. In *Smart Polymers and Their Applications* (pp. 237–270).
58. Ahn, S.-K., Kasi, R. M., Kim, S.-C., Sharma, N., & Zhou, Y. (2008). Stimuli-responsive polymer gels. *Soft Matter, 4*(6), 1151–1157.

59. Langer, R., & Peppas, N. A. (2003). Advances in biomaterials, drug delivery, and bionanotechnology. *AIChE Journal, 49*(12), 2990–3006.
60. Galaev, I., & Mattiasson, B. (2003). *Stimulus responsive surfaces. Possible implications for bioseparation, cell cultivation and drug delivery.*
61. Crespy, D., & Rossi, R. M. (2007). Temperature-responsive polymers with LCST in the physiological range and their applications in textiles. *Polymer International, 56*(12), 1461–1468.
62. Heskins, M., & Guillet, J. E. (1968). Solution properties of poly (N-isopropylacrylamide). *Journal of Macromolecular Science—Chemistry, 2*(8), 1441–1455.
63. Wang, T., Turhan, M., & Gunasekaran, S. (2004). Selected properties of pH-sensitive, biodegradable chitosan–poly (vinyl alcohol) hydrogel. *Polymer International, 53*(7), 911–918.
64. Bonina, P., Petrova, T., & Manolova, N. (2004). PH-sensitive hydrogels composed of chitosan and polyacrylamide–preparation and properties. *Journal of Bioactive and Compatible Polymers, 19*(2), 101–116.
65. Grancarić, A. M., et al. (2018). Conductive polymers for smart textile applications. *Journal of Industrial Textiles, 48*(3), 612–642.
66. Gu, J. F., Gorgutsa, S., & Skorobogatiy, M. (2010). Soft capacitor fibers using conductive polymers for electronic textiles. *Smart Materials and Structures, 19*(11), Article 115006.
67. Onggar, T., Kruppke, I., & Cherif, C. (2020). Techniques and processes for the realization of electrically conducting textile materials from intrinsically conducting polymers and their application potential. *Polymers, 12*(12), 2867.
68. Mia, R., & Sultana, S. (2020). Fabrication and properties of silver nanowires (AgNWs) functionalized fabric. *SN Applied Sciences, 2*(12), 2052. https://doi.org/10.1007/s42452-020-03845-1
69. Zannat, A., Uddin, M. N., Mahmud, S. T., Prithu, P. S. S., & Mia, R. (2023). Textile-based soft robotics for physically challenged individuals. *Journal of Materials Science, 58*, 12491–12536. https://doi.org/10.1007/s10853-023-08799-4
70. Alhashmi Alamer, F., Badawi, N. M., Alodhayb, A., Okasha, R. M., & Kattan, N. A. (2020). Effect of dopant on the conductivity and stability of three different cotton fabrics impregnated with PEDOT: PSS. *Cellulose, 27*, 531–543.
71. Tseghai, G. B., Mengistie, D. A., Malengier, B., Fante, K. A., & Van Langenhove, L. (2020). PEDOT: PSS-based conductive textiles and their applications. *Sensors, 20*(7), 1881. https://doi.org/10.3390/s20071881
72. Roemer, M. (2004). *Elektrochemo-und mikromechanisches Verhalten elektronisch leitfähiger Polymere.*
73. Crimmann, P. (2003). *Grenzflächenmodifizierung und-analyse im Polypropylen-Kupfer-Verbund*. Halle (Saale), Univ., Diss.
74. Saidman, S. (2003). Influence of anion and pH on the electrochemical behaviour of polypyrrole synthesised in alkaline media. *Electrochimica acta, 48*(12), 1719–1726.
75. Chauvet, O., et al. (1994). Magnetic and transport properties of polypyrrole doped with polyanions. *Synthetic metals, 63*(2), 115–119.
76. Bredas, J. L., & Street, G. B. (1985). Polarons, bipolarons, and solitons in conducting polymers. *Accounts of chemical research, 18*(10), 309–315.
77. Perez, P. N. G., Sino, P. A. L., Herrera, M. U., & Tapia, A. K. G. (2018). Conducting properties of polyaniline emeraldine salt on paper in the low-frequency region. In *Materials Science Forum* (vol. 923, pp. 72–76). Trans Tech Publications.
78. Manding, K. V. S., Lopez, J. E. S., & Tapia, A. K. G. (2018). AC conduction in polyaniline Emeraldine salt-graphite pellets. In *Materials Science Forum* (vol. 923, pp. 66–71). Trans Tech Publications.
79. Mäkelä, T. (2008). *Towards printed electronic devices. Large-scale processing methods for conducting polyaniline.*
80. Marr, I., Stöcker, T., & Moos, R. (2013). F3-resistives Gasdosimeter auf Basis von PEDOT: PSS zur Detektion von NO und NO2. *Tagungsband,* 317–320.

81. Lange, U., & Mirsky, V. M. (2011). Chemiresistors based on conducting polymers: A review on measurement techniques. *Analytica Chimica Acta, 687*(2), 105–113.
82. Mia, R., Habib, S., Assiri, M. A., Liu, H., & Mahmud, S. (2023). Coloration and antibacterial treatment of aramid fabric: A comprehensive review. *Surface Innovations, 12*(3–4), 130–146. https://doi.org/10.1680/jsuin.23.00065
83. Machado, M. V., Leite, K. R., de Araujo, E. S., Zatta, L., & Marangoni, R. (2023). Biocomposites based on alginate–chitosan–collagen and nimesulide/layered double hydroxides: Preparation and characterization. *Cuadernos de Educación y Desarrollo, 15*(12), 16493–16510.
84. Sanjay, M., Siengchin, S., Pruncu, C. I., Jawaid, M., Kumar, T. S. M., & Rajini, N. (2019). Biomedical applications of polymer/layered double hydroxide bionanocomposites. In *Nanostructured Polymer Composites for Biomedical Applications* (pp. 315–322). Elsevier.
85. He, J., Wei, M., Li, B., Kang, Y., Evans, D. G., & Duan, X. (2006). Preparation of layered double hydroxides. *Layered Double Hydroxides,* 89–119.
86. Cavani, F., Trifirò, F., & Vaccari, A. (1991). Hydrotalcite-type anionic clays: Preparation, properties and applications. *Catalysis today, 11*(2), 173–301.
87. Thostenson, E. T., Ren, Z., & Chou, T.-W. (2001). Advances in the science and technology of carbon nanotubes and their composites: A review. *Composites Science and Technology, 61*(13), 1899–1912.
88. Ko, F. K., Khan, S., Ali, A., Gogotsi, Y., Naguib, N., & Yang, G. (2002). Structure and properties of carbon nanotube reinforced nanocomposites. In *43rd AIAA/ASME/ASCE/AHS/ASC Structures, Structural Dynamics, and Materials Conference* (p. 1426).
89. Park, C., et al. (2001). Polymer-single wall carbon nanotube composites for potential spacecraft applications. *MRS Online Proceedings Library (OPL), 706*, Z3. 30.1.
90. Shao, C., Kim, H.-Y., Gong, J., Ding, B., Lee, D.-R., & Park, S.-J. (2003). Fiber mats of poly (vinyl alcohol)/silica composite via electrospinning. *Materials Letters, 57*(9–10), 1579–1584.
91. Fong, H., Liu, W., Wang, C.-S., & Vaia, R. A. (2002). Generation of electrospun fibers of nylon 6 and nylon 6-montmorillonite nanocomposite. *Polymer, 43*(3), 775–780.
92. Dai, H., Gong, J., Kim, H., & Lee, D. (2002). A novel method for preparing ultra-fine alumina-borate oxide fibres via an electrospinning technique. *Nanotechnology, 13*(5), 674.
93. Zhang, Y., & Fan, X. (2010). Surface modification of cotton fabrics by transesterification with ion-paired subtilisin Carlsberg in solvents. *Cellulose, 17*, 903–911.
94. Cai, S.-B., Huang, Z.-H., Zhang, X.-Q., Cao, Z.-J., Zhou, M.-H., & Hong, F. (2011). Identification of a keratinase-producing bacterial strain and enzymatic study for its improvement on shrink resistance and tensile strength of wool-and polyester-blended fabric. *Applied Biochemistry and Biotechnology, 163*, 112–126.
95. Shen, J., & Smith, E. (2015). Enzymatic treatments for sustainable textile processing. *Sustainable Apparel,* 119–133.
96. Matama, T., & Cavaco-Paulo, A. (2010). Enzymatic modification of polyacrylonitrile and cellulose acetate fibres for textile and other applications. In *Advances in Textile Biotechnology* (pp. 98–131). Elsevier.
97. Awad, G. E., Abd El Aty, A. A., Shehata, A. N., Hassan, M. E., & Elnashar, M. M. (2016). Covalent immobilization of microbial naringinase using novel thermally stable biopolymer for hydrolysis of naringin. *3 Biotech, 6*, 1–10.
98. Agostinelli, E., et al. (2007). Polyketone polymer: A new support for direct enzyme immobilization. *Journal of Biotechnology, 127*(4), 670–678.
99. Homaei, A. A., Sariri, R., Vianello, F., & Stevanato, R. (2013). Enzyme immobilization: An update. *Journal of Chemical Biology, 6*, 185–205.
100. Bai, Y., et al. (2012). Identification of an acidic α-amylase from Alicyclobacillus sp. A4 and assessment of its application in the starch industry. *Food Chemistry, 131*(4), 1473–1478.
101. Schückel, J., Matura, A., & Van Pee, K.-H. (2011). One-copper laccase-related enzyme from Marasmius sp.: Purification, characterization and bleaching of textile dyes. *Enzyme and Microbial Technology, 48*(3), 278–284.

102. Li, F.-Y., Xing, Y.-J., & Ding, X. (2007). Immobilization of papain on cotton fabric by sol–gel method. *Enzyme and Microbial Technology, 40*(7), 1692–1697.
103. Singh, A., Lee, Y., & Dressick, W. J. (2004). Self-cleaning fabrics for decontamination of organophosphorous pesticides and related chemical agents. *Advanced Materials, 16*(23–24), 2112–2115.
104. Yan, M., Ge, J., Liu, Z., & Ouyang, P. (2006). Encapsulation of single enzyme in nanogel with enhanced biocatalytic activity and stability. *Journal of the American Chemical Society, 128*(34), 11008–11009.
105. Shishoo, R. (2007). *Plasma technologies for textiles*. Elsevier.
106. Bhat, N., Netravali, A., Gore, A., Sathianarayanan, M., Arolkar, G., & Deshmukh, R. (2011). Surface modification of cotton fabrics using plasma technology. *Textile Research Journal, 81*(10), 1014–1026.
107. Morent, R., De Geyter, N., Leys, C., Gengembre, L., & Payen, E. (2007). Surface modification of non-woven textiles using a dielectric barrier discharge operating in air, helium and argon at medium pressure. *Textile Research Journal, 77*(7), 471–488.
108. Verschuren, J., Kiekens, P., & Leys, C. (2007). Textile-specific properties that influence plasma treatment, effect creation and effect characterization. *Textile Research Journal, 77*(10), 727–733.
109. Guo, L., Campagne, C., Perwuelz, A., & Leroux, F. (2009). Zeta potential and surface physico-chemical properties of atmospheric air-plasma-treated polyester fabrics. *Textile Research Journal, 79*(15), 1371–1377.
110. Horrocks, A. (2019). Smart flame retardant textile coatings and laminates. In *Smart Textile Coatings and Laminates* (pp. 205–236). Elsevier
111. Horrocks, A., Nazaré, S., Masood, R., Kandola, B., & Price, D. (2011). Surface modification of fabrics for improved flash-fire resistance using atmospheric pressure plasma in the presence of a functionalized clay and polysiloxane. *Polymers for Advanced Technologies, 22*(1), 22–29.
112. Alongi, J., Tata, J., & Frache, A. (2011). Hydrotalcite and nanometric silica as finishing additives to enhance the thermal stability and flame retardancy of cotton. *Cellulose, 18*, 179–190.
113. Malkov, G. S., & Fisher, E. R. (2010). Pulsed plasma enhanced chemical vapor deposition of poly (allyl alcohol) onto natural fibers. *Plasma Processes and Polymers, 7*(8), 695–707.
114. Szymanowski, H., Sobczyk, A., Gazicki-Lipman, M., Jakubowski, W., & Klimek, L. (2005). Plasma enhanced CVD deposition of titanium oxide for biomedical applications. *Surface and Coatings Technology, 200*(1–4), 1036–1040.
115. Jaroszewski, M., Pospieszna, J., & Ziaja, J. (2010). Dielectric properties of polypropylene fabrics with carbon plasma coatings for applications in the technique of electromagnetic field shielding. *Journal of Non-Crystalline Solids, 356*(11–17), 625–628.
116. Rosace, G., Canton, R., & Colleoni, C. (2010). Plasma enhanced CVD of SiOxCyHz thin film on different textile fabrics: Influence of exposure time on the abrasion resistance and mechanical properties. *Applied Surface Science, 256*(8), 2509–2516.
117. Shahidi, S., Ghoranneviss, M., Moazzenchi, B., Rashidi, A., & Mirjalili, M. (2007). Investigation of antibacterial activity on cotton fabrics with cold plasma in the presence of a magnetic field. *Plasma Processes and Polymers, 4*(S1), S1098–S1103.
118. Shahidi, S., et al. (2010). Investigation of metal absorption and antibacterial activity on cotton fabric modified by low temperature plasma. *Cellulose, 17*, 627–634.
119. Shahidi, S., & Ghoranneviss, M. (2012). Comparison between oxygen and nitrogen plasma treatment on adhesion properties and antibacterial activity of metal coated polypropylene fabrics. *Fibers and Polymers, 13*, 971–978.
120. Ghoranneviss, M., Moazzenchi, B., Shahidi, S., Anvari, A., & Rashidi, A. (2006). Decolorization of denim fabrics with cold plasmas in the presence of magnetic fields. *Plasma Processes and Polymers, 3*(3), 316–321.
121. Ghoranneviss, M., Shahidi, S., Moazzenchi, B., Anvari, A., Rashidi, A., & Hosseini, H. (2007). Comparison between decolorization of denim fabrics with oxygen and argon glow discharge. *Surface and Coatings Technology, 201*(9–11), 4926–4930.

122. Shahidi, S., Ghoranneviss, M., Moazzenchi, B., Rashidi, A., & Dorranian, D. (2007). Effect of using cold plasma on dyeing properties of polypropylene fabrics. *Fibers and Polymers, 8*, 123–129.
123. Shahidi, S., Moazzenchi, B., Ghoranneviss, M., & Azizi, S. (2013). Investigation on dyeability of polypropylene fabrics grafted with chitosan after plasma modification. *The European Physical Journal-Applied Physics, 62*(1), 10801.
124. Shahidi, S., & Ghoranneviss, M. (2011). Investigation on dye ability and antibacterial activity of nanolayer platinum coated polyester fabric using DC magnetron sputtering. *Progress in Organic Coatings, 70*(4), 300–303.
125. Shahidi, S., Ghoranneviss, M., Moazzenchi, B., Rashidi, A., & Dorranian, D. (2013). Study of surface modification of wool fabrics using low temperature plasma. In *Proceedings of the 3rd International Conference on the Frontiers of Plasma Physics and Technology (PC/5099).*
126. Shahidi, S., Ghoranneviss, M., Moazzenchi, B., Anvari, A., & Rashidi, A. (2007). Aluminum coatings on cotton fabrics with low temperature plasma of argon and oxygen. *Surface and Coatings Technology, 201*(9–11), 5646–5650.
127. Shahidi, S., Rashidi, A., Ghoranneviss, M., Anvari, A., & Wiener, J. (2010). Plasma effects on anti-felting properties of wool fabrics. *Surface and Coatings Technology, 205*, S349–S354.
128. Ghoranneviss, M., & Shahidi, S. (2014). Flame retardant properties of plasma pretreated/ metallic salt loaded cotton fabric before and after direct dyeing. *Journal of Fusion Energy, 33*, 119–124.
129. Alongi, J., Carosio, F., Frache, A., & Malucelli, G. (2013). Layer by Layer coatings assembled through dipping, vertical or horizontal spray for cotton flame retardancy. *Carbohydrate Polymers, 92*(1), 114–119.
130. Michel, M., Toniazzo, V., Ruch, D., & Ball, V. (2012). Deposition mechanisms in layer-by-layer or step-by-step deposition methods: From elastic and impermeable films to soft membranes with ion exchange properties. *ISRN Materials Science.*
131. Bertrand, P., Jonas, A., Laschewsky, A., & Legras, R. (2000). Ultrathin polymer coatings by complexation of polyelectrolytes at interfaces: Suitable materials, structure and properties. *Macromolecular Rapid Communications, 21*(7), 319–348.
132. Borges, J., & Mano, J. F. (2014). Molecular interactions driving the layer-by-layer assembly of multilayers. *Chemical Reviews, 114*(18), 8883–8942.
133. Li, Z., Tan, B. H., Lin, T., & He, C. (2016). Recent advances in stereocomplexation of enantiomeric PLA-based copolymers and applications. *Progress in Polymer Science, 62*, 22–72.
134. Stockton, W., & Rubner, M. (1997). Molecular-level processing of conjugated polymers. 4. Layer-by-layer manipulation of polyaniline via hydrogen-bonding interactions. *Macromolecules, 30*(9), 2717–2725.
135. Benjamin, I., Hong, H., Avny, Y., Davidov, D., & Neumann, R. (1998). Poly (phenylenevinylene) analogs with ring substituted polar side chains and their use in the formation of hydrogen bonding based self-assembled multilayers. *Journal of Materials Chemistry, 8*(4), 919–924.
136. Tieke, B., El-Hashani, A., Toutianoush, A., & Fendt, A. (2008). Multilayered films based on macrocyclic polyamines, calixarenes and cyclodextrins and transport properties of the corresponding membranes. *Thin Solid Films, 516*(24), 8814–8820.
137. Shimazaki, Y., Ito, S., & Tsutsumi, N. (2000). Adsorption-induced second harmonic generation from the layer-by-layer deposited ultrathin film based on the charge-transfer interaction. *Langmuir, 16*(24), 9478–9482.
138. Shimazaki, Y., Nakamura, R., Ito, S., & Yamamoto, M. (2001). Molecular weight dependence of alternate adsorption through charge-transfer interaction. *Langmuir, 17*(3), 953–956.
139. Uğur, ŞS., Sarıışık, M., Aktaş, A. H., Uçar, M. Ç., & Erden, E. (2010). Modifying of cotton fabric surface with nano-ZnO multilayer films by layer-by-layer deposition method. *Nanoscale Research Letters, 5*, 1204–1210.
140. Ugur, ŞS., Sariişik, M., & Aktaş, A. H. (2010). The fabrication of nanocomposite thin films with TiO2 nanoparticles by the layer-by-layer deposition method for multifunctional cotton fabrics. *Nanotechnology, 21*(32), Article 325603.

141. Wang, Q., & Hauser, P. (2010). Developing a novel UV protection process for cotton based on layer-by-layer self-assembly. *Carbohydrate Polymers, 81*(2), 491–496.
142. Cai, G., & Wilkie, C. A. (2018). Recent advances on montmorillonite-based fire retardant composites in polymer matrices. *The Chemistry and Physics of Engineering Materials,* 339–356.
143. Lee, H. C., & Lee, S. (2024). Flame-retardant behavior and nanocrystal synthesis through formation of multilayers of poly (acrylic acid) and zinc phosphate on cotton fabric. *ACS omega.*
144. Xiong, R., et al. (2015). Ultrarobust transparent cellulose nanocrystal-graphene membranes with high electrical conductivity. *Advanced Materials (Deerfield Beach, Fla.), 28*(7), 1501–1509.
145. Li, F., Biagioni, P., Finazzi, M., Tavazzi, S., & Piergiovanni, L. (2013). Tunable green oxygen barrier through layer-by-layer self-assembly of chitosan and cellulose nanocrystals. *Carbohydrate Polymers, 92*(2), 2128–2134.
146. Mao, Y., Wang, D., & Fu, S. (2022). Layer-by-layer self-assembled nanocoatings of Mxene and P, N-co-doped cellulose nanocrystals onto cotton fabrics for significantly reducing fire hazards and shielding electromagnetic interference. *Composites Part A: Applied Science and Manufacturing, 153*, Article 106751.
147. Carosio, F., Alongi, J., & Malucelli, G. (2011). α-Zirconium phosphate-based nanoarchitectures on polyester fabrics through layer-by-layer assembly. *Journal of Materials Chemistry, 21*(28), 10370–10376.
148. Carosio, F., Laufer, G., Alongi, J., Camino, G., & Grunlan, J. C. (2011). Layer-by-layer assembly of silica-based flame retardant thin film on PET fabric. *Polymer Degradation and Stability, 96*(5), 745–750.
149. Krogman, K. C., Zacharia, N. S., Grillo, D. M., & Hammond, P. T. (2008). Photocatalytic layer-by-layer coatings for degradation of acutely toxic agents. *Chemistry of Materials, 20*(5), 1924–1930.
150. Dubas, S. T., Kumlangdudsana, P., & Potiyaraj, P. (2006). Layer-by-layer deposition of antimicrobial silver nanoparticles on textile fibers. *Colloids and Surfaces A: Physicochemical and Engineering Aspects, 289*(1–3), 105–109.
151. Maddalena, L., Indias, J. M., Bettotti, P., Scarpa, M., & Carosio, F. (2024). Cellulose nanocrystals polyelectrolyte complexes as flame retardant treatment for cotton fabrics. *Polymer Degradation and Stability, 220*, Article 110646.
152. Pakdel, E., Fang, J., Fang, J., Sun, L., Wang, X., & Wang, X. (2018). Nanocoatings for smart textiles. In *Smart Textiles: Wearable Nanotechnology* (pp. 247–300).
153. Kathirvelu, S., D'souza, L., & Dhurai, B. (2009). *UV protection finishing of textiles using ZnO nanoparticles.*
154. Ismail, W. N. W. (2016). Sol–gel technology for innovative fabric finishing—a review. *Journal of Sol-Gel Science and Technology, 78*, 698–707.
155. Tomšič, B., et al. (2008). Sol–gel coating of cellulose fibres with antimicrobial and repellent properties. *Journal of Sol-Gel Science and Technology, 47*, 44–57.
156. Pakdel, E., Wang, J., Allardyce, B. J., Rajkhowa, R., & Wang, X. (2016). Functionality of nano and 3D-microhierarchical TiO2 particles as coagulants for sericin extraction from the silk degumming wastewater. *Separation and Purification Technology, 170*, 92–101.
157. Nguyen-Tri, P., et al. (2019). Recent progress in the preparation, properties and applications of superhydrophobic nano-based coatings and surfaces: A review. *Progress in Organic Coatings, 132*, 235–256.
158. Zeng, C., Wang, H., Zhou, H., & Lin, T. (2015). Self-cleaning, superhydrophobic cotton fabrics with excellent washing durability, solvent resistance and chemical stability prepared from an SU-8 derived surface coating. *Rsc Advances, 5*(75), 61044–61050.
159. Taurino, R., Fabbri, E., Messori, M., Pilati, F., Pospiech, D., & Synytska, A. (2008). Facile preparation of superhydrophobic coatings by sol–gel processes. *Journal of Colloid and Interface Science, 325*(1), 149–156.

160. Shi, Y., Wang, Y., Feng, X., Yue, G., & Yang, W. (2012). Fabrication of superhydrophobicity on cotton fabric by sol–gel. *Applied Surface Science, 258*(20), 8134–8138.
161. Mahltig, B. (2016). Smart hydrophobic and soil-repellent protective composite coatings for textiles and leather. In *Smart Composite Coatings and Membranes* (pp. 261–292). Elsevier.
162. Parvinzadeh Gashti, M. (2014). Nanocomposite coatings: State of the art approach in textile finishing. *Textile Science & Engineering, 4*(2).
163. Montazer, M., & Pakdel, E. (2011). Functionality of nano titanium dioxide on textiles with future aspects: Focus on wool. *Journal of Photochemistry and Photobiology C: Photochemistry Reviews, 12*(4), 293–303.
164. Gashti, M. P., & Almasian, A. (2013). Citric acid/ZrO2 nanocomposite inducing thermal barrier and self-cleaning properties on protein fibers. *Composites Part B: Engineering, 52*, 340–349.
165. Alimohammadi, F., Parvinzadeh Gashti, M., & Shamei, A. (2013). Functional cellulose fibers via polycarboxylic acid/carbon nanotube composite coating. *Journal of Coatings Technology and Research, 10*, 123–132.
166. Daoud, W. A., Leung, S., Tung, W. S., Xin, J. H., Cheuk, K., & Qi, K. (2008). Self-cleaning keratins. *Chemistry of Materials, 20*(4), 1242–1244.
167. Meilert, K., Laub, D., & Kiwi, J. (2005). Photocatalytic self-cleaning of modified cotton textiles by TiO2 clusters attached by chemical spacers. *Journal of Molecular Catalysis A: Chemical, 237*(1–2), 101–108.
168. Hsu, P.-C., et al. (2015). Personal thermal management by metallic nanowire-coated textile. *Nano Letters, 15*(1), 365–371.
169. Kayabaşı, G., Özen, Ö., & Yılmaz, D. (2022). A novel yarn spinning method for fabricating conductive and nanofiber-coated hybrid yarns. *Journal of Industrial Textiles, 51*(6), 953–976.
170. Gashti, M. P., Alimohammadi, F., Song, G., & Kiumarsi, A. (2012). Characterization of nanocomposite coatings on textiles: A brief review on microscopic technology. *Current Microscopy Contributions to Advances in Science and Technology, 2*, 1424–1437.
171. Wang, H., Wang, J., Hong, J., Wei, Q., Gao, W., & Zhu, Z. (2007). Preparation and characterization of silver nanocomposite textile. *Journal of Coatings Technology and Research, 4*, 101–106.
172. Shim, J.-J., Yates, M. Z., & Johnston, K. P. (1999). Polymer coatings by rapid expansion of suspensions in supercritical carbon dioxide. *Industrial & Engineering Chemistry Research, 38*(10), 3655–3662.
173. Hay, J., & Khan, A. (2002). Review environmentally friendly coatings using carbon dioxide as the carrier medium. *Journal of Materials Science, 37*, 4743–4752.
174. Abate, M. T., Ferri, A., Guan, J., Chen, G., Ferreira, J. A., & Nierstrasz, V. (2019). Single-step disperse dyeing and antimicrobial functionalization of polyester fabric with chitosan and derivative in supercritical carbon dioxide. *The Journal of Supercritical Fluids, 147*, 231–240.
175. Iwai, Y., Sameshima, S., Yonezawa, S., & Katayama, S. (2015). Fabrication of conductive cotton by electroless plating method with supercritical carbon dioxide. *The Journal of Supercritical Fluids, 100*, 46–51.
176. Mohamed, A. L., Er-Rafik, M., & Moller, M. (2013). Supercritical carbon dioxide assisted silicon based finishing of cellulosic fabric: A novel approach. *Carbohydrate Polymers, 98*(1), 1095–1107.
177. Baba, T., Hirogaki, K., Tabata, I., Okubayashi, S., Hisada, K., & Hori, T. (2010). Impregnation of chitin/chitosan into polyester fabric using supercritical carbon dioxide. *Sen'i Gakkaishi, 66*(3), 63–69.
178. Xu, Y.-Y., Zheng, L.-J., Ye, F., Qian, Y.-F., Yan, J., & Xiong, X.-Q. (2015). Water/oil repellent property of polyester fabrics after supercritical carbon dioxide finishing. *Thermal Science, 19*(4), 1273–1277.
179. Eren, H. A., Yiğit, İ., Eren, S., & Avinc, O. (2020). Sustainable textile processing with zero water utilization using super critical carbon dioxide technology. In *Sustainability in the Textile and Apparel Industries: Production Process Sustainability* (pp. 179–196).

180. De Smet, D., Weydts, D., & Vanneste, M. (2015). Environmentally friendly fabric finishes. In *Sustainable Apparel* (pp. 3–33). Elsevier.
181. Islam, M. T., & Asaduzzaman, S. (2019). Environmentally-friendly textile finishing. *Textiles and Clothing,* 101–129.
182. Schindler, W. D., & Hauser, P. J. (2004). *Chemical finishing of textiles.* Elsevier.
183. Welch, C. M. (1992). Formaldehyde-free durable-press finishes. *Review of Progress in Coloration and Related Topics, 22*(1), 32–41.
184. Ibrahim, N., Abo-Shosha, M., Elnagdy, E., & Gaffar, M. (2002). Eco-friendly durable press finishing of cellulose-containing fabrics. *Journal of Applied Polymer Science, 84*(12), 2243–2253.
185. Raza, Z. A., Siddque, A., & Mohsin, M. (2017). Multi-performance enhancement of knitted wool fabric with citric acid: An eco-compatible cross-linker. *Journal of Natural Fibers, 14*(6), 887–896.
186. Mohsin, M., Farooq, U., Ramzan, N., Rasheed, A., Ahmad, S., & Ahsan, M. (2014). Softener impact on environment friendly low and zero formaldehyde cross-linker performance for cotton. *Industria Textila, 65*(3), 134–139.
187. Hashem, M., Ibrahim, N. A., El-Shafei, A., Refaie, R., & Hauser, P. (2009). An eco-friendly–novel approach for attaining wrinkle–free/soft-hand cotton fabric. *Carbohydrate Polymers, 78*(4), 690–703.
188. Choi, B., et al. (2018). Highly conductive fiber with waterproof and self-cleaning properties for textile electronics. *ACS Applied Materials & Interfaces, 10*(42), 36094–36101.
189. Das, S., Kumar, S., Samal, S. K., Mohanty, S., & Nayak, S. K. (2018). A review on superhydrophobic polymer nanocoatings: Recent development and applications. *Industrial & Engineering Chemistry Research, 57*(8), 2727–2745.
190. Xue, C.-H., Zhang, L., Wei, P., & Jia, S.-T. (2016). Fabrication of superhydrophobic cotton textiles with flame retardancy. *Cellulose, 23*, 1471–1480.
191. Xue, C.-H., Fan, Q.-Q., Guo, X.-J., An, Q.-F., & Jia, S.-T. (2019). Fabrication of superhydrophobic cotton fabrics by grafting of POSS-based polymers on fibers. *Applied Surface Science, 465*, 241–248.
192. Cai, R., et al. (2018). Environmentally friendly super-water-repellent fabrics prepared from water-based suspensions. *ACS Applied Materials & Interfaces, 10*(18), 15346–15351.
193. Wen, O. Y., Tohir, M. Z. M., Yeaw, T. C. S., Razak, M. A., Zainuddin, H. S., & Hamid, M. R. A. (2023). Fire-resistant and flame-retardant surface finishing of polymers and textiles: A state-of-the-art review. *Progress in Organic Coatings, 175*, Article 107330.
194. Yusuf, M. (2018). A review on flame retardant textile finishing: Current and future trends. *Current Smart Materials, 3*(2), 99–108.
195. Rosace, G., Migani, V., Guido, E., & Colleoni, C. (2015). Flame retardant finishing for textiles. *Flame Retardants: Polymer Blends, Composites and Nanocomposites,* 209–246.
196. Morris, H., & Murray, R. (2020). Medical textiles. *Textile Progress, 52*(1–2), 1–127.
197. Iyigundogdu, Z. U., Demir, O., Asutay, A. B., & Sahin, F. (2017). Developing novel antimicrobial and antiviral textile products. *Applied Biochemistry and Biotechnology, 181*, 1155–1166.
198. Nadi, A., Boukhriss, A., Bentis, A., Jabrane, E., & Gmouh, S. (2018). Evolution in the surface modification of textiles: A review. *Textile Progress, 50*(2), 67–108.
199. Butola, B., & Mohammad, F. (2016). Silver nanomaterials as future colorants and potential antimicrobial agents for natural and synthetic textile materials. *RSC Advances, 6*(50), 44232–44247.
200. Jeremiah, S. S., Miyakawa, K., Morita, T., Yamaoka, Y., & Ryo, A. (2020). Potent antiviral effect of silver nanoparticles on SARS-CoV-2. *Biochemical and Biophysical Research Communications, 533*(1), 195–200.
201. Román, L. E., Gomez, E. D., Solís, J. L., & Gómez, M. M. (2020). Antibacterial cotton fabric functionalized with copper oxide nanoparticles. *Molecules, 25*(24), 5802.

202. Sadalage, P. S., Nimbalkar, M. S., Sharma, K. K. K., Patil, P. S., & Pawar, K. D. (2020). Sustainable approach to almond skin mediated synthesis of tunable selenium microstructures for coating cotton fabric to impart specific antibacterial activity. *Journal of Colloid and Interface Science, 569*, 346–357.
203. Kim, B.-M., Minh, N. V., Choi, H.-Y., & Kim, W.-G. (2019). Coralmycin derivatives with potent anti-gram negative activity produced by the myxobacteria corallococcus coralloides M23. *Molecules, 24*(7), 1390.
204. Pandimurugan, R., & Thambidurai, S. (2017). UV protection and antibacterial properties of seaweed capped ZnO nanoparticles coated cotton fabrics. *International Journal of Biological Macromolecules, 105*, 788–795.
205. Singh, N., & Sahu, O. (2019). Sustainable cyclodextrin in textile applications. In *The Impact and Prospects of Green Chemistry for Textile Technology* (pp. 83–105). Elsevier
206. Gulati, R., Sharma, S., & Sharma, R. K. (2022). Antimicrobial textile: Recent developments and functional perspective. *Polymer Bulletin, 79*(8), 5747–5771.
207. Sunthornvarabhas, J., et al. (2019). Utilization of lignin extracts from sugarcane bagasse as bio-based antimicrobial fabrics. *Sugar Tech, 21*, 355–363.
208. Benltoufa, S., Miled, W., Trad, M., Slama, R. B., & Fayala, F. (2020). Chitosan hydrogel-coated cellulosic fabric for medical end-use: Antibacterial properties, basic mechanical and comfort properties. *Carbohydrate Polymers, 227*, Article 115352.
209. Mavrić, Z., Tomšič, B., & Simončič, B. (2018). Recent advances in the ultraviolet protection finishing of textiles. *Tekstilec, 61*(3).
210. Ranjan Das, B. (2010). UV radiation protective clothing. *The Open Textile Journal, 3*(1).
211. Sankaran, A., Kamboj, A., Samant, L., & Jose, S. (2021). Synthetic and natural UV protective agents for textile finishing. *Innovative and Emerging Technologies for Textile Dyeing and Finishing,* 301–324.
212. Laperre, J., et al. (2001). Determination of the ultraviolet protection factor of textile materials: Measurement precision. *Photodermatology, Photoimmunology & Photomedicine, 17*(5), 223–229.
213. Tsuzuki, T., & Wang, X. (2010). Nanoparticle coatings for UV protective textiles. *Research Journal of Textile and Apparel, 14*(2), 9–20.
214. Grancarić, A. M., Tarbuk, A., & Botteri, L. (2014). Light conversion and scattering in UV protective textiles. *AUTEX Research Journal, 14*(4), 247–258.
215. Das, B., Ishtiaque, S., Rengasamy, R., Hati, S., & Kumar, A. (2010). Ultraviolet absorbers for textiles. *Research Journal of Textile and Apparel, 14*(1), 42–52.
216. Mikame, K., et al. (2021). Natural organic ultraviolet absorbers from lignin. *ACS Sustainable Chemistry & Engineering, 9*(49), 16651–16658.
217. Yousif, E., El-Hiti, G. A., Haddad, R., & Balakit, A. A. (2015). Photochemical stability and photostabilizing efficiency of poly (methyl methacrylate) based on 2-(6-methoxynaphthalen-2-yl) propanoate metal ion complexes. *Polymers, 7*(6), 1005–1019.
218. Broasca, G., Borcia, G., Dumitrascu, N., & Vrinceanu, N. (2013). Characterization of ZnO coated polyester fabrics for UV protection. *Applied Surface Science, 279*, 272–278.
219. Ojstršek, A., et al. (2021). Metallisation of textiles and protection of conductive layers: An overview of application techniques. *Sensors, 21*(10), 3508.
220. Ojstršek, A., Jug, L., & Plohl, O. (2022). A review of electro conductive textiles utilizing the dip-coating technique: Their functionality, durability and sustainability. *Polymers, 14*(21), 4713. https://doi.org/10.3390/polym14214713
221. Stoppa, M., & Chiolerio, A. (2014). Wearable electronics and smart textiles: A critical review. *Sensors, 14*(7), 11957–11992. https://doi.org/10.3390/s140711957
222. Wang, F. (2018). Moisture absorption and transport through textiles. In *Engineering of High-Performance Textiles* (pp. 247–275). Elsevier.
223. Ivanova, N., & Zaretskaya, A. (2010). Simple treatment of cotton textile to impart high water repellent properties. *Applied Surface Science, 257*(5), 1800–1803.
224. Islam, M. S., & van de Ven, T. G. (2021). Cotton-based flame-retardant textiles: A review. *BioResources, 16*(2), 4354.

225. Kundu, C. K., Li, Z., Song, L., & Hu, Y. (2020). An overview of fire retardant treatments for synthetic textiles: From traditional approaches to recent applications. *European Polymer Journal, 137*, Article 109911.
226. Maity, S., & Chatterjee, A. (2015). Conductive polymer based electro-conductive textiles for novel applications. *Technical Textiles, 1*, E16–E18.

Chapter 8
Bio-Derived Polymers for Eco-Friendly Textile Finishing

Zehra Yildiz, Oguz Eryilmaz, Berivan Ozer, Betul Nur Kus, Pelin Yetman, Muslum Demircioglu, and Erhan Sancak

Abstract The textile industry can be accepted as a significant contributor to the global economy and has been experiencing a paradigm change recently towards environmentally friendly and sustainable processes. Alternative approaches have been suggested by academics, researchers, and industry professionals due to the worries about the environmental effects of conventional textile finishing stages. In the textile business, finishing treatments are essential for improving the final product's antibacterial, flame-retardant, antistatic, oleophobic, and other qualities. The conventional finishing processes often involve the usage of petrochemicals, excess amounts of water, and volatile organic compounds (VOCs) that are all accepted as environmental challenges due to their non-biodegradable nature, and possible damage to the ecosystem and human health. Bio-polymers and bio-derived compounds have become increasingly attractive for environmentally friendly textile finishing as a substitute for petroleum-based chemicals as they are reusable and biodegradable. This chapter examines the application of bio-derived polymers in textile finishing stages, emphasizing how they have the potential to transform the sector by providing environmentally friendly solutions without any loss on the performance and functionality of the end product. Numerous bio-polymers, including starch, cellulose, chitosan, protein, etc., are utilized in textile finishing applications for a variety of objectives according to the literature. For example, chitosan, which is produced by the deacetylation of chitin, which is present in sea animal shells, possesses an inherent antimicrobial property and is known for its versatility in enhancing antibacterial fabric properties. Starch, a carbohydrate polymer, offers a sustainable option for improving fabric stiffness and wrinkle resistance. Cellulose, extracted from plant sources, can be utilized for its excellent moisture management and breathability. Enzymes are being used for bio-polishing effects on textiles to enhance the luster

Z. Yildiz (✉) · O. Eryilmaz · B. Ozer · B. N. Kus · P. Yetman · M. Demircioglu · E. Sancak
Department of Textile Engineering, Marmara University, Istanbul, Turkey
e-mail: zehra.yildiz@marmara.edu.tr

O. Eryilmaz
e-mail: oguz.eryilmaz@marmara.edu.tr

E. Sancak
e-mail: esancak@marmara.edu.tr

M. Shahid et al. (eds.), *Advancements in Textile Finishing*,
https://doi.org/10.1007/978-981-96-6385-9_8

and softness of the fabrics. Proteins such as soy protein and collagen have gained attention for their ability to impart softness and durability to textiles. Essential oils with a variety of benefits, including pain and tension relief, relaxation, and refreshing qualities, such as orange, vanilla, and jasmine, are utilized to complete fragrances. Besides, epoxidized soybean oil (ESBO) is a notable bio-derived alternative to traditional finishing agents generated from petrochemicals, helping to drive the industry's green shift. After the chemical modification of ESBO, photocurable formulations can be derived by using additives, bio-based oligomers, and photoinitiators and can be used to give fabrics certain qualities during the finishing phases. The usage of photocuring in textile finishing contributes to sustainability by minimizing waste along with water and VOC. The key step in the mentioned process involves exposing the treated textile to UV light, triggering a polymerization reaction in the ESBO-based formulation. This reaction results in a cross-linked, robust, and sustainable finish on the textile.

Keywords Bio-polymer · Textile finishing · Cellulose · Sustainability · Enzyme · Vegetable oils

8.1 Definition of Bio-Derived Polymers

Bio-polymers are produced from synthetic monomers such as polycaprolactone (PCL), polyester-amides (PEA), and aliphatic or aromatic co-polyesters (PBSA, PBAT), as well as from natural sources like proteins (e.g., chitosan, collagen), lipids (e.g., wax, fatty acids), and polysaccharides derived from plants or animals (e.g., starch, cellulose, lignin) [1]. These large-structure polymers consist of monomeric units connected by covalent bonds, with biological macromolecules forming their basis. Bio-polymers are non-toxic, renewable, and biodegradable [2]. When broken down by environmental microorganisms, they rapidly decompose into carbon dioxide and water, making them more environmentally friendly than synthetic polymers [3–5]. Bio-polymers are organic substances that form naturally from the life periods of bacteria, fungi, mammals, and green plants. In addition to bio-polymers made from animal proteins like collagen, gelatin, wool, and silk, bio-polymers also comprise polysaccharides like cellulose, starch, and carbohydrate polymers made by bacteria and fungus [6]. One such product is BioSteel™, a protein fiber made from goat milk that has undergone genetic modification to include genes for spider silk. According to reports, the polymers have ten times the strength of steel for the same weight [7]. The importance of bio-derived polymers can be explained in some points below:

- Bio-polymers are materials derived from renewable resources that are both biocompatible and biodegradable [8].
- Due to their special nontoxicity and biodegradability characteristics, bio-polymers have more uses in technology, medical equipment, energy, food packaging, and other industries than synthetic polymers [9].

- Improved characteristics of bio-polymers, such as antibacterial, UV protection, flame retardancy, and water resistance, can enhance their practical applicability and market expansion [9].
- Biologically treated biowaste is applied to improve soil carbon levels while protecting the soil through the use of bio-polymers [10].
- Biodegradable bio-polymers (BDPs) offer a high degree of environmental protection by preventing or minimizing the effects of waste from industries on the environment [10].
- The use of BDPs guarantees the internal market's operation and aids in preventing trade barriers, competition distortion, and limitations both inside and outside the European Union [10].

8.2 Usage of Bio-Derived Polymers in Textile Industry

Industries are looking for creative ways to lessen their ecological impact as worries about climate change and the degradation of the environment become increasingly urgent. The textile sector uses a lot of chemicals and as a result, releases harmful substances into the environment [11]. The development and usage of natural polymers present a way to decrease plastic waste and create more sustainable, environmentally friendly alternatives [12]. Thanks to the rising sustainability awareness, re-evaluation of classical applications with the awareness of green innovation and taking remedial reactions are on the agenda of the textile industry. Biological macromolecules provide the basis for bio-derived polymers. They're non-toxic, renewable, and biodegradable [2]. Besides renewable resources, bio-derived polymers have inherent antibacterial activity, biodegradability, and biocompatibility. As a result, they are perfect for usage in many different industries, including paper coatings, automotive, textiles, medicine, and ophthalmology [10]. Smart polymers, known as biodegradable polymers, are being used in many different kinds of sectors, including food, agriculture, healthcare, tissue culture, and smart textiles [13]. The use of bio-derived polymers is such an approach that it is also gaining popularity in the textile industry (Fig. 8.1).

A wide range of products and industries use biodegradable polymers, such as automotive parts, glass fiber agents, erosion control, toys, tubes, medical supplies, fast food containers, paper coating, mulch films for lawn and garden, textiles, erosion control, and biologically based resins [13]. In polymers, natural resources are the source of a particular type of materials named bio-polymers. Advantages of bio-derived polymers are the sustainability of bio-derived polymers, their biodegradability and their sourcing from renewable resources are only a few of their benefits. Bio-derived polymers are biodegradable materials that are a compelling substitute for fossil-based polymers [8]. Bio-derived polymers reduce reliance on petrochemicals by minimizing chemical dependency while offering customizable properties to the final product. Their abundance, renewability, biodegradability, and ease of modification have made them a focal point of research and development in sustainable materials. As new techniques for producing and functionalizing nano-adsorbents have

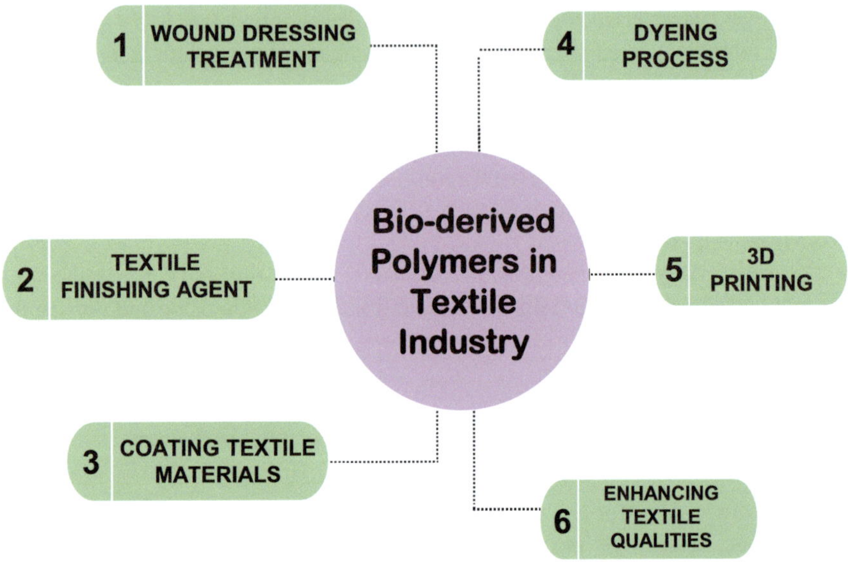

Fig. 8.1 Usage of bio-derived polymers in the textile industry

emerged, they have become more popular in the removal of dyes, toxic substances, oil spills, and excess pesticides. With the application of bio-polymers, particle coagulation-flocculation in wastewater has been successfully achieved. Their potential in producing biohydrogen and combating desertification has shown promise. Bio-polymer-based electrolytes have been commercialized for fuel cell and rechargeable battery applications, showcasing the versatility of bio-polymers in reducing pollution and promoting environmental sustainability. These applications highlight their potential to contribute significantly to greener technologies [14]. Intelligent bio-polymers can adapt to their surroundings and respond to stimuli, making them ideal for textile industries [13]. Because bio-based materials are non-toxic, biodegradable, and biocompatible, they are very competitive when it comes to producing hydrophobic or superhydrophobic coatings. However, these coatings cannot be directly applied using plants or animals, and modifying their wetting behavior from superhydrophilic to superhydrophobic requires appropriate modifications [15]. Bio-polymers have demonstrated significant potential in producing biohydrogen and combating desertification. Additionally, bio-polymer-based electrolytes have been commercialized for fuel cells and rechargeable batteries, highlighting their versatile role in reducing pollution and protecting the environment through sustainable applications. In automobiles, using bio-derived polymers lowers weight, increases impact and heat resistance, and increases fuel efficiency. Manufacturing large quantities and composites research have reduced costs and increased demand across several sectors, including the automobile industry [16]. There are several benefits that biocomposites offer for industries in automobiles. Because they are portable, they require less fuel and produce less greenhouse gas emissions [14, 17]. Biocomposites are superior to non-renewable composites in terms of their acoustic and thermal qualities, making them ideal for interior automobile components [18]. A bio-derived polymer's synthesis may start in one area of the cell and then proceed in another as needed. Plants and algae are rich sources of bio-derived polymers that accumulate in organic ecosystems. Nutrient agar and alginin, for instance, are isolated from various brown algae or seaweeds, such as the genus Gelidium red algae. hyaluronic acid, which is taken out of newborns' umbilical cords, is one exception. Purified enzymes can be used to create certain bio-polymers in vitro under conditions devoid of cells [19]. Antimicrobial textile treatments [20], which include natural and synthetic ingredients, including triclosan, quaternary ammonium, polybiguanides, N-halamines, and chitosan, have been used to preserve textiles since the time of the ancient Egyptians [21] (Table 8.1).

Table 8.1 Bio-polymers for textile finishing

Bio-polymer	Origin	Characteristics	References
Cellulose	Sugar molecules	Printing paste, coating, backfilling, stiffening, and anti-stretching processes	[22]
Starch	Plant substance	Used as sizing	[23–26]
Chitosan	Exoskeleton of crustaceans	Deodorizing, antistatic, dyeing, and antibacterial properties	[19, 27]
Lignin	Woody plants and their cell walls	Active packaging, antioxidant agent, antimicrobial agent, and supercapacitor	[28, 29]
Keratin	Extracted from horns, claws, and hooves	Superior mechanical properties, poor density, hydrophobic characteristics, flame retardant, sound absorption, and affordability	[23, 30]
Pectin	Primary cell walls of most plants	Increasing tensile strength, elongation at break, enhancing coagulation, and as a thickening or gelling agent	[6, 31–33]
Sericin	Silk	Antibacterial, antioxidant, anticancer, UV-resistant, and moisture-absorbing qualities	[22, 34]
Collagen	Extracted from horns, claws, and hooves	Wound dressing treatment, reproducible, hydrating, hydrophilic, and biodegradable, and is used in sponges, films, and membranes	[10, 35, 36]
Soybean	Soybean cake	Accessibility, availability, potential for use, low rate, biodegradability, and thermoplastic characteristics	[37]
Alginate	Brown seaweed	Affordability and accessibility, biocompatibility, wound-healing potential, strong ion-exchange capacity, high moisture adsorption, and thickening agent in textile printing	[38]

8.3 Usage of Cellulosic Materials in Textile Finishing

The mechanical and barrier properties of biocomposites can be improved by incorporating cross-linking agents and renewable nanofillers derived from polysaccharide biopolymers. Their high surface area, nanoscale size, biodegradability, and reactivity make them suitable for integration into polymer chains or fabric structures. Due to their low cost, biodegradability, and nanoscale dimensions, they present a compelling alternative to synthetic polymers [39].

8.3.1 *Cellulose*

An organic substance with a lengthy chain of joined sugar molecules is called cellulose. This linear natural polymer is made up of repeated units of *D-anhydroglucose* ($C_6H_{11}O_5$) connected by 1,4-b-D-glycosidic bonds at carbon 1 (C-1) and carbon 4 carbon locations [40]. The formation of the polymeric structure starts when two glucose molecules come together to create the repeating dimeric unit of the cellobiose molecule. Long and stiff microfibrils are formed as a result of cellulose's structure. Additionally, the many hydroxyl groups found in glucose rings aid in chemical modification activities, enabling the creation of cellulose derivatives with various characteristics [41]. A study aimed at effectively managing waste and utilizing groundnut husk and sugarcane bagasse successfully extracted cellulose. This extracted cellulose was then applied to leather finishing to enhance low-quality leathers, particularly those with grain damage. Without altering its appearance, the cellulose provided a refined finish to the final leather product [9]. Cellulose was utilized in the preparation of printing paste, coating, backfilling, stiffening, and anti-stretching processes for textiles [22]. Sumesh et al. concluded that demonstrating the thermal and mechanical responsiveness of composites made of natural fibers was enhanced by the use of cellulose made from peanut oil cake as a filler, resulting in good mechanical property assessments [18]. Phosphorylated cellulose was produced in research using 1,3-dimethyl imidazolium methyl phosphite, proving its dispersibility and suitability for cellulose spinning. In comparison to pure lyocell fibers, this structural affinity enhanced the mechanical strength of lyocell fibers, producing flame retardant fibers with superior flame retardancy, a lower peak heat release rate, and a higher breaking strength [42].

8.3.2 *Starch*

Nature starch is utilized as an inexpensive, easily accessible source of industrial materials since it is the most frequent plant substance, following cellulose [26]. Starch, an inexpensive, easily obtainable polymer, is commonly used in the textile sector for warp sizing, particularly in cotton fabrics. However, it struggles with stability due to factors like temperature fluctuations, large molecular sizes, rigidity, and susceptibility to rot and microbial degradation. Chemical modifications like etherification, grafting, cross-linking, oxidation, and hydrolysis are employed to address these drawbacks, depending on the intended application of the starch product [24, 25]. Starch is a renewable polysaccharide derived from various sources and can supplement or replace fossil fuel-based polymers. Its key advantages include absorbency, biodegradability, biocompatibility, and non-toxicity, making it environmentally safe. Beyond agriculture, starch finds uses in culinary, medicinal, food, paper, chemical, textile, and pharmaceutical sectors. Its industrial applications are expanding, particularly in textile stiffening agents, adhesives, non-food fillers, and packaging additives [43]. Natural dyes may increase color strength, create materials that are antibacterial, and have fastness qualities against light, sweat, and washing, according to a study on chitosan and gelatinous starch. This environmentally friendly, multipurpose, nontoxic finish has been improved for use with printing using natural dyes on many kinds of fabrics [44]. Flame retardants based on starch are used in building materials, coatings, and textiles. Ammonium starch phosphate (ASTP), a novel phosphorus-based flame retardant, was successfully cotton fibers that have been grafted using P-O-C covalent linkages to provide high and long-lasting cotton that is flame retardant textiles (LOI: 45.2%, even after 50 laundry cycles) [42].

8.3.3 Chitosan

It is primarily created by alkaline deacetylation of chitin and is composed of glucosamine and N-acetylglucosamine units linked by 1–4 glucosidic links. The process of removing acetyl groups (CH_3-CO) from chitin's linear molecular chain leaves a reactive NH_3 group, which is then converted into chitosan $[(C_6H_{11}O_4N)_n]$. Chitosan is a natural, biodegradable material ideal for antibacterial textile finishing due to its non-toxic properties, bacterial resistance, and solubility in diluted acetic acid [21, 45]. Both chitin and chitosan are polysaccharides structurally similar to cellulose, available in various molecular weights, purities, and crystallinities. However, the presence of an acetamide group, containing nitrogen, distinguishes cellulose from chitosan and chitin [46]. Chitin, a by-product of the shrimp canning industry, constitutes about 40% of the waste produced during shrimp processing. Much of this waste is discarded into the ocean, so repurposing it directly benefits the environment by reducing chitin deposits in coastal areas [47]. Chitosan, derived from chitin, is an excellent material for antimicrobial coatings due to its ability to absorb nutrients required by bacteria for growth, reduce enzymatic activity, and a high water-binding capacity [48]. The study searches into how the type of textile construction and treatment materials (Chitosan and AgNPs) affect the mechanical and physical characteristics of cellulosic fabrics. It finds that the concentration of chitosan has a major impact on fabric properties, which may be useful for traffic and medical workers [49]. In addition to its antibacterial properties, chitosan can be used to enhance the dyeing process and provide fabrics with an antistatic and deodorizing finish [21]. Chitosan is a versatile textile finishing chemical with deodorizing, antistatic, dyeing, and antibacterial properties. It stiffens fabric surfaces by producing a coating, but its effectiveness is limited to higher concentrations. The primary disadvantage is that it only inhibits microbes at high concentrations, limiting air permeability and creating a coating on the fabric's surface [27]. Chitosan is a multipurpose treatment that improves fabrics' antibacterial qualities and color strength. Because it offers color fastness that is equal to commercial samples, it is also utilized in textile printing [44]. In a study, utilizing glyoxal as a precursor, chitosan was modified with sodium pyrophosphate and melamine, two sources of phosphorus and nitrogen, to produce green, multipurpose chemical agents. The flammability, antibacterial activity, and UV protection of textiles treated with modified chitosan were assessed. Significant bacterial inhibition, good UV protection (UPF > 100), improved fire retardancy (LOI > 30), and satisfactory mechanical qualities were also displayed by the treated fabrics [42, 50]. Octadecylamine (ODA) was chosen for the superhydrophobic coating in this work due to its long hydrocarbon chain, which reduces surface energy and repels liquid droplets. Glutaraldehyde (GA), a crosslinking agent, was used to covalently bind ODA to chitosan. When applied to hydrophilic textile fibers like cotton and polyester, this coating can transform PU sponges into a superhydrophobic state, providing excellent self-cleaning properties [15]. The other study coated the surfaces of polyamide 66 fabric with flame-retardant coatings made of bio-derived polysaccharides, such as sodium alginate and chitosan. Subsequently, inorganic salts comprising iron, nickel, and boron were added to these coatings. The outcomes demonstrated that the flame retardant performance of the fabric samples treated with metal-salt and LbL-assembly was superior [51].

8.3.4 Lignin

Lignin, a structural component found in the cell walls of woody plants, plays a key role in providing strength and forming vascular networks that transport water via hydrophobic interactions [29]. It also protects plants from insects and microbes. Lignin's antibacterial properties stem from its methoxy groups and phenolic monomers, which can inhibit bacterial growth by disrupting their cell walls [52–55]. The growing demand for natural packaging has lifted the need for biopolymers, particularly lignin, which acts as an antioxidant and radical scavenger. It is increasingly used in sunscreens, textiles, packaging, varnishes, and composites due to its UV-blocking properties [28, 56]. Lignin's dark color presents difficulties for dye dispersion applications, but despite certain drawbacks because of its dark color, it has been employed in many studies to meet consumer demand for biobased nonwoven processing alternatives. These studies have shown that lignin can save costs and lead to innovations in nonwoven coloration technologies [57]. Harshvardhan et al. successfully produced natural vanillin using natural bacterial groups, reducing lignocellulose in bamboo chips without chemical treatment. Hosoya et al.

synthesized vanillin using soft lignin and tetrabutylammonium ions, offering superior heat, chemical resistance, and mechanical strength. Vanillin and its derivatives could be a potential class of biobased aromatic chemicals from lignin [58]. In their study, Zimniewska assessed how well nano lignin blocked UV rays in linen fabric. After ultrasonically treating kraft lignin to produce nanofibers, various amounts of silicon emulsions were applied to the nanofibers. Using a Sola screen device, the nano lignin consolidated dry linen fabric's UV protection factor (UPF) was calculated. The use of nano lignin in linen fabric improved its UV protection characteristics [56]. In a different study, lignin sulfate was linked together with polyvinyl alcohol (PVA) to create lignosamine after amino groups were grafted onto it to increase its antibacterial properties. By using in situ reduction with silver ions, a polymer hydrogel with silver nanoparticles was produced, and it demonstrated superior antibacterial properties against *S. aureus* and *E. Coli* [59].

8.4 Usage of Protein Based Substances in Textile Finishing

Proteins are essential components of our daily lives. They are some of the basic elements that comprise body tissue. Proteins are made up of separate amino acids that can be conjugated thanks to peptide bonds. Amino bonds occur when molecules of carboxylic acid and amino acid react to form peptide bonds. Proteins are lengthy macromolecules that are joined by polypeptide bonds [23]. Polypeptides, often known as proteins, are vital biomolecules present in every living organism. These structures can be primary, secondary, tertiary, or quaternary, and they are made up of peptide bonds connecting up to 20 amino acids. Primary structures are linear sequences; secondary structures are three-dimensional configurations of the backbone; tertiary structures are complete proteins; and quaternary structures are proteins that have more than one polypeptide chain [3].

8.4.1 Keratin

The name "keratin," which first appeared about 1850, refers to the substance that makes up hard tissues like animal hooves and horns. The term "kera" is derived from the word "horn." Keratin is a tiny protein with a consistent size that is insoluble and extremely stable [32]. Keratin is an abundant protein bio-polymer found in the epithelial cells of higher vertebrates. It consists of approximately 90% proteins, 2–5% sulfur, 15–18% nitrogen, 3.2% mineral elements, and 1.27% fat. Key amino acids required for its structure include cysteine, glycine, proline, and serine. Keratin is the main structural protein present in hair, horn, and feathers. It is commonly classified as a cysteine sulfur-crosslinked fibrous protein. Wool is a notable source of keratin [3, 60]. Organic solvent-insoluble keratin fibers that are renewable, non-abrasive, and biodegradable may be obtained from chicken feathers. They are the best option for high structural reinforcement in polymer composites because of their excellent mechanical qualities, low density, hydrophobic behavior, sound absorption, flame retardant, and cost [30]. Furthermore, enzymes that break down proteins cannot readily break them down. Chemical and mechanical strength are provided by the high cysteine content seen in keratin [33].

8.4.2 Pectin

Glycoside linkages hold *D-galactopyranosyluronic* acids together to form pectin, a linear polysaccharide. It is an essential part of citrus by-products and may be found in the cell walls of plant tissue. It is found in the primary cell walls of most plants, particularly non-woody parts, besides hemicelluloses and cellulose [6, 31, 60]. A linear polymer of 1,4-linked d-galacturonic acid with some monomer groups combined with a methyl group is the fundamental structure of pectin which is used as a thickening or gelling agent [60]. A study investigates the use of natural thickeners derived from fruit peels, specifically orange and pomegranate peels, focusing on pectin in printing paste. The pectin was treated with hexamine to enhance its storage stability and tolerance to high pH, making it suitable for use in printing paste for cotton fabrics. The results show that pectin-based thickeners increase printed materials' elongation and tensile strength at break. Additionally, the modified pectin develops and modifies a thin coating in the microstructure of the fabric filling up surface voids and enhancing elongation and tensile strength at break [61]. Also, pectin enhances coagulation, which is important in reducing bleeding [6].

8.4.3 Sericin

Sericin is a naturally occurring, environmentally benign, protein-based, and biodegradable bio-polymer that results from the silk industry. It comprises 20%–30% sericin, a water-soluble globular protein matrix, and 70–80% fibroin. Due to the presence of hydroxyl groups, such as those in serine and aspartic acid, sericin is hydrophilic. It acts as a gum, binding the fibroin filaments together, and is removed during the degumming process to restore the silky sheen [22]. Research is focusing on sericin because of its potential as a bio-polymer and moisturizer. It is well-known that sericin has anti-oxidant, UV-absorbing, and moisture-absorbing qualities. Restoring sericin from degumming solutions can enhance the hygroscopic characteristics of synthetic fibers and produce a bio-polymer that benefits the environment. Sericin is a valuable finishing agent in textile, cosmetic, and biomedical polymeric products due to its biological features, which include antibacterial, antioxidant, anticancer, UV-resistant, and moisture-absorbing qualities [22, 34].

8.4.4 Collagen

Collagen, an animal protein, is the main structural component of bone and skin. It is derived from marine sources like fish skin and scale fish and shares biochemical and physical properties with collagen from pigs and cows. Collagen is reproducible, hydrating, hydrophilic, and biodegradable, and is used in sponges, films, and membranes. Its biomaterial properties make it commercially effective, offering benefits like structural, chemical, physical, and immunological characteristics [10, 36]. Collagen can be utilized to produce bio-ink for 3D printing, enabling the development of novel materials. Due to their superior cell-activating properties, collagen-based materials have been identified as a promising alternative for fabricating ideal mimetic bio-inks for tissue regeneration. In a paper published in 2012, Marques et al. discussed the latest advancements in 3D printing with collagen-based materials for hard tissue engineering, including the development of substrates for bone and cartilage regeneration [35]. The study by Busra et al. outlines the manufacturing processes used to produce various types of collagen substrates for tissue engineering applications. Additionally, collagen is employed in wound treatment therapies. Pallaske et al. reviewed its significance in wound dressing treatments, highlighting collagen's crucial role in wound healing where collagen has been shown to have an important part in wound healing, and as a result, several products based on this substance have been produced [35].

8.4.5 Soybean

A promising plasticizer for molding into polymer products and films is soy protein, a renewable bio-polymer with a 40–55% protein concentration. The environmental danger associated with polymer items is substantial since they do not decompose in the environment. In order to tackle this issue, soybean polymers have been created, which differ from conventional polymers in terms of accessibility, availability, potential for use, low rate, biodegradability, and thermoplastic characteristics. In the plastics industry, these factors have helped to make them popular [37]. The development of soy protein films aims to reduce pollution levels in the environment and promote a sustainable textile sector. However, their application is complicated by poor water stability and limited adhesive strength. A study looked into how sodium dodecyl sulfonate affects soy protein unfolding and how nano-chitosan reinforces unfolded soy protein adhesives. Comparing the results to unmodified soy protein, water stability, and tensile strength both increased by 85% and 157%, respectively [62]. Additionally, the dry and wet strengths of pulp/viscose wet-laid nonwovens were enhanced by modified soy protein. Natural soy protein (SP) consists of various amino acids and can be easily isolated. Due to its biodegradability, biocompatibility, availability, and renewable nature, soy protein has been widely used in textiles and apparel, culinary applications, healthcare, tissue engineering, and the cosmetic industry [63].

8.4.6 Alginate

Alginic acid, found in brown seaweed, is composed of two uronic acids with a distinct structure which are β-D-mannuronic acid and α-L-guluronic acid. Ionic interactions between alginic acid and divalent cations can result in the formation of persistent gels. Apart from its conventional use as an agent for thickening in printing on textiles, alginate possesses unique characteristics including affordability and accessibility, biocompatibility, wound-healing potential, strong ion-exchange capacity, and high moisture adsorption [38]. Alginate has been shown to have strong gelling characteristics, great biocompatibility, and remarkable ease of processing and modification, which have made it a viable biomaterial for a variety of uses, such as medication delivery systems and wound healing goods and structures for the engineering of tissues. Alginate has been a highly promising option for textile antibacterial finishing in recent times [64]. Because of their strength and flexibility, alginates—polymers containing divalent cations including calcium, magnesium, and sodium—are employed in the food and pharmaceutical sectors. To prepare sodium alginate films, certain brown seaweeds are evaluated for bioactive, antibacterial, and antioxidant compounds. Research is being conducted to create textile textiles coated in biodegradable alginate film for use in wound healing [65]. Hospitals need to practice good textile hygiene since a lot of materials might harbor dangerous germs that stick to their surfaces and cause diseases. To stick securely to fibers and avoid washing out, researchers are investigating antibacterial coatings for textiles. Polyunsaturated hydroxyls (PUDs) are candidates that mix with natural resources to enhance biocompatibility and antibacterial action. Chitosan, cyclodextrin, and collagen are used as chain extenders in the synthesis of these PUDs. In the work, aliphatic isocyanate-based PUDs were made using sodium alginate, a naturally occurring antimicrobial polymer, and dimethylol propionic acid (DMPA), an in emulsifier. It is advised that these alginate-based dispersions be considered viable options for textile antibacterial finishing [66]. In a study, the antibacterial action of cotton that has been coated fibers has been demonstrated to enhance when copper NPs are present, either in the form of calcium alginate or using alkali hydrolyzed silver nitrate alginate solution. In a different investigation, even after many washing cycles, the modified polyester fabric containing alginate and TiO2 NPs demonstrated a respectable level of UV protection efficacy and antibacterial activity [67]. To impart antibacterial properties to alginate/cellulose fibers (SACF), Varaprasad et al. modified them with nano zinc oxide. The results of the antibacterial tests demonstrated that the modified fibers exhibited strong bacterial killing capabilities, while the SACF alone showed no inhibitory zone [68].

8.5 Enzymes in the Textile Industry

In the textile industry (Fig. 8.2), enzymatic treatments have been recognized as non-toxic, energy-efficient, and stereospecific solutions.

Enzyme acceptance as biocatalysts in several areas has been remarkable because of their green chemistry and substrate selectivity. Although enzymes may replace harsh chemicals and save energy and water, they are widely used in the textile processing sector [69]. Several textile processing processes, including biostoning (of denim), degumming silk, bleaching and shrink-proofing wool, desizing, scouring, bleaching, and biopolishing cotton, as well as handling textile effluents, can benefit from the usage of enzymes [69–71]. Enzymes are biodegradable and serve as replacements for hazardous chemicals, helping to reduce effluent load. Additionally, they operate under very mild conditions, making processes involving fiber degradation safer. Furthermore, due to their specific actions, enzymes do not compromise quality. Since they don't require expensive chemicals or auxiliaries and use less steam energy, enzymatic processes are cost-effective. As a catalyst, enzymes allow a reaction to occur and then return to its original state for reuse [72]. Nonpolluting, nontoxic, and biodegradable enzymes are used in finishing applications for thermal regulation, scent therapy, fragrance finish, biological agents, freshening, insect repellant, protection from UV rays, and static-resistant reasons [73]. Common enzymes used in the textile industry include amylases, peroxidases, cellulases, catalases, and laccase. They also take part in several activities, clean starchy soils, and break down excessive hydrogen peroxide. Environmentally friendly enzymes can revitalize the economy

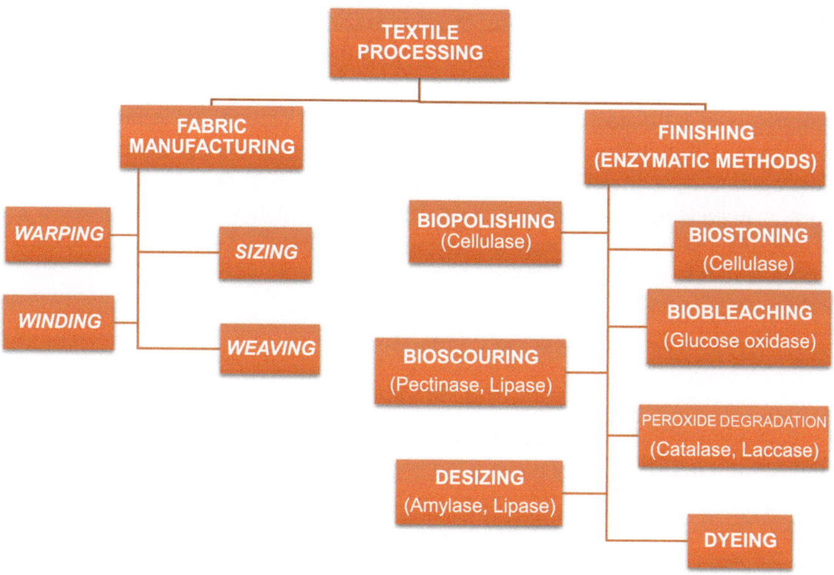

Fig. 8.2 Enzymes used in textile processing

of underdeveloped countries [74]. Enzymes such as catalase for removing excess oxidants, pectinase for bioscouring, amylase for desizing, and cellulase for mercerizing and biopolishing contribute to improving the fabric's quality, strength, and texture [75].

8.5.1 Pectinase

Pectinases are a novel class of enzymes utilized in the textile and fruit industries. As essential enzymes in industrial biotechnology, they are widely used in the food and beverage production sectors [76]. Pectinase is used in textile chemical processing for a pretreatment known as bio-scouring, which helps to make cotton fabric more absorbent and ready for additional coloring [77]. Since pectinases can depolymerize pectin and break it down into low molecular weight water-soluble oligomers, they seem to be the most effective enzymes for achieving the best scouring outcomes. This enhances the textile material's brightness and absorption [78]. Pectinases, amylases, lipases, cellulases, and hemicellulases have been used together to safely and environmentally friendly remove impurities from cotton fibers, replacing the previously used toxic caustic soda. A novel method called *bio-bleaching* uses certain enzymes to extract non-cellulosic contaminants from fibers. For this reason, pectinases have been utilized without causing any detrimental side effects on the breakdown of cellulose [76]. In order to make the surface of raw cotton more hydrophilic, a different and environmentally friendly technique called *bioscouring* uses certain enzymes to remove noncellulosic *impurities* [78]. Bioscouring involves using specific enzymes to target noncellulosic contaminants. During the bioscouring process, some of the naturally occurring colors associated with these noncellulosic substances may be removed from the fiber. One other benefit of this method is that, in addition to being more ecologically friendly and energy-efficient, the enzymes employed in bioscouring do not harm the cellulose backbone, significantly reducing damage to fibers. In addition to the pectinases' compatibility with pH and temperature, factors such as residual pectin, water adsorption, whiteness, and the required treatment period are taken into account when selecting pectinases for bioscouring [79].

8.5.2 Cellulase

The most often utilized enzyme in the wet processing of textiles is cellulase. Industrial cellulases are non-uniformly constituted complexes of many cellulases, cellobiase, and related enzymes with an average molecular weight ranging from 10,000 to 4,00,000. The enzyme system known as cellulase is made up of several components, such as cellobiohydrolase (CBHS) cellobiases, and endoglucalases (EGs). Complete cellulase systems are best for cleaning and removal effects, but EG or EG rich preparations are best for aging and heart arrest of fiber surfaces. The enzyme cellulase is specifically employed to scour cotton textiles. Its major purpose is to make contact with the primary wall by penetrating the fiber strand's outer coat, or cuticle [80]. Bio-polishing is a safe, permanent, and non-toxic treatment that smooths the surface of cellulosic fabrics like cotton, viscose, and linen by eliminating superficial microfibrils through cellulase enzyme hydrolysis. It prevents pill formation, clears fuzz, and can be performed at any stage of wet processing, especially after bleaching [73, 80, 81]. Cellulase-based denim washing is a popular environmentally friendly method for giving cotton textures and denim a nice appearance and non-abrasiveness. When used appropriately, cellulase has several benefits, such as being less damaging to the environment, lowering wear and tear on equipment, enhancing clothing quality, inflicting less harm on clothing without sacrificing fabric strength, and increasing the load of clothing in the machine [81].

8.5.3 Amylase

Amylase is an extremely adaptable enzyme that finds extensive use in both industrial and laboratory settings [82]. Microorganisms, plants, and mammals all manufacture *α-amylases*, a type of enzymes. Companies that use starch in their manufacturing processes benefit from amylases because they can hydrolyze or break down starch molecules into polymers that contain glucose units [83]. *α-amylase* is used in the textile industry for desizing. Before fabric is made, starch, a sizing agent, is applied to the yarn to ensure a quick and safe weaving process. During weaving, starch functions as a strengthening factor by preventing warp yarn breaking. In the textile finishing business, starch on woven fabrics is removed by a wet procedure. α-amylase breaks down starch used in weaving into water-soluble components during the desizing process. But the fibers are unharmed [82]. Desizing helps colors and chemicals better penetrate into the cloth by removing the sizing material from the yarn. Chemical desizing weakens the cloth, which increases wastewater treatment costs and produces hazardous effluents. Using amylase for enzymatic desizing offers several benefits including the production of non-toxic byproducts, and decreased water and energy usage. However, the main challenge lies in obtaining low-cost, consistently high-quality enzymes that are stable at high temperatures and pH levels. In this regard, producing amylase in-house using textile industry effluents could present an appealing alternative to traditional chemical desizing methods [75]. Research indicated that the optimal temperature for *α-amylase* activity was 80 °C. Enzyme activity increased significantly at this temperature ($p < 0.05$). 64% of the original activity remained when the temperature hit 90° C. Compared to previous research, the ideal temperature for *α-amylase* generated from *Trichoderma pseudokoningii* was 50 °C, but amylase produced from *A. awamori* and *Laceyella sacchari* had respective optimal temperatures of 55 and 70 °C. According to the study's findings, *α-amylase* activity is most effectively activated around 80 °C, making it more suitable for usage in the textile industry [75].

8.5.4 Laccase

Laccases, also called benzenediol: oxygen oxidoreductases (EC 1.10.3.2), are multi-copper oxidases that help reduce molecular oxygen to water and oxidize aromatic thiols, anilines, and substituted phenols to their respective radicals in just one electron. Laccases are of great importance to the pulp and paper, beverage processing, textile dye decolorization, sugar beet pectin gelation, and baking sectors [84]. Green catalyst options are highly sought after for dye production due to the large global demand for dyes and the difficult conditions needed for their creation. Consequently, researchers have focused on eco-friendly alternatives, like bio-based options, to minimize environmental harm when making dyes. Plenty of studies have been done in the past few years on green catalysts, such as enzymes, and safe oxidants, like oxygen, for the production of dyes. In particular, there is a lot of promise for using laccases in enzymatic processes [85]. It is interesting to develop laccase-based strategies since the laccase enzyme can break down a variety of colors. Immobilized laccases have been used in place of laccases due to the industrial usage of laccases' limitations in decolorizing textile wastewaters, including the enzyme's nonreusability. These immobilized solutions are not only more affordable, simple to use, and reusable, but they also provide greater flexibility. Because laccases oxidize aromatic molecules, they can also aid in the formation of color. These oxidizing techniques result in industrial hues like phenoxazine and azo dyes. Heteropolymeric dyes are additional types of polymeric dyes that are used to color cotton and wool textiles. Laccases are also used in laundry detergents because they are capable of removing odors during cloth production. Laccase-mediated systems could also reduce wool shrinkage [86].

8.5.5 Peroxidase

Peroxidases (EC 1.11.1.7) are widely distributed in nature, and various species, including bacteria, plants, and mammals, produce enzymes [87]. Peroxidases may be used to bioremediate wastewater that contains phenols, cresols, and chlorinated phenols; they can also be used to remove peroxide from food and industrial waste, breakdown textile colors, and generate bio pulp and bio bleach for the paper industry. Peroxide may also be extracted from food and industrial waste. Rhodamine dyes, which can be eliminated by peroxidase but are resistant to conventional bleaching, are the reason for the bright color of process water from textile factories. This process has the potential to reduce environmental pollution. The heme-containing glycoproteins known as peroxidases (EC1.11.1.7; hydrogen-peroxide oxidoreductases) catalyze the oxidation of many substrates in the presence of hydrogen peroxide, including aromatic amines, phenols, polyaromatic hydrocarbons (PAHs), etc. [88].

8.5.6 Catalase

A class of enzymes known as catalases has gained recognition as an essential biocatalyst in the industrial setting owing to their remarkable stability and high catalytic rate. The main application of this enzyme is in textile bleaching before dyeing, which has been carried out extensively for many years. New techniques continue to make use of their strong catalytic ability to digest hydrogen peroxide, while catalases' (per)oxidase activity opens up new options [89]. Because hydrogen peroxide has a harsh effect on the environment, catalase was also utilized to remove it from the bleaching effluents of the textile industry. Furthermore, because catalase causes hydrogen peroxide (H_2O_2) to break down into water and oxygen, it is employed in the bioremediation process where H_2O_2 is employed as an oxygen source under aerobic circumstances [90]. The wastewater from textile bleaching was treated using a catalase-peroxidase (CP) derived from the recently identified *Bacillus SF* [91].

8.6 Essential and Vegetable Oils in Textile Finishing

8.6.1 Essential Oils

Since textile manufacturing utilizes a lot of chemicals, it discharges toxic materials into the environment. Thanks to the rising sustainability awareness, re-evaluation of classical applications with the awareness of green innovation and taking remedial reactions are on the agenda of the textile industry. In various research, the main starting factors for alternative solutions are evaluated as natural materials, plant ingredients, and plant extracts. The usage, cost-effectiveness, and minimal side effects of essential oils compared to chemicals promote their adoption as a sustainable trend. Essential oils derived from various parts of plants, such as seeds, roots, leaves, flowers, and more—examples include basil, cinnamon, marigold, lavender, rose, vetiver, jasmine, and lemongrass—are utilized in functional finishes like fragrance and insect repellent treatments on a variety of surfaces. Fragrance finishing refreshes the mood of people by helping to eliminate negative psychological states such as stress, anxiety, and fatigue. In addition, various studies have shown that textile materials scent finished with essential oils bring calmness and reduce pain. With fragrance finishing, we can obtain antibacterial and antiviral structures up to a certain ratio. Such scent-finished textile materials are widely used in both domestic and industrial applications [73]. Additionally, scent-finished fabrics are being studied for their potential to combat airborne illnesses, purify the air, and save the environment. Furthermore, scented masks and handkerchiefs smell good and aid in protection. To lessen tension, scent coatings have also been placed on businessmen's jackets and ties. Through nano-coating or microcapsule padding, they are also applied to socks, athletic socks, and home furnishings to provide a pleasant aroma and offer protection against bacteria, germs, viruses, and fungi (Fig. 8.3).

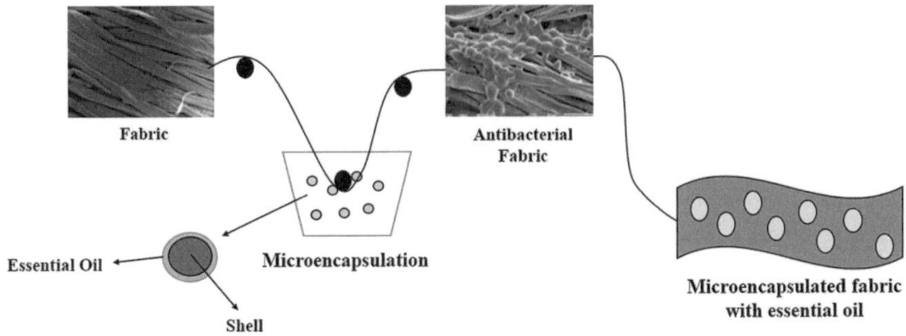

Fig. 8.3 Microencapsulation of essential oil for antibacterial textiles

8.6.1.1 Orange

Utilization of oranges as raw material for fresh produce, juice, and concentrate production generates a significant amount of residue (approximately 50–60% of fresh oranges), predominantly comprising peel, mostly discarded as waste. The waste resulting from orange consumption presents a significant opportunity for advanced transformation in the bio-based economy. Limonene is the primary volatile oil component of orange peel. Due to its limonene content, orange peel has clinical applications as an antidiabetic, mosquito repellent, and antioxidant. Performance fabrics utilize this essential oil to enhance the final product appearance on fabric surfaces with nanoparticle structures, thereby improving fabric versatility. In a study, Sliced orange peels into little pieces after washing. The use of ultrasound-assisted extraction technology was used to extract bioactive components from wet orange peels. The orange essential oil was used to treat cellulose fabric, giving it antibacterial, antioxidant, UV protection, and photocatalytic cleaning qualities. Additionally, treatment with microencapsulated orange essential oil extract, obtained via spray drying, imparted insect-repellent properties to linen fabric surfaces [92]. In another study, the acquisition of antimicrobial barriers as an effective method has been suggested due to the continuous contact of textiles with microorganisms. It has been demonstrated that orange oil functionalizes textile materials in terms of antimicrobial properties. Studies have demonstrated the antifungal [14–16] and insecticidal [13] qualities of orange oil, which can be added to cotton textiles to help prevent the microbiological degradation process that develops during use and can result in major functional, hygienic, and aesthetic issues as well as textile deterioration, discoloration, and odor [93].

8.6.1.2 Vanilla Oil

Vanilla oil is an aromatic chemical used in textile manufacturing that is extracted or distilled from vanilla beans or seeds. It is applied to textiles, clothing, and household textiles like bedding and curtains, to give them a lovely vanilla aroma. In order to add scent and improve the overall sensory experience of textile goods, vanilla oil is frequently added to fabric softeners, laundry detergents, or fabric sprays. It has a lingering, natural scent that may make you feel cozy and warm. Furthermore, vanilla oil might possess certain antibacterial qualities that help keep textiles fresh by preventing the growth of germs that cause odors. Although years of applying fabric conditioners to give textiles a scent boost, the results are fleeting. Fragrances can stay on clothing for a considerable portion of their lifespan because of microencapsulation technology. With a shelf life of three to five years, companies have produced fabrics that release natural scents for up to twenty-five washing cycles. s. In one study, rosemary, jasmine, vanilla, and lavender essential oils were compared for fragrance finishing. The results show that the fragrance finish provided by vanilla essential oil lasts up to 5–10 wash cycles [94]. Textile structures treated with microencapsulation methods using essential oils such as sesame oil, aloe vera, and vanilla have exhibited increased wash durability, pleasant and long-lasting scent, and better antimicrobial properties compared to sesame oil and aloe vera [95]. In an additional investigation, cotton fabricators were treated with polysulfone microcapsules containing vanillin. The goal of the study was to examine the microcapsules' antibacterial qualities and thermal stability against Staphylococcus aureus bacteria, which led to the inhibition of bacterial growth in specific environmental circumstances [96].

8.6.1.3 Lemon Oil

Lemon oil can impart a pleasant lemon scent to textile products, providing a sense of cleanliness, while also possessing antimicrobial properties, thus helping to inhibit microbial growth and prolong the freshness of the products. Additionally, lemon oil can be utilized for stain removal due to its acidic nature, particularly effective in removing oil or food-based stains. Furthermore, fabrics infused with lemon oil may serve as an ideal option in aromatherapy applications, imparting a sense of vitality and well-being. Overall, lemon oil can be employed to add a natural and effective scent, cleanliness, and freshness to textile products. Different metal nanoparticles encapsulated in bio-polymer material can be used in proteinic and cellulosic fabrics with or without lemon oil in order to create textile material with antibacterial, fragrant, and mosquito-repelling properties [73, 97]. One of the studies in the literature aimed to create fragrant fabrics and textiles that repel mosquitoes by directly applying a unique nanoparticle made of natural polymers like pectin and gelatin to the fabric while it is infused with lemon oil. This was achieved through a mechanical homogenization mixing process [98]. A study demonstrated that the essential oil derived from citrus lemons had a higher and longer-lasting antibacterial ability and that the green lemon peel extract had stronger antibacterial activity against *Escherichia coli* and *Staphylococcus aureus* microbes when cotton fabric treated with essential oils from orange, black, and green lemon peels was compared [99]. Cellulosic organic textiles such as viscose and protein organic textiles such as wool are suitable for providing ambient conditions such as moisture, temperature, and oxygen that attract microorganisms. As a result, they can cause unpleasant odors and allergic reactions. To strengthen wool and viscose textiles' resistance to microorganisms, lemon peel oil was made and applied. It was encapsulated in nanoemulsion and nano clay. Results from antimicrobial investigations showed that textiles treated with lemon peel oil and its derivatives significantly outperformed untreated fabrics against the *Staphylococcus aureus* bacteria [100].

8.6.1.4 Lavender Oil

Lavender oil, derived from the lavender plant, is a natural essential oil known for its soothing aroma and various therapeutic properties. In textile finishing, lavender oil can be utilized in several ways to enhance the properties of textile products. One common application is to impart a pleasant lavender scent to fabrics, providing a sense of relaxation and freshness. Lavender oil can be incorporated into fabric softeners, detergents, or fabric sprays used during the finishing process to infuse the textiles with its aromatic properties. Additionally, lavender oil is known for its antimicrobial and insect-repelling properties, making it a valuable additive in textiles, especially for bedding, towels, and other items prone to bacterial growth or pests. Furthermore, lavender oil may contribute to the overall well-being of individuals through aromatherapy, promoting relaxation and improving sleep quality when infused into textiles such as bedding or sleepwear. Overall, lavender oil offers multiple benefits in textile finishing, including imparting a pleasant scent, providing antimicrobial protection, and enhancing the overall sensory experience of textile products. A study demonstrates the use of polyurethane shell capsules to release lavender oil from cotton fabrics. The capsules, combined with environmentally friendly propylene glycol and isocyanate, effectively release lavender oil and reduce environmental impact. The capsules lose 30–50% of their weight after 30 days and are sensitive to mechanical stimuli. These textiles can be utilized as transient or throwaway textiles, which will help with anxiety, palliative care, and sleep issues [101]. In a lavender oil-based study, ultrafine nanoparticles of a copolymer loaded with lavender oil were applied to cotton, cotton/polyester, and polyester-elastane fabrics. The results showed that the fragrance of the fabrics impregnated with copolymer nanoparticles lasted for up to 25 washes and an average of 60 days if stored at room temperature. The study's findings suggest that the long-lasting odor emission from fabrics is caused by the ease with which the small nanoparticles used in this study enter and remain in the pores of the fabric fibers [102].

8.6.1.5 Rose Oil

Rose oil, obtained through distillation from rose flowers, is a natural oil known for its pleasant fragrance and various therapeutic properties. In the textile industry, rose oil can be utilized for various purposes. For instance, it can be incorporated into fabric softeners, detergents, or textile sprays to impart a delightful rose scent to textiles. Additionally, rose oil possesses antimicrobial and insect-repelling properties, contributing to the hygiene and protection of textile products. Rose oil is also widely used in aromatherapy and can promote relaxation and enhance sleep quality when added to bedding or sleepwear. Therefore, rose oil provides both pleasant fragrance and health benefits to textile products. Rose essential oil is a particularly effective oil for dry or aging skin. It has been revealed by researchers to contain various therapeutic compounds such as antimicrobial and anti-inflammatory compounds known to promote healing. As a result, rose essential oil improves skin texture and tone and can help in the treatment management of skin conditions such as psoriasis, and dermatitis. Using padding and exhaustion processes, commercial microcapsules containing rose essential oil were applied to cotton and cotton/polyester knitted materials. The peak areas of the remaining chemicals on the samples significantly decreased after being exposed to sweat solution, suggesting the establishment of a releasing propensity. Overall findings showed that functionalized treatments only required one washing cycle; nevertheless, the number of microcapsules dramatically dropped after 9000 abrasion cycles. All experimental versions exhibited an antibacterial effect against *Staphylococcus aureus* bacteria [103].

8.6.2 Vegetable Oil

Vegetable oil is an abundant hydrocarbon that occurs naturally in a wide variety of compositions and is found in natural plants, seeds, and fruits. Because of their countless uses in the biofuel, food, cosmetics, paint and coating, textile, biofuel, and plastic industries, vegetable oils have garnered enormous attention in this new era. Vegetable oil has been the subject of numerous studies aimed at improving its extraction and utilization in key applications such binders, polymerizations, and lubricants [104].

8.6.2.1 Soybean Oil

Soybean oil is a versatile bio-based alternative in the textile industry, offering eco-friendly benefits. It improves dyeing properties, imparts softness, smoothness, and wrinkle resistance, and acts as a biodegradable lubricant in textile production processes. In addition, it offers a sustainable substitute for conventional scaling agents that come from petrochemical sources. Pretreating them with soybean oil and applying non-thermal plasma to their surface offers a novel way to increase the dyeing color strength of para-aramid materials. One study showed that the results demonstrated promising characteristics, as non-thermal plasma surface treatments alone did not significantly improve the dyeing of para-aramid fabric with dispersion dye. However, the outcomes of the staged experiments indicated that the color strength of the dye is enhanced when soybean oil is pretreated before being treated with non-thermal plasma [105]. Castor and palm oils, among other vegetable oils, have been adjusted to contain fluorine through the use of soybean oil, a crucial component of soybeans. The results of the investigation demonstrated the exceptional water resistance of cotton textiles treated with modified vegetable oils. After five washing cycles, all treated fabrics exhibited good water-resistant durability, with those treated with modified soybean oil showing the highest level of durability [106]. In a research effort aimed at developing a methodology for fabricating laminated textiles with reduced environmental impact, a formulation incorporating soybean oil and UV-curing technology was developed for a bio-based coating. This formulation involved the synthesis of an acrylated epoxy-esterified soybean oil oligomer, which was subsequently applied to cotton fabrics using UV-curing. Pretreatment procedures using pectinase, sodium hydroxide, and a commercial detergent were used before this coating technique to improve properties like wettability, durability, and abrasion resistance [107]. A new method for raising the surface roughness of cotton fabric is presented in a study targeted at producing a superhydrophobic surface. This involves covering the fabric with a thermoset made from epoxidized soybean oil, altering it with stearic acid, and etching it with a solvent that contains choline chloride and oxalic acid. As a result, coatings give cotton fabric remarkable oil–water separation capabilities, strong self-cleaning properties, and durable superhydrophobic properties [108]. In a different investigation, epoxidized soybean oil (ESBO) and a phosphorus-containing specialized monomer (Sipomer PAM 200) reacted to create an oligomer produced from soybean oil. The resultant oligomer was then combined with lignocellulosic material (Platanus orientalis (PO) leaf powder), applied on cotton fabrics as a coating material to be utilized as a lightweight tarpaulin, and UV-cured (Fig. 8.4b and c). To strengthen the connection between the oligomer and PO powder, UV light, and enzymes have been used to modify the surfaces of the powder. Due to its reduced weight loss with abrasion outcomes, enzyme-treated PO powder has been proven to be more successful in terms of bonding performance [109]. The ring opening reaction between the ESBO and AA has been used to create a photocurable oligomer for this purpose (Fig. 8.4a). Next, cotton fabrics treated with a water-repellent finish (WRF) agent were coated with coating formulations including the produced acrylated ESBO (AESBO) oligomer. Spray and absorbency tests have proved that the WRF on the fabric by itself is not durable enough against several washings. So, the hydrophobic character of the fabrics with only WRF gradually decreases after 10 ECE washings. Whereas the fabrics with both WRF and AESBO coating layers still show hydrophobic character even after 10 ECE washings. In conclusion, a bio-based photocurable coating layer can be employed to improve the washing fastness of textile finishing agents [110].

Cotton fabric coatings with flame-resisting qualities have made use of the photocurable oligomer following the addition of phosphorus groups to the ESBO polymer backbone. Researchers have looked into how the type of reactive diluent affects the cotton fabrics' mechanical, wettability, and thermal properties. The findings demonstrate that the reactive diluents' aromatic, aliphatic, flexible, rigid, and other functional groups had an impact on a variety of fabric properties, including wettability, mechanical qualities, thermal transitions, drapeability of cotton fabrics, and weight loss of abrasion-resistant fabrics. Figure 8.5 shows the scheme of the

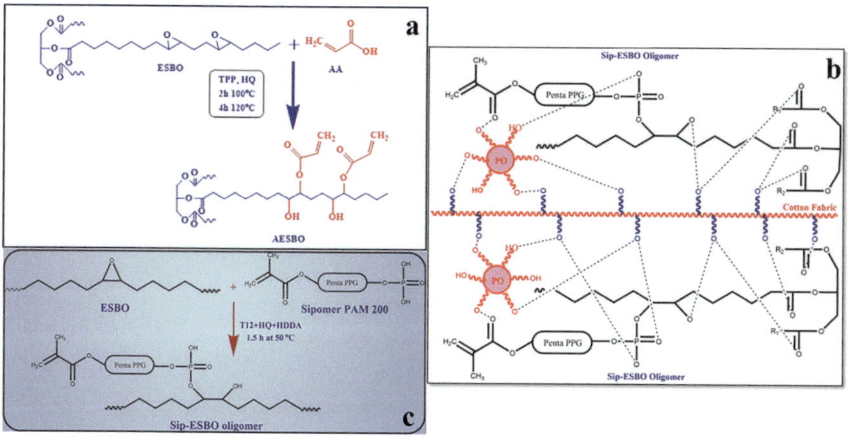

Fig. 8.4 Synthesis scheme of the ESBO oligomer

coated and uncoated textiles using various types of reactive diluent and LOI test results [111].

In a different investigation, epoxidized soybean oil (ESBO) and acrylic acid (AA) combined to form an oligomer known as acrylated ESBO (AESBO). The isocyanate groups of a 2,4-toluene diisocyanate-2-hydroxyethyl methacrylate (TDI-HEMA) adduct were then treated with this oligomer. TDI was selected as an isocyanate source due to its widespread use in the global polyurethane industry, while HEMA provided the hydroxyl and acrylate groups essential for UV curing and polyurethane production. Additionally, the oligomer's flexibility is enhanced by HEMA's methyl side group. The coating compositions included the TDI-HEMA-modified AESBO oligomer, various reactive diluents, and a photoinitiator. Following spray coating, these compositions were applied to aluminum plates and allowed to cure under a UV lamp. The results showed that FTIR spectra identified the formation of carbamate ester groups upon interacting with the TDI-HEMA adduct and also suggested the presence of acrylic functional groups in the AESBO oligomer backbone. The addition of reactive diluents increased the crosslinking density, shifting Tg values to higher temperatures. Through an increase in crosslinking density, this addition improved adhesion. By modifying the polymer network's free volume and molecular mobility, the quantity and kind of functional groups in the reactive diluents (aliphatic, aromatic, etc.) are essential for defining coating quality and performance [112].

8.6.2.2 Flaxseed Oil

Made from the seeds of the Linum usitatissimum plant, flaxseed oil is a vegetable oil that is high in antioxidants, lignans, and alpha-linolenic acid (ALA). In the field of textile processing, flaxseed oil has various applications. It can be utilized to provide

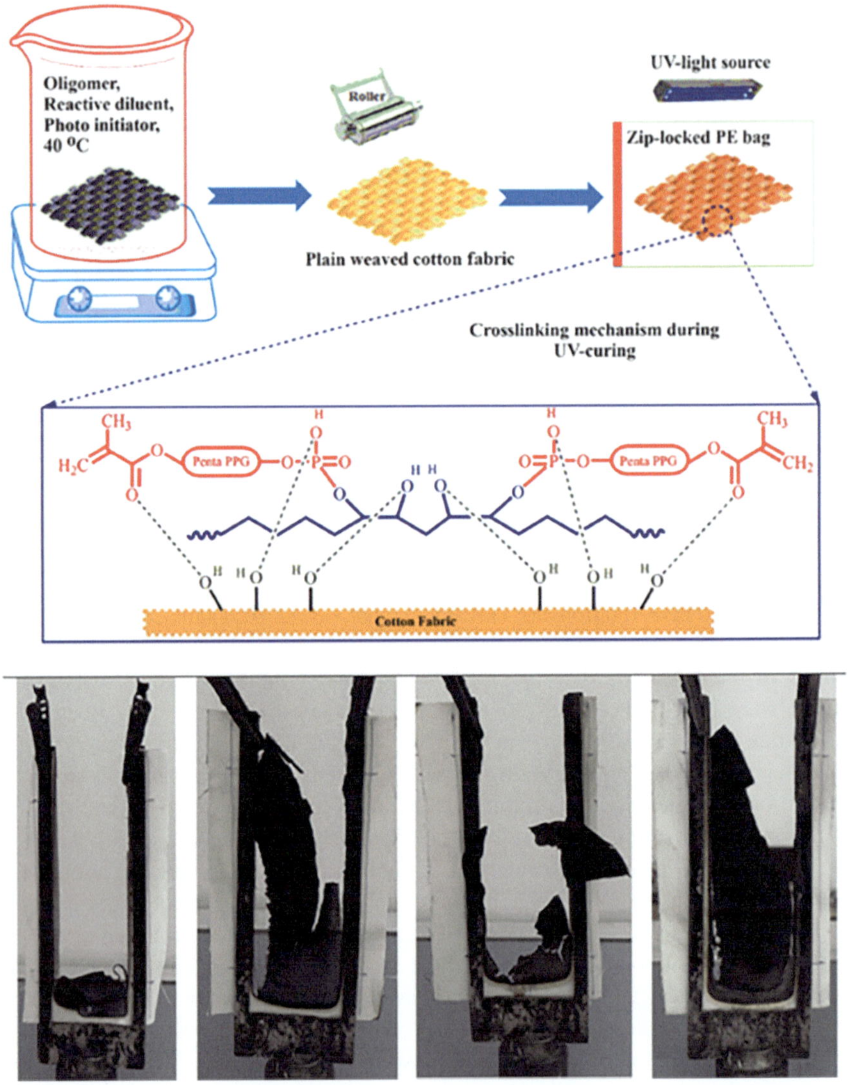

Fig. 8.5 Schematic of the coating process and digital images of coated fabric after the LOI test. Reprinted with permission from Ref [111], Copyright Elsevier, 2023

waterproofing, soften fabrics, enhance color, impart antimicrobial properties, and serve as an environmentally friendly alternative. Studies have indicated that flaxseed oil can be utilized in the production of oil-based bio-polymer resins, which are biodegradable composites. The process of turning flaxseed oil into epoxy oil, creating acrylated epoxidized flaxseed oil, combining it with styrene, and curing it are all part of a study looking at several production methods for nonwoven flax mats and biobased resin matrix systems. According to the study, composites made with flax reinforcement have non-cytotoxic, platelet-anti-aggregative, and bacteriostatic properties. The application of flax fiber as reinforcement in polymer matrix composites is becoming more widespread, and thermoplastic and thermosetting polymers may eventually be replaced by it [113]. The practical evaluation of linseed oil's antifungal abilities was accomplished by the application of the agar disk diffusion method. Antimicrobial activity results showed that there were no germs on the fabric's surface. Furthermore, comparing the outcomes of physical testing revealed no discernible difference between the treated and untreated materials. As a result, the canvas cotton fabric treated with linseed oil retained its physical characteristics unaffected by the antimicrobial treatment. Thus, it has been determined that canvas material treated with linseed oil is appropriate for use in bags and shoes [114]. A study that revealed how to use flaxseed oil effectively produced an acrylated flaxseed oil-beeswax emulsion system that was free of surfactants and solvents. The objective of this technology is to provide transparent, moisture-resistant, and environmentally friendly coatings that can be applied industrially and become more affordable when exposed to ultraviolet radiation. The coatings showed the appropriate levels of thermal resistance, hydrophobicity, transparency, and compostability [115].

8.6.2.3 Teatree Oil

Melaleuca alternifolia leaves are used to make tea tree oil, which is well known for having strong antibacterial and antifungal qualities. In textile finishing, tea tree oil can be incorporated into fabric treatments to impart antibacterial and antifungal properties, thus enhancing the hygiene and longevity of textile products. Additionally, tea tree oil can be utilized to create fragranced textiles, providing a natural and refreshing scent. Its application in textile finishing requires careful consideration of concentration, compatibility with different fabric types, and adherence to safety regulations to ensure effective and safe use. Tea tree oil/ethyl cellulose microcapsules were used in a study to manufacture disposable antibacterial cotton fabrics. *Escherichia coli* and *Staphylococcus* aureus were the pathogens against which the antibacterial properties of the microcapsules were evaluated after they were described and applied to cotton fabric samples. Based on the available information, the samples exhibit antibacterial properties [116]. Encapsulating tea tree oil in PVA preserved its antibacterial activity on textile surfaces. Tea tree oil is known to be effective against fungal infections and has been recommended for use in acne and skin disorders. Using the cushioning method, tea tree oil-containing capsules were applied to textile surfaces. The study produced textile surfaces with strong antibacterial activity against *Staphylococcus aureus* and *Escherichia coli* [117].

8.7 Conclusions and Future Prospects

The exploration and application of bio-derived polymers in textile finishing represent a significant step toward the adoption of sustainable and environmentally friendly practices in the textile sector. This chapter has covered a variety of bio-polymers, including proteins, chitosan, starch, and cellulose, and how they can enhance the antibacterial activity, stiffness, wrinkle resistance, moisture management, and longevity of textiles. The incorporation of fragrance finishes through essential oils and photocurable formulations using ESBO highlights the promise of these bio-based alternatives to conventional petrochemical-based finishing agents. Research results highlighted that bio-derived polymers not only meet the functional requirements of textile finishing but also contribute to environmental sustainability by being renewable, biodegradable, and reducing the application of dangerous substances. The succesful application of these bio-polymers in textile finishing processes demonstrates their viability as sustainable alternatives, ensuring that the textile industry can achieve its performance goals while mitigating environmental impact. Due to continuous research and innovation, bio-derived polymers have strong future possibilities in textile finishing. Important areas of research and development include integrating with cutting-edge technologies like nanotechnology and smart textiles, scaling up laboratory overcomes to industrial applications, and enhancing functionalization to improve textile qualities with multifunctional coatings. Assessing environmental effects and reducing waste need life cycle assessment (LCA) evaluations and the development of a circular economy. Demand for eco-friendly textiles can also be driven by raising consumer knowledge through educational initiatives and certifications.

Funding This study was financially supported by the TUBITAK IRA-SME project with the grant number "122N399".

Declaration The authors declare that they have no known competing financial interests or personal relationships that could have appeared to influence the work reported in this chapter. Generative AI and AI-assisted technologies have been used in the writing process to improve the readability and language of the work only. After using this tool, the authors reviewed and edited the content as needed and take full responsibility for the content of the publication.

References

1. Patti, A., & Acierno, D. (2022). Towards the sustainability of the plastic industry through biopolymers: Properties and potential applications to the textiles world. *Polymers, 14*(4), 692.
2. Rani, S., Kadam, V., Rose, N. M., Jose, S., Yadav, S., & Shakyawar, D. (2020). Wheat starch, gum arabic and chitosan biopolymer treatment of wool fabric for improved shrink resistance finishing. *International Journal of Biological Macromolecules, 163*, 1044–1052.
3. Mowafi, S., El-Sayed, H., & Abou, T. M. (2018). Utilization of proteinic biopolymers: Current status and future prospects. *Journal of Textiles, Coloration and Polymer Science, 15*(1), 15–31.

4. Yildiz, Z., & Eryilmaz, O. (2023) 12-Preimpregnated natural fiber preforms. In M. Midani, T. Hamouda, A. H. Hassanin, & A. -F. M. Seyam (Eds.), *Multiscale textile preforms and structures for natural fiber composites* (pp. 327–40). Woodhead Publishing.
5. Eryilmaz, O., Kocak, E. D., & Sancak, E. (2023). 8-braided natural fiber preforms. In M. Midani, T. Hamouda, A. H. Hassanin, & A. F. M. Seyam (Eds), *Multiscale textile preforms and structures for natural fiber composites* (pp. 221–37). Woodhead Publishing.
6. Yadav, P., Yadav, H., Shah, V.G., Shah, G., & Dhaka, G. (2015). Biomedical biopolymers, their origin and evolution in biomedical sciences: A systematic review. *Journal of Clinical and Diagnostic Research, 9*(9), Ze21–5.
7. Padinjakkara, A., Thankappan, A., Jr Souza, F. G., & Thomas, S. (2018). *Biopolymers and biomaterials.* CRC Press.
8. Phiri, R., Rangappa, S. M., Siengchin, S., Oladijo, O. P., & Dhakal, H. N. (2023). Development of sustainable biopolymer-based composites for lightweight applications from agricultural waste biomass: A review. *Advanced Industrial and Engineering Polymer Research.*
9. Sukumaran, N. P., & Gopi, S. (2021). Overview of biopolymers: Resources, demands, sustainability, and life cycle assessment modeling and simulation. In *Biopolymers and their industrial applications* (pp. 1–19). Elsevier.
10. Rajeswari, A., Christy, E. J. S., & Pius, A. (2021). Biopolymer blends and composites: Processing technologies and their properties for industrial applications. In *Biopolymers and their industrial applications* (pp. 105–47). Elsevier.
11. Yildiz, Z., Kartal, I., Kocak, E. D., Ozer, B., Kus, B. N., & Eryilmaz, O. (2024). Sustainable fabric printing by using pre-consumed cellulosic textile wastes: The effect of waste particle content. *Journal of Cleaner Production, 448*, 141635.
12. Flórez, M., Cazón, P., & Vázquez, M. (2023). Selected biopolymers' processing and their applications: A review. *Polymers, 15*(3), 641.
13. Younes, B. (2017). Classification, characterization, and the production processes of biopolymers used in the textiles industry. *The Journal of the Textile Institute, 108*(5), 674–682.
14. Kanmani, P., Aravind, J., Kamaraj, M., Sureshbabu, P., & Karthikeyan, S. (2017). Environmental applications of chitosan and cellulosic biopolymers: A comprehensive outlook. *Bioresource Technology, 242*, 295–303.
15. Kaushal, N., & Singh, A. K. (2023). Advancement in utilization of bio-based materials including cellulose, lignin, chitosan for bio-inspired surface coatings with special wetting behavior: A review on fabrication and applications. *International Journal of Biological Macromolecules, 246*, 125709.
16. Eryilmaz, O., & Sancak, E. (2021). Effect of silane coupling treatments on mechanical properties of epoxy based high-strength carbon fiber regular (2 x 2) braided fabric composites. *Polymer Composites, 42*(12), 6455–6466.
17. Eryilmaz, O., Oz, M. E., Jois, K. C., Sackmann, J., & Gries, T. (2024). FEA and experimental ultimate burst pressure analysis of type IV composite pressure vessels manufactured by robot-assisted radial braiding technique. *International Journal of Hydrogen Energy, 50*, 597–612.
18. Trivedi, A. K., Gupta, M., & Singh, H. (2023). *Advanced industrial and engineering polymer research.* Elsevier.
19. Aggarwal, J., Sharma, S., Kamyab, H., & Kumar, A. (2020). The realm of biopolymers and their usage: An overview. *Journal of Environmental Treatment Techniques, 8*(2), 1005–1016.
20. Eryılmaz, O., Ateş, P. S., Ünal, İ, Üstündağ, Ü. V., Bay, S., Alturfan, A. A., et al. (2018). Evaluation of the interaction between proliferation, oxidant–antioxidant status, Wnt pathway, and apoptosis in zebrafish embryos exposed to silver nanoparticles used in textile industry. *Journal of Biochemical and Molecular Toxicology, 32*(1), e22015.
21. Kamel, M. Y., & Hassabo, A. G. (2021). Anti-microbial finishing for natural textile fabrics. *Journal of Textiles, Coloration and Polymer Science, 18*(2):83–95.

22. Jajpura, L., Rangi, A., & Khandual, A. (2020). 10-Natural finishes, technologies and recent developments. In R. Nayak (Ed.), *Sustainable technologies for fashion and textiles* (pp. 209–229). Woodhead Publishing.
23. Thomas, S., Gopi, S., & Amalraj, A. (2020). *Biopolymers and their industrial applications: From plant, animal, and marine sources, to functional products*. Elsevier.
24. Hebeish, A., Aly, A. A., El-Shafei, A. M., & Zaghloul, S. (2008). Innovative starch derivatives as textile auxiliaries for application in sizing, finishing and flocculation. *Starch-Stärke, 60*(2), 97–109.
25. Hashim, S. O. (2020). Starch-modifying enzymes. In G. Mamo & B. Mattiasson (Eds.), *Alkaliphiles in biotechnology* (pp. 221–244). Springer International Publishing.
26. Teli, M., Rohera, P., Sheikh, J., & Singhal, R. (2009). Application of germinated maize starch in textile printing. *Carbohydrate polymers, 75*(4), 599–603.
27. Joshia, M., Ali, S. W., & Purwar, R. (2009). Ecofriendly antimicrobial finishing of textiles using bioactive agents based on natural products. *Indian Journal of Fibre & Textile Research, 34*, 295–304.
28. Solihat, N. N., Sari, F. P., Falah, F., Ismayati, M., Lubis, M. A. R., Fatriasari, W., et al. (2021). Lignin as an active biomaterial: A review. *Jurnal Sylva Lestari, 9*(1), 1–22.
29. Chung, H., & Washburn, N. R. (2013). Chemistry of lignin-based materials. *Green Materials, 1*(3), 137–160.
30. Flores-Hernández, C. G., Colín-Cruz, A., Velasco-Santos, C., Castaño, V. M., Rivera-Armenta, J. L., Almendarez-Camarillo, A., et al. (2014). All green composites from fully renewable biopolymers: Chitosan-starch reinforced with keratin from feathers. *Polymers, 6*(3), 686–705.
31. Ebrahim, S. A., Othman, H., Mosaad, M. M., & Hassabo, A. A. (2022). Valuable observation on Pectin as an eco-friendly material for valuable utilisation in textile industry. *Egyptian Journal of Chemistry, 65*(4):555–68.
32. Saha, S., Arshad, M., Zubair, M., & Ullah, A. (2019). Keratin as a biopolymer. In *Keratin as a Protein Biopolymer: Extraction from Waste Biomass and Application* (pp. 163–85).
33. Chilakamarry, C. R., Mahmood, S., Saffe, S. N. B. M., Arifin, M. A.B., Gupta, A., Sikkandar, M. Y., et al. (2021). Extraction and application of keratin from natural resources: A review. *3 Biotech, 11*(5), 220.
34. Saha, J., Mondal, M., Sheikh, M. K., & Habib, M. A. (2019). Extraction, structural and functional properties of silk sericin biopolymer from Bombyx mori silk cocoon waste. *Journal of Textile Science and Engineering, 9*(2).
35. Sionkowska, A. (2021). Collagen blended with natural polymers: Recent advances and trends. *Progress in Polymer Science, 122*, 101452.
36. Hassabo, A. G., Othman, H., Bakr, M., & Zayed, M. (2022). Collagen biopolymer in textile wet processing. *Journal of Textiles, Coloration and Polymer Science, 19*(2), 111–124.
37. Srivastava, A., Srivasatva, A. K., Singh, A., Singh, P., Verma, S., Vats, M., et al. (2022). Biopolymers as renewable polymeric materials for sustainable development-an overview. *Polimery, 67*(5), 185–196.
38. Heliopoulos, N. S., Papageorgiou, S. K., Galeou, A., Favvas, E. P., Katsaros, F. K., & Stamatakis, K. (2013). Effect of copper and copper alginate treatment on wool fabric. Study of textile and antibacterial properties. *Surface and Coatings Technology, 235*, 24–31.
39. Mostafa, K., Ameen, H., Morsy, M., El-ebiassy, A., El-Sanabary, A., Adel, M., et al. (2019). Production of high-performance textiles via pioneering strengthening approach using starch nanoparticles. *Journal of Industrial Textiles, 50*(3), 278–92.
40. Motaung, T. E., & Linganiso, L. Z. (2018). Critical review on agrowaste cellulose applications for biopolymers. *International Journal of Plastics Technology, 22*(2), 185–216.

41. da Silva, C. J. G., de Medeiros, A. D. L. M., de Amorim, J. D. P., do Nascimento, H. A., Converti, A., & Costa, A. F. S., et al. (2021). Bacterial cellulose biotextiles for the future of sustainable fashion: A review. *Environmental Chemistry Letters, 19*(4), 2967–80.
42. Kumar, M., & Barbhai, M. D. (2023). Sustainable fire safety solutions: Bioactive natural polysaccharides and secondary metabolites as innovative fire retardants for textiles. *Emergency Management Science and Technology, 3*(1). https://doi.org/10.48130/EMST-2023-0017.
43. Gamage, A., Liyanapathiranage, A., Manamperi, A., Gunathilake, C., Mani, S., Merah, O., et al. (2022). Applications of starch biopolymers for a sustainable modern agriculture. *Sustainability, 14*(10), 6085.
44. Abdou, E. S., El-Hennawi, H. M., & Ahmed, K. A. (2013). Preparation of novel chitosan-starch blends as thickening agent and their application in textile printing. *Journal of Chemistry, 2013*, 595810.
45. Shahid ul, I., Shahid, M., & Mohammad, F. (2013). Green chemistry approaches to develop antimicrobial textiles based on sustainable biopolymers—a review. *Industrial & Engineering Chemistry Research, 52*(15), 5245–60.
46. Grover, A., Arya, N., Rukhaya, S., & Bisht, D. (2022). *Textile's green finishing and their standards: A great initiative towards eco-friendliness*
47. Goswami, P., & O'Haire, T. (2016). 3-developments in the use of green (biodegradable), recycled and biopolymer materials in technical nonwovens. In G. Kellie (Ed.), *Advances in Technical Nonwovens* (pp. 97–114). Woodhead Publishing.
48. Zubair, M., & Ullah, A. (2020). Recent advances in protein derived bionanocomposites for food packaging applications. *Critical Reviews in Food Science and Nutrition, 60*(3), 406–434.
49. Ramadan, M. A., Taha, G. M., & EL-Mohr, W. Z. E.-A. (2020) Antimicrobial and UV protection finishing of Polysaccharide-based textiles using biopolymer and AgNPs. *Egyptian Journal of Chemistry, 63*(7), 2707–16
50. Patankar, K. C., Biranje, S., Pawar, A., Maiti, S., Shahid, M., More, S., et al. (2022). Fabrication of chitosan-based finishing agent for flame-retardant, UV-protective, and antibacterial cotton fabrics. *Materials Today Communications, 33*, 104637.
51. Kundu, C. K., Hossen, M. T., Islam, T., Mollick, S., Song, L., & Hu, Y. (2022). Flame retardant coatings from bio-derived chitosan, sodium alginate, and metal salts for polyamide 66 textiles. *ACS Omega, 7*(35), 30841–30848.
52. Try, P., Nissa Nurfajrin, S., Widya, F., & Deded, S. N. (2021). Natural and synthetic antimicrobials agent for textile: A review. *Indonesian Journal of Industrial Research, 16*(2), 33–48.
53. Eryılmaz, O., & Ovalı, S. (2024). Investigation and analysis of new fiber from allium fistulosum L. (Scallion) plant's tassel and its suitability for fiber-reinforced composites. *Uludağ Üniversitesi Mühendislik Fakültesi Dergisi, 29*(1), 51–66.
54. Ovalı, S., & Eryılmaz, O. (2024). Physical and chemical properties of a new cellulose fiber extracted from the Mentha pulegium L. (Pennyroyal) Plant's Stem. *Çukurova Üniversitesi Mühendislik Fakültesi Dergisi, 39*(1), 211–20.
55. Ovalı, S., Eryılmaz, O., & Uyanık, S. (2024). Exploring the potential of sustainable natural cellulosic fiber from Sorghum bicolor (Sorghum vulgare var. technicus) stem for textile and composite applications. *Cellulose, 31, 3289–302.*
56. Raman, A., Sankar, A., S. D. A, Anilkumar, A., & Saritha, A. (2022). Insights into the sustainable development of lignin-based textiles for functional applications. *Macromolecular Materials and Engineering, 307*(8), 2200114.
57. Gavin Gaynor, J., Szlek, D. B., Kwon, S., Tiller, P. S., Byington, M. S., & Argyropoulos, D. S. (2022). Lignin use in nonwovens: A review. *BioResources, 17*(2).
58. Tian, W., Huang, K., Zhu, C., Sun, Z., Shao, L., Hu, M., et al. (2022). Recent progress in biobased synthetic textile fibers. *Frontiers in Materials, 9*. https://doi.org/10.3389/fmats.2022.1098590
59. Li, K., Zhong, W., Li, P., Ren, J., Jiang, K., & Wu, W. (2023). Antibacterial mechanism of lignin and lignin-based antimicrobial materials in different fields. *International Journal of Biological Macromolecules, 252*, 126281.

60. Goyal, S., Dotter, M., Diestelhorst, E., Storck, J. L., Ehrmann, A., & Mahltig, B. (2022). Extraction of keratin from wool and its use as biopolymer in film formation and in electrospinning for composite material processing. *Journal of Engineered Fibers and Fabrics*. https://doi.org/10.1177/1558925022109049
61. Ebrahim, S. A., Othman, H. A., Mosaad, M. M., & Hassabo, A. G. (2023). Eco-friendly natural thickener (Pectin) extracted from fruit peels for valuable utilization in textile printing as a thickening agent. *Textiles, 3*(1), 26–49.
62. Xu, X., Hu, W., Ke, Q., Liu, H., Li, J., & Zhao, Y. (2020). Bio-adhesives from unfolded soy protein reinforced by nano-chitosan for sustainable textile industry. *Textile Research Journal, 90*(9–10), 1094–1101.
63. Wang, Y., Wang, Q., Zhu, Y., Shen, Y., Cheng, S., Zheng, H., et al. (2021). Structure and properties of oxycellulose fabric crosslinked with soy protein. *Carbohydrate Polymers, 257*, 117548.
64. Li, J., He, J., & Huang, Y. (2017). Role of alginate in antibacterial finishing of textiles. *International Journal of Biological Macromolecules, 94*, 466–473.
65. Janarthanan, M., & Senthil, K. M. (2019). Extraction of alginate from brown seaweeds and evolution of bioactive alginate film coated textile fabrics for wound healing application. *Journal of Industrial Textiles, 49*(3), 328–351.
66. Misbah, Bhatti, I. A., Zia, K. M., Bhatti, H. N., & Shahid, M. (2021). Synthesis, biological efficiency evaluation and application of sodium alginate-based polyurethane dispersions using cycloaliphatic isocyanate, as antibacterial textile coating. *Journal of Industrial Textiles, 50*(10), 1625–42.
67. Sadeghi-Kiakhani, M., Safapour, S., Habibzadeh, S. A., & Tehrani-Bagha, A. R. (2021). Grafting of wool with alginate biopolymer/nano ag as a clean antimicrobial and antioxidant agent: Characterization and natural dyeing studies. *Journal of Polymers and the Environment, 29*(8), 2639–2649.
68. Zhang, Y., Li, T.-T., Shiu, B.-C., Lin, J.-H., & Lou, C.-W. (2022). Multifunctional sodium Alginate@ urushiol fiber with targeted Antibacterial, acid corrosion resistance and flame retardant properties for personal protection based on wet spinning. *Applied Surface Science, 584*, 152573.
69. Madhu, A., & Chakraborty, J. N. (2017). Developments in application of enzymes for textile processing. *Journal of Cleaner Production, 145*, 114–133.
70. Roy Choudhury, A. K. (2020). 5-enzyme applications in textile chemical processing. In R. Nayak (Ed.), *Sustainable technologies for fashion and textiles* (pp. 91–115). Woodhead Publishing.
71. Gautam, R. L., Bharadwaj, A. K., Kumar, S., & Naraian, R. (2023). Chapter 14-microbial enzymes for the variable applications of textile industry processing. In V. K. Gupta (Ed.), *Valorization of biomass to bioproducts* (pp. 297–321). Elsevier.
72. Jegannathan, K. R., & Nielsen, P. H. (2013). Environmental assessment of enzyme use in industrial production–a literature review. *Journal of Cleaner Production, 42*, 228–240.
73. Natarajan, G., Rajan, T. P., & Das, S. (2022). Application of sustainable textile finishing using natural biomolecules. *Journal of Natural Fibers, 19*(11), 4350–4367.
74. Niyonzima, F. N., More, V. S., Nsanganwimana, F., Rao, A. S., Nair, A., Anantharaju, K. S., et al. (2023). Chapter 23-microbial enzymes used in textile industry. In G. Brahmachari (Ed.), *Biotechnology of microbial enzymes (second edition)* (pp. 649–684). Academic Press.
75. Kalia, S., Bhattacharya, A., Prajapati, S. K., & Malik, A. (2021). Utilization of starch effluent from a textile industry as a fungal growth supplement for enhanced α-amylase production for industrial application. *Chemosphere, 279*, 130554.
76. Pasha, K. M., Anuradha, P., & Subbarao, D. (2013). Applications of pectinases in industrial sector. *International Journal of pure and Applied sciences and Technology, 16*(1), 89.
77. Aggarwal, R., Dutta, T., & Sheikh, J. (2020). Extraction of pectinase from Candida isolated from textile mill effluent and its application in bio-scouring of cotton. *Sustainable Chemistry and Pharmacy, 17*, 100291.

78. Ahlawat, S., Dhiman, S. S., Battan, B., Mandhan, R. P., & Sharma, J. (2009). Pectinase production by Bacillus subtilis and its potential application in biopreparation of cotton and micropoly fabric. *Process Biochemistry, 44*(5), 521–526.
79. Hoondal, G., Tiwari, R., Tewari, R., Dahiya, N., & Beg, Q. (2002). Microbial alkaline pectinases and their industrial applications: A review. *Applied microbiology and biotechnology, 59*, 409–418.
80. Shah, S. (2013). Chemistry and application of cellulase in textile wet processing. *Research Journal of Engineering Sciences ISSN, 2278*, 9472.
81. Korsa, G., Konwarh, R., Masi, C., Ayele, A., & Haile, S. (2023). Microbial cellulase production and its potential application for textile industries. *Annals of Microbiology, 73*(1), 13.
82. Farooq, M. A., Ali, S., Hassan, A., Tahir, H. M., Mumtaz, S., & Mumtaz, S. (2021). Biosynthesis and industrial applications of α-amylase: A review. *Archives of Microbiology, 203*(4), 1281–1292.
83. Ahmad, M., Isah, U., Raubilu, I., Muhammad, S., & Ibrahim, D. (2019). An overview of the enzyme: Amylase and its industrial potentials. *Bayero Journal of Pure and Applied Sciences, 12*(1), 352–358.
84. Khlifi, R., Belbahri, L., Woodward, S., Ellouz, M., Dhouib, A., Sayadi, S., et al. (2010). Decolourization and detoxification of textile industry wastewater by the laccase-mediator system. *Journal of Hazardous Materials, 175*(1), 802–808.
85. Pezzella, C., Giacobbe, S., Giacobelli, V. G., Guarino, L., Kylic, S., Sener, M., et al. (2016). Green routes towards industrial textile dyeing: A laccase based approach. *Journal of Molecular Catalysis B: Enzymatic, 134*, 274–279.
86. Khatami, S. H., Vakili, O., Movahedpour, A., Ghesmati, Z., Ghasemi, H., & Taheri-Anganeh, M. (2022). Laccase: Various types and applications. *Biotechnology and Applied Biochemistry, 69*(6), 2658–2672.
87. Bansal, N., & Kanwar, S. S. (2013). Peroxidase(s) in environment protection. *The Scientific World Journal, 2013*, 714639.
88. Kalsoom, U., Bhatti, H. N., & Asgher, M. (2015). Characterization of plant peroxidases and their potential for degradation of dyes: A review. *Applied Biochemistry and Biotechnology, 176*(6), 1529–1550.
89. Lončar, N., & Fraaije, M. W. (2015). Catalases as biocatalysts in technical applications: Current state and perspectives. *Applied Microbiology and Biotechnology, 99*(8), 3351–3357.
90. Kaushal, J., Mehandia, S., Singh, G., Raina, A., & Arya, S. K. (2018). Catalase enzyme: Application in bioremediation and food industry. *Biocatalysis and Agricultural Biotechnology, 16*, 192–199.
91. Fruhwirth, G., Paar, A., Gudelj, M., Cavaco-Paulo, A., Robra, K. H., & Gübitz, G. (2002). An immobilised catalase peroxidase from the alkalothermophilic Bacillus SF for the treatment of textile-bleaching effluents. *Applied Microbiology and Biotechnology, 60*(3), 313–319.
92. Rehan, M., Abdel-Wahed, N. A., Farouk, A., & El-Zawahry, M. M. (2018). Extraction of valuable compounds from orange peel waste for advanced functionalization of cellulosic surfaces. *ACS Sustainable Chemistry & Engineering, 6*(5), 5911–5928.
93. Iordache, O., Cozea, A., Vărzaru, E., Stoica, E., Platon, C., Rodino, S., et al. (2016). Antimicrobial activity of textiles treated with rosemary and orange essential oils against a selection of pathogenic fungi. *Scientific Bulletin Series F Biotechnologies, 20.*
94. Fouda, M. M., & Fahmy, H. (2011). Multifunctional finish and cotton cellulose fabric. *Carbohydrate polymers, 86*(2), 625–629.
95. Nelson, G. (2002). Application of microencapsulation in textiles. *International Journal of Pharmaceutics, 242*(1), 55–62.
96. Panisello, C., Peña, B., Gilabert Oriol, G., Constantí, M., Gumí, T., & Garcia-Valls, R. (2013). Polysulfone/vanillin microcapsules for antibacterial and aromatic finishing of fabrics. *Industrial & Engineering Chemistry Research, 52*(29), 9995–10003.

97. Vankar, P. S., & Shukla, D. (2012). Biosynthesis of silver nanoparticles using lemon leaves extract and its application for antimicrobial finish on fabric. *Applied Nanoscience, 2*(2), 163–168.
98. Mohamed, A. L., El-Zawahry, M., Hassabo, A. G., & Abd, E. -A. (2023). Encapsulated lemon oil and metal nanoparticles in biopolymer for multifunctional finishing of cotton and wool fabrics. *Industrial Crops and Products, 204*, 117373.
99. Wolela, A. D. (2020). Antibacterial finishing of cotton textiles with extract of citrus fruit peels. *Current Trends in Fashion Technology and Textile Engineering, 6*(1), 1–7.
100. Ismail, S., & Hebeish, A. (2022). Preparation and application of lemon peel oil (Citrus aurantifolia) to improve microbial resistance of wool and viscose fabrics. *Egyptian Journal of Chemistry, 65*(8), 511–521.
101. Özsevinç, A., & Alkan, C. (2023). Polyurethane shell medicinal lavender release microcapsules for textile materials: An environmentally friendly preparation. *Industrial Crops and Products, 192*, 116131.
102. Saade, H., de León-Gómez, R. D., Treviño, M., Rodríguez-Fernández, O., Soriano-Corral, F., Borjas, J., et al. (2021). Performance of cotton, cotton-polyester, and polyester-elastane fabrics impregnated with ultrafine polymeric nanoparticles loaded with lavender oil. *Journal of Nanomaterials, 2021*, 1–10.
103. Chirilă, L., Popescu, A., Chiriac, L., Roxana, R., Constantınescu, E., Chelaru, C., et al. (2018). *Functional finishing of textiles using bioactive agents based on natural products.* ICAMS.
104. Farooq, S., & Ngaini, Z. (2024). *Introduction to vegetable oils. Vegetable oil-based composites: Processing, properties and applications* (pp. 1–20). Springer.
105. Morris, M., Ye, X. P., & Doona, C. J. (2021). Soybean oil and nonthermal plasma pretreatment to dye para-aramid woven fabrics with disperse dye using a glycerol-based dye bath. *Journal of the American Oil Chemists' Society, 98*(4), 463–473.
106. Ren, Q., & Zhao, T. (2010). Synthesis and application of modified vegetable oils in water-repellent finishing of cotton fabrics. *Carbohydrate Polymers, 80*(2), 381–386.
107. Yıldız, Z. (2022). Usage of UV-curable soybean oil based coating formulations for pretreated cotton fabrics. *Textile and Apparel, 32*(3), 232–42.
108. Xu, Q., Wang, X., Wang, P., Cheng, L., Wan, Y., & Wang, Z. (2022). Sustainable and super-hydrophobic coating from epoxidized soybean oil and stearic acid on cotton fabric etched by deep eutectic solvent. *Materials Today Chemistry, 26*, 101211.
109. Yildiz, Z. (2022). Bio-based photocurable coatings reinforced with platanus orientalis (Chinar Tree) leaf powder for lightweight Tarpaulin. In *3rd International Conference of Innovative Textiles.* Tekirdag Namik Kemal University.
110. Yildiz, Z., Onen, H. A., & Gungor, A. (2022). Soybean oil based UV-curable coating formulations for improvement of washing fastness of waterproof cotton fabrics. In *3rd International Conference of Innovative Textiles.* Tekirdag Namik Kemal University.
111. Yildiz, Z. (2023). Photocurable soybean oil based phosphorus containing coatings for cotton fabrics: The influence of reactive diluents. *Progress in Organic Coatings, 174*, 107255.
112. Kus, B. N., Ozer, B., Yetman, P., Demircioglu, M., Yildiza, Z., Eryilmaz, O., & Sancak, E. (2024). Synthesis and application of soybean oil based photocurable polyurethane acrylates for metal coating. In *14th International Fiber and Polymer Research Symposium.*
113. Shadhin, M., & Shuvo, I. I. (2019). Biopolymer flax (Linum Usitatissimum L.) and its prospects in biodegradable composite fabrication-a short review. *Journal of Textile Engineering and Fashion Technology, 5*(5), 267–74.
114. Harlapur, S. F., Harlapur, S., & Shridhar, A. (2018). Reveal on the effect of linseed oil on fabrics: A competent antimicrobial agent. *Materials Today: Proceedings, 5*(10), 22638–22642.
115. Vijayan, S. P., Aparna, S., & Sahoo, S. K. (2023). Effect of beeswax on hydrophobicity, moisture resistance and transparency of UV curable linseed oil based coating for compostable paper packaging. *Industrial Crops and Products, 197*, 116645.

116. Beşen, B. S. (2020). Tea tree oil/ethyl cellulose microcapsule loaded antimicrobial textiles. *AATCC Journal of Research, 7*(2), 1–6.
117. Beşen, B. S. (2019). Production of disposable antibacterial textiles via application of tea tree oil encapsulated into different wall materials. *Fibers and Polymers, 20*, 2587–2593.

Chapter 9
Microencapsulation in Textile Finishing: Challenges and Recent Trends

Tonmoy Saha and Shariful Islam

Abstract Microencapsulation is an innovative finishing technique primarily used in food and drug industries, but recently, it has received considerable attention in the field of textile finishing. This process involves formation of capsules consisting of core materials and polymeric shell materials. Core materials play a key role in imparting functionalization to the textile substrate. There are numerous methods, such as spray drying, in situ polymerization, and coacervation, to produce microcapsules. In the field of textile engineering, the application of microcapsules can be achieved by different post-treatment processes or during the spinning stage. Core materials from the microcapsules can be released due to physical action with the help of different stimuli such as temperature. Microencapsulation brings lots of benefits for the user by incorporating core substances that can work against different environmental conditions. However, it also has some limitations, considering the use of polymeric materials and application methods. This chapter focuses on the morphology of microcapsules, methods to produce and apply them to the textile substrate, release mechanism, and application field in the textile sector.

Keywords Microencapsulation · Microcapsules · Core materials · Shell materials · Textile finishing

T. Saha (✉)
Department of Textile Engineering, National Institute of Textile Engineering and Research, Dhaka, Bangladesh
e-mail: tonmoybutex23@gmail.com

S. Islam
Department of Bioproducts and Biosystems, School of Chemical Engineering, Aalto University, Espoo, Finland
e-mail: shariful.islam@alumni.aalto.fi

M. Shahid et al. (eds.), *Advancements in Textile Finishing*,
https://doi.org/10.1007/978-981-96-6385-9_9

9.1 Introduction

The idea of using microcapsules in textile technology was first introduced in the early 1970s. Since then, most of the literature on microencapsulation for textile use has been in the form of patents [1]. With the progression into the twenty-first century, the textile industry is seeing an increasing number of commercial uses for microencapsulation, especially in regions like Western Europe, Japan, and North America [2]. Microencapsulation, a modern technique in the textile industry, involves creating small capsules with active ingredients within a coating shell of polymeric material. There are several ways to fabricate these microcapsules, such as spray drying, in situ polymerization, coacervation, etc. After manufacturing these microcapsules, they are applied to the textile substrate for further functionalization. These microcapsules are introduced via standard finishing methods like padding and open possibilities in various fields such as healthcare, beauty, and functional textiles. These capsules are added to fabrics for long-lasting effects and controlled release of compounds, enhancing stability [3]. Microencapsulation in textile finishing embeds microcapsules into textile fibers to impart specific traits or release active substances under certain conditions. The release of these capsules onto the substrate depends on factors such as external pressure, temperature, etc. Thus, the microencapsulation process in textile finishing encompasses a variety of stages, such as the formation of the microcapsules, their incorporation into the fabric, and a fixation procedure to guarantee their proper attachment. The encapsulated active ingredients can subsequently be discharged at once or in a regulated way, either in reaction to particular stimuli or over a period of time [4]. The active components in the microcapsules can range from essential oils to medications, with their formulation and efficacy depending on the carrier polymer, active component, and encapsulation technique. Research has been conducted on the integration of moisturizers, essential oils, fragrances, and leuco dyes into fabrics through the process of microencapsulation [5]. Additionally, the encapsulation of anti-microbial substances is becoming increasingly prevalent in the production of sportswear and medical textiles [6]. Thus, this technique holds the promise to bring about a transformation in the textile sector, providing improved functionality, enhanced comfort for the user, and even ecological advantages. Nevertheless, it also poses certain challenges, such as environmental degradation of microcapsules, ensuring they maintain their durability through multiple cycles of washing and usage [7]. In this book chapter, we will discuss the morphology of microcapsules, how they formed, the advantages and limitations of different techniques, methods of fabrication onto the textiles, how they release, and what benefits they offer.

9.2 What is Microencapsulation?

Microencapsulation is a process where thin polymeric layers are deposited onto liquid droplets of very small diameter or tiny solid particles. Microscopic capsules can be formed with diverse features obtained by encasing droplets of liquid, solid particles, and gas molecules covered by shell material. Thus, a microcapsule produced from the microencapsulation has two main components. One is core material, which is responsible for functionalization, and the other is shell material, which provides support for the core materials. The main reason for microencapsulation is listed below [8].

- Isolating components that don't mix well
- Transforming liquids into solids that flow freely
- Enhancing stability (shielding the encapsulated substances from oxidation or deactivation due to environmental reactions)
- Concealing the smell, taste, and function of the encapsulated substances
- Safeguarding the surrounding environment
- Regulated dispensation of active ingredients (prolonged or postponed release)
- Specific delivery of the encapsulated substance

9.3 Microcapsule

A microcapsule is a tiny, spherical structure that encloses a core material inside a protective shell. Its size ranges from micrometers to millimeters. With different functional components, the active substance, also known as the core materials, can range from gaseous to liquid and solid, three different states of matter. The outer shell controls the release of the core by enclosing and isolating it [9] (Fig. 9.1).

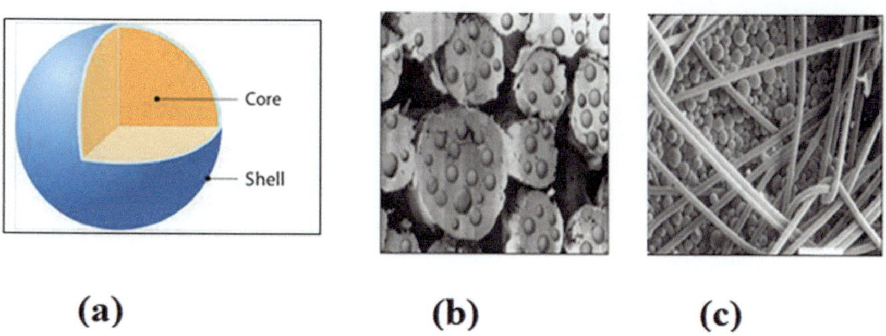

Fig. 9.1 **a** Morphology of microcapsules (Adapted from [10]). **b** Coating of PCM microcapsules on the surface and **c** implanted within fiber (Adapted from [11])

9.4 Morphology of Microcapsules

The physical attributes such as shape. size, dimensions, and form of microcapsules are referred to as their morphology. These attributes are crucial in defining their characteristics and efficacy across a range of uses. The encapsulation technique, the material selection, and the microcapsules' intended use all affect their morphology. Figure 9.2 illustrates the various forms that microcapsules can take.

- **Mono-Core Microcapsules**: These microcapsules are composed of a singular core substance that is enveloped by a defensive casing. The casing functions to segregate and manage the discharge of the substance within the core.
- **Poly core Microcapsules**: Poly core microcapsules consist of more than one core material in one shell. Different active substances can be found in each core, which offers a platform for combining multiple functions into a single microcapsule.
- **Multi-Shell Microcapsules**: In multi-shell microcapsules, there are multiple layers of protective shells surrounding the core material. Each shell layer can have distinct properties, influencing the release kinetics and stability of the encapsulated substance.
- **Matrix Microcapsules**: The core material of matrix microcapsules is dispersed or embedded within a matrix-like structure, usually composed of polymers, rather than having a distinct core and shell structure.
- **Irregular microcapsules**: This is used to describe microcapsules that don't have a regular or uniform spherical shape. Irregular microcapsules can have non-uniform or uneven shapes, in contrast to traditional microcapsules, which are typically spherical.

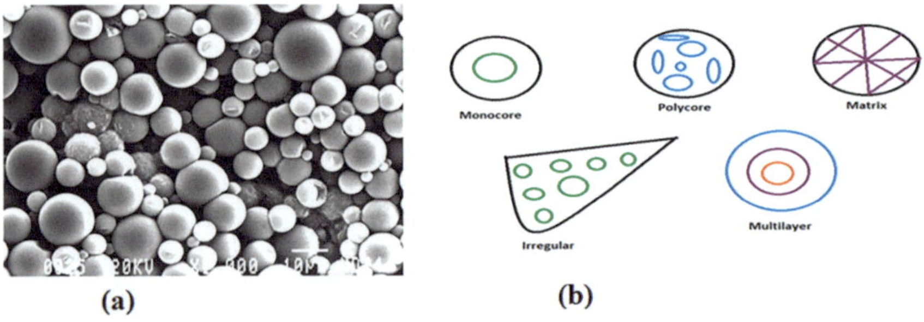

Fig. 9.2 **a** microcapsules under a microscope [12] and **b** Different morphology of microcapsules (Adapted and re-drawn from [13])

Table 9.1 Materials used to produce microcapsule

Materials	Materials type	References
Core materials	Fragrances	[15]
	Lecco dyes	[16]
	Essential oils	[17]
	Phase change materials	[18]
	Vitamins, moisturizing agents, depilatory agents, coloring agents, bleaching agents	[19, 20]
Shell materials	Synthetic based shell materials such as melamine–formaldehyde, polyurethane, polyamides, polyester	[21, 22]
	Bio-based polymeric shell materials for instance acetophthalate, chitosan, alginate, gum Arabic, gelatin	[23]
Cross linking agents	Carboxymethyl cellulose, tannic acid, citric acid, glutaraldehyde, genipin	[24]

9.5 Materials Used for the Microencapsulation

The components needed to create microcapsules vary depending on the intended use and the shell and core characteristics needed (Table 9.1). Whether liquid or solid, the core material is customized to the particular purpose of the microcapsule—for example, to encapsulate essential oils, fragrances, or bio-based compounds. Microcapsules can be used in the textile industry to give textiles features like phase-change capabilities, fragrance, or antimicrobial qualities. Phase-change materials for thermoregulation, antimicrobial agents to stop the growth of microorganisms, and aromatic compounds for fragrance could make up the core material. Polymers compatible with textiles and robust enough to endure the rigors of manufacturing and subsequent use are frequently used in the selection of shell materials [14].

9.6 Methodologies of Preparing Microcapsules

Numerous methods have been suggested for the fabrication of microcapsules. The method chosen is influenced by the core release rate and the thickness, penetrability, solubility, and physical properties of the shell. Microencapsulation methods can be categorized into three types: those based on physical principles, those involving physicochemical interactions, and those purely chemical in nature [23] (Fig. 9.3).

Each of these methods possesses its own set of benefits and drawbacks, which are outlined in Table 9.2 Spray drying and coacervation are the techniques most utilized to create microcapsules among these methods. A brief discussion of these two methods is provided below.

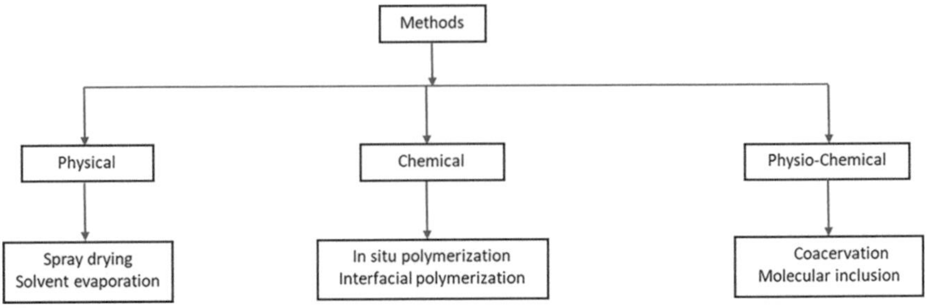

Fig. 9.3 Methods of microencapsulation that are frequently employed in enhancing textile functionality (Re-drawn from [24])

Table 9.2 Benefits and limitations of microencapsulation methods [26]

Method	Benefits	Limitations
Spray drying	– Quick and east – High efficiency – Possibility to obtain sterile product	– Creation of capsules of a larger size – Material loss because of active ingredient particles adhering to the drying chamber's walls
Coacervation	– Process is simple, fast and reproducible – Small size capsules – Suitable for aqueous and anhydrous environment	– Clustering of droplets because of solvent evaporation – The harmful effects of the solvent utilized
Polymerization	– By incorporating surfactants in minimal amounts, the dimensions of the microcapsules can be manipulated	– The reaction is challenging to regulate – Extensive cleansing is necessary to eliminate monomers, organic solvents, and surfactants

9.6.1 Spray Drying Method

Spray drying method for producing microcapsules is commonly used in different industries. The following are the fundamental procedures for creating microcapsules with a spray dryer: To create the microencapsulated product, (1) the core materials are dispersed into a polymeric solution and sprayed into a heated chamber, and (2) heated air is then given to evaporate the solvent from the coated material. This method quickly turns dried powder into a liquid or slurry using hot gas. After the core material is sprayed and dried, it dissolves in the polymer as a melt or solution, and the shell material hardens to form a microcapsule. The main benefit of this approach is the duration time of the dryer for which it can handle delicate textiles [25] (Fig. 9.4).

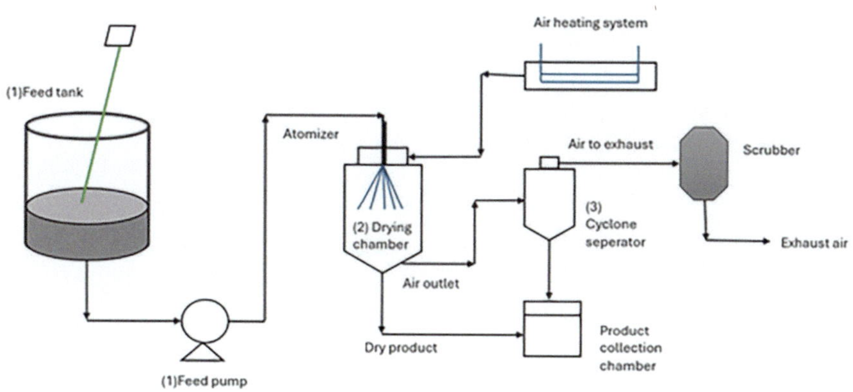

Fig. 9.4 Spray drying technology. (1) Emulsification, (2) atomization and dispersion; (3) particle collection (Re-drawn from [26])

9.6.2 In Situ Polymerization

Within the polymerization mixture takes place a polymer chemistry process known as in situ polymerization. Polymer nanocomposites are formed because of the direct synthesis of unstable oligomers in the reaction mixture. Separating these oligomers is a difficult task. The method begins with an initiating step and continues through several polymerization steps to produce a hybrid structure consisting of nanoparticles and polymer molecules (Fig. 9.5).

Amine-aldehyde resins are most used as polymeric shell materials for in situ polymerization process. For instance, researchers encapsulated citronella oil as core materials and used melamine–formaldehyde as protective material. The release of the capsules is performed by external pressure and partial release of core materials. The morphology is illustrated in Fig. 9.6.

Functional textile areas such as photochromic materials, and microcapsules are getting popular day by day due to their unique properties and controlled mechanism

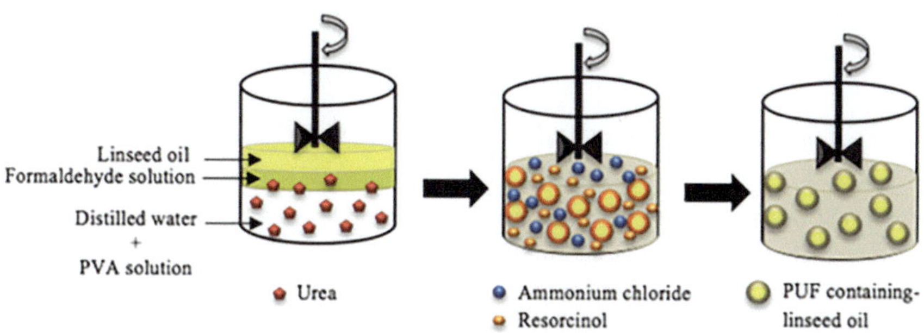

Fig. 9.5 In-situ polymerization process (Adapted from [27])

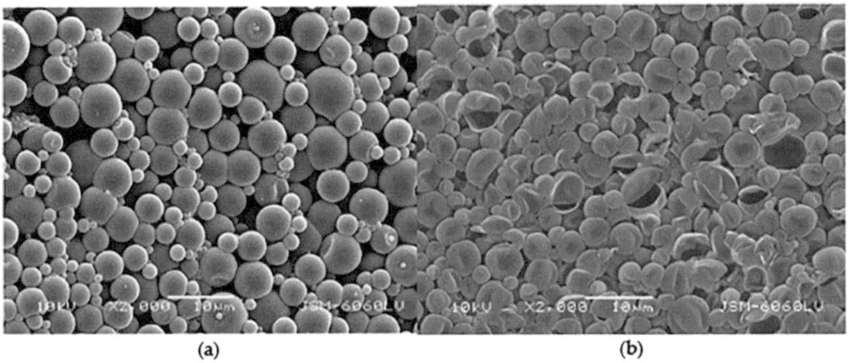

Fig. 9.6 Microcapsules formed of citronella oil and melamine formaldehyde **a** intact microcapsules before subjected to mechanical action **b** Activation of microcapsules by pressure (Adapted from [24])

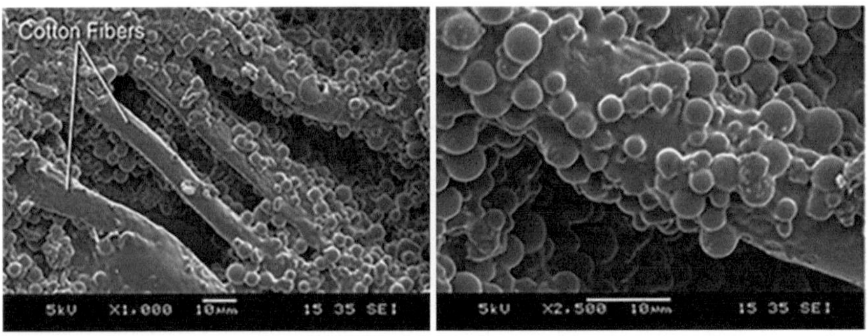

Fig. 9.7 Morphology of microcapsules attached on cotton fibers (Adapted from [28])

for uniform distribution of nanomaterials into the target materials. Ethyl cellulose and melamine-urea–formaldehyde were employed as polymers in in situ polymerization techniques recently to microencapsulate photochromic pigments. Photomicrographs of microcapsules were attached to the cotton fibers illustrated in Fig. 9.7.

9.6.3 Coacervation Method

When one or more hydrocolloid phases are deposited on an active substance that is either suspended or emulsified in the same reaction media, the process known as coacervation results in the formation of a microcapsule with a core–shell structure. The resulting capsule boasts a high load capacity and may be ejected efficiently. Three steps make up the coacervation-phase separation method of microencapsulation,

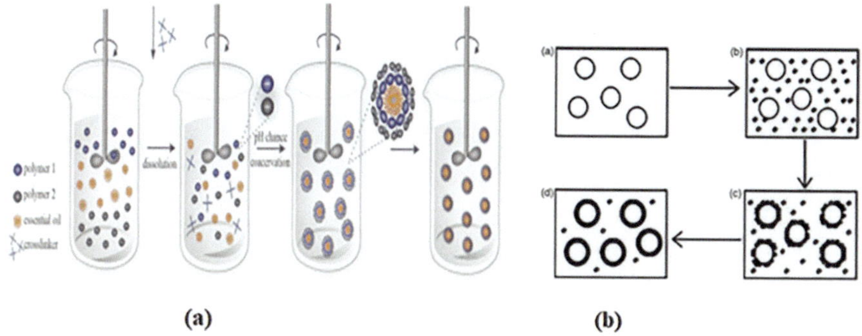

Fig. 9.8 **a** Schematic diagram of coacervation process (Adapted from [29]). **b** Diagrammatic depiction of the coacervation process-[**a** Dispersal of core material in shell polymer solution; **b** separation of coacervate from the solution; **c** enveloping of core material by coacervate microdroplets; **d** fusion of coacervate to form a continuous shell around core particles (Adapted from [30])

which is carried out with continuous agitation: the formation of three immiscible chemical phases, coating deposition, and coating rigidization.

A liquid manufacturing vehicle phase, a core material phase, and a coating material phase are the three immiscible chemical phases that develop in the first step. To create the three phases, the core material is dissolved in the liquid production vehicle phase, which serves as a solvent for the coating polymer (Fig. 9.8).

Step 2 of the process is applying the layer of liquid polymer on the basic core material. To do this, the core material and the coating material are physically combined in the fabrication vehicle under supervision. The liquid polymer coating is deposited around the core material when the polymer is absorbed at the contact that arises between the phase of liquid and the core material. To ensure the effectiveness of the coating, this adsorption phenomenon is required.

The third step in the process entails the hardening of the layering, typically accomplished through cross-linking agents, thermal modification or desolvation methods, resulting in the formation of a microcapsule [31].

9.7 Method of Applying Microcapsules on Textile Surface

There are many different approaches to applying microcapsules mechanically to the fabric at different stages of textile processing. Incorporating capsules within polymer during the spinning stage is one of the examples [32]. Secondly, microcapsules can also be applied to fabrics via the immersion procedure. Unlike the padding approach, this method does not include passing the material through squeeze rollers [23]. It involves immersion of capsules impregnated in contact with fabric. Again, the printing method can be useful to impart microcapsules onto the surface of textiles; in this process, tiny capsules are combined with a compatible binder material to prepare

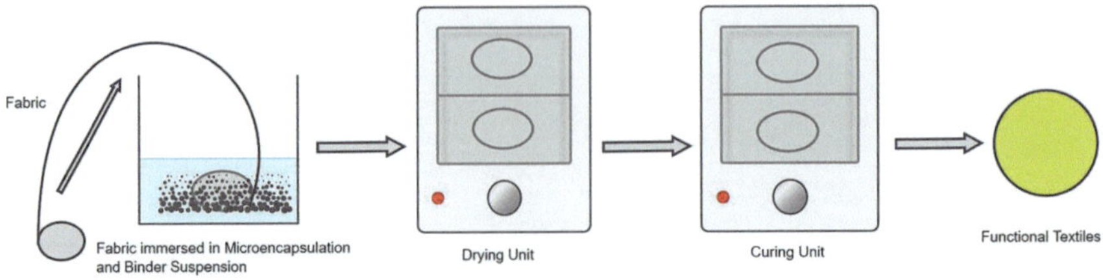

Fig. 9.9 Process of applying microcapsules to textile carriers by impregnation (Adapted from [1])

the paste required for the printing process. There are several processes to apply those microcapsules onto the surface of the fabric, such as screen, sublimation transfer, and inkjet techniques [33]. The spraying method is also an alternative for applying microcapsules on textiles. Microcapsules are sprayed on the surface of the material in a closed chamber by using a spray nozzle, followed by heating at 150 °C. The main objective of the heating process is to stabilize the microcapsules within the textile substrate [5]. Microcapsules can be applied to the textile surface by means of air knife or rod coater as well [34]. Lastly, the most popular method, known as the padding method, can be used to impart microcapsules into the textile substrate. In this method, the fabric is padded with a solution containing microcapsules and squeezed between rollers to pick the solution (Fig. 9.9).

9.8 Release Mechanism

The core material can be released in a targeted, regulated, or sustained manner thanks to various release mechanisms present in encapsulated materials. Three mechanisms exist for the release of core material from the capsules: physical burst in the wall, dissolution of the wall material in the changing stimuli, or melting of the capsule wall due to temperature change. Less well-known processes include biodegradation and low wall erosion (ablation). Among the uses for these long-lasting textile materials are medical textiles, cosmetics, deodorants, perfumes, and insect repellents. It is also possible to categorize the release process (high or low) using the migratory molecular weight (Fig. 9.10).

The choice of release mechanism depends on factors such as the nature of the shell material, the intended application, and the desired release profile. There are several ways to release microcapsules onto the textile substrate. They are briefly tabulated in the Table 9.3.

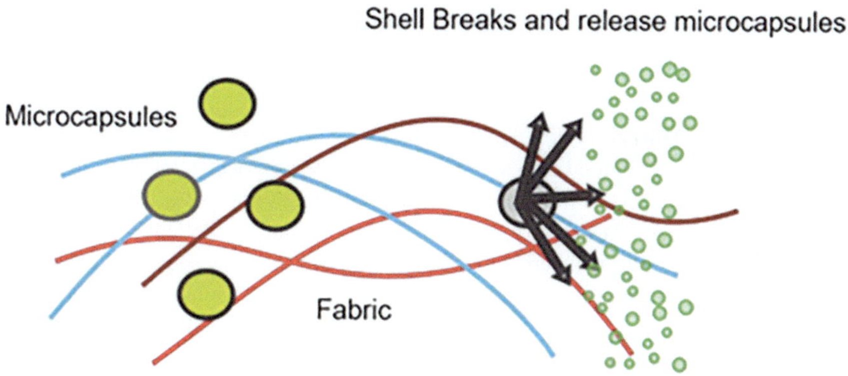

Fig. 9.10 Release mechanism of microencapsulation (Adapted from [10])

Table 9.3 Different release mechanism of microencapsule [10]

Types of mechanism	Main principle
External and internal pressure	The primary idea is to regulate how much material is released from the microcapsule when pressure or mechanical force is applied. This process results in the mechanical breaking of the microcapsules. For instance, if the core shell contains substances that, when exposed to specific stimuli (such as radiation), change into gaseous products, as is the case when light synthetic leather is made, this could potentially lead to the microcapsule wall breaking [35]
Microcapsule walls abrasion	This involves the physical wearing down or frictional disruption of the microcapsule shell, leading to the controlled release of the encapsulated core material. This wall abrasion release is very popular in the case of fragrance and scent release
Temperature changes	Changes in temperature can help release of core material. It involves two different release process: one is temperature-sensitive, meaning that the wall expands and descends when the critical temperature is reached. Another is fused, meaning that when the temperature rises, the wall dissolves
Chemical reactions	Chemical-encompassing additives are added to textile cleaning or washing solutions and which, because of changes in pH or chemical composition, release their contents during the wash cycle
Enzymatic degradation	Enzymatic degradation mainly occurs when capsules are subjected to an environment, that creates enzyme to degrade shell under different stimuli's. [36]

9.9 Applications of Microencapsulation Technique in Textiles

Phase Change Materials (PCMs): These high-performance materials of phase change behavior have gained considerable attention. Microencapsulation is one of the finest techniques for producing these kinds of materials. For instance, core substances for PCM, such as paraffin, can be placed inside and shielded by shell through microencapsulation. These microcapsules can efficiently trap and release heat during the phase transition and regulate the thermal conditions. The application of PCMs is widely used in textiles, construction, and other industries where the incorporation of such materials improves the efficiency of the final products [37].

Antimicrobial Textiles: The encapsulation of antimicrobial agents within microcapsules is a critical aspect in developing antimicrobial textiles. The antimicrobial agents in these capsules release gradually, resulting in a controlled and sustained release over time. When integrated into textiles, like linens or clothes, these microcapsules give the fabric strong, long-lasting antimicrobial qualities. Antimicrobial textiles can help stop the spread of infections in healthcare settings, which makes this application especially useful. For example, researchers experimented with citric acid; named as eco-binding agent during microencapsulation to impart antimicrobial properties to cellulosic fibers [38].

Fire Retardant textiles: Fire-resistant agents can be released gradually thanks to microcapsules found in textiles that resist flames. These capsules create a barrier of defense that retards or stops the spread of flames. Microencapsulation is essential for improving safety and fire resistance in home textiles, industrial workwear, and firefighting gear. With the help of this technology, textiles can now be given fire-retardant qualities with consistency. Paraffin can be microencapsulated as phase-changing materials to reduce flammability and function as a fire retardant [39].

Anti-mosquito finish: This technology offers a dependable way to give textiles insect-repelling qualities. For example, researchers used sodium alginate, Acacia arabica, and Moringa oleifera gum can be used as shell materials and as a core material, they choose combination of different essential oils to create mosquito repellent microcapsules through the ionic gelation process. The Moringa oleifera gum-made microcapsules maintained 60% of their mosquito-repelling power even after 30 washing cycles [40] (Fig. 9.11).

Medical textiles: Therapeutic agents are released by microcapsules in medical textiles to provide regulated treatment. By forming a targeted delivery system, they facilitate the process of healing. Microencapsulation is crucial for wound dressings and medical textiles because it guarantees targeted and long-lasting medical benefits. With the help of this technology, textiles can be given therapeutic qualities for use in medical applications in a dependable manner [41, 42].

Cosmetic textiles: Cosmetic textiles with microcapsules release skincare ingredients for regulated application. They give the skin long-lasting benefits by forming a layer that enhances its beauty. Microencapsulation is crucial for sportswear and regular apparel because it guarantees long-lasting cosmetic effects. With the help of

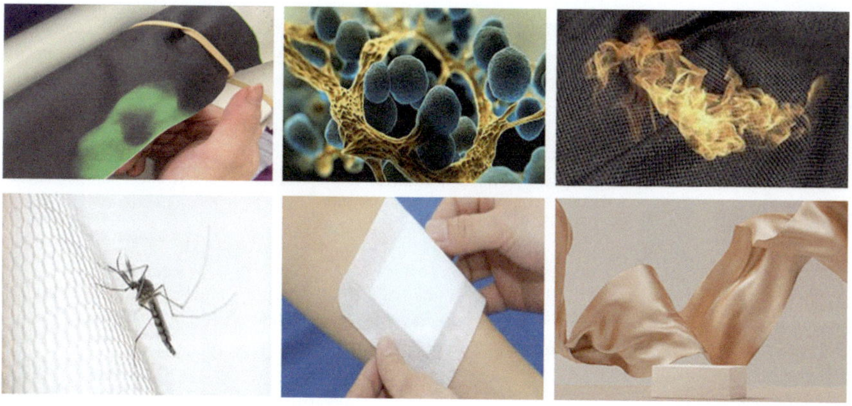

Fig. 9.11 Application of microencapsule in textile finishing

this technology, skincare qualities can be reliably added to textiles for use in cosmetic applications [43].

9.10 Conclusion

There are several ways to functionalize textile materials. Still, the limited endurance and resistance to different mechanical actions such as hot washing, rubbing fastness, and generation of textile effluent are major barriers to their widespread use. These factors are vital for textile maintenance and sustained functionality. Again, most of the microencapsulation process requires polymeric shell materials, which further contribute to generating pollutants in the form of microplastic. So, the main objective should be to produce biodegradable, environmentally benign microcapsules that will not hurt the environment as much as traditional synthetic shell materials, which take a long time to decompose and can cause major environmental problems. To lower the number of microcapsules lost to wastewater during washing, stronger adhesion must be created between textile fibers and microcapsules. The quest for bio-based functional textiles embedded with green microcapsules is one of the most significant hurdles in textile functionalization research. Microencapsulation technology is a viable strategy for imparting functional properties in textile applications. However, this field is in its infancy and remains largely untapped. Further research should investigate the potential of novel eco-friendly materials for coating compositions with improved technological characteristics. Alternatively, the biodegradable textiles' functional groups and the functional groups on the microcapsule shell could form covalent bonds, strengthening the microcapsule's adherence to the fiber.

Declaration The authors declare that they have no known competing financial interests or personal relationships that could have appeared to influence the work reported in this chapter. Generative

AI and AI-assisted technologies have been used in the writing process to improve the readability and language of the work only. After using this tool, the authors carefully reviewed and revised the content to ensure accuracy and quality. The authors take full responsibility for the final version of this publication.

References

1. Bojana, B. P., & Marica, S. (2016). Microencapsulation technology and applications in added-value functional textiles. *Physical Sciences Reviews, 1*(1), 20150003.
2. Nelson, G. (2001). Microencapsulation in textile finishing. *Review of Progress in Coloration and related Topics, 31*, 57–64.
3. Shrimali, K., & Dedhia, E. (2015). Microencapsulation for textile finishing. *IOSR Journal of Polymer and Textile Engineering, 2*(2), 2348–3181.
4. Van Parys, M. (2006). Smart textiles using microencapsulation technology. *Functional Coatings: By Polymer Microencapsulation*, 221–258.
5. Dedhia, A. T. D. E. (2021). An overview of microencapsulation technology in the application of aroma and antibacterial finishes. *International Journal of Home Science, 7*(1), 34–39.
6. Soroh, A. (2019). *Novel eco-friendly antimicrobial coatings for use in healthcare and sport textiles*. De Montfort University.
7. Peng, X., et al. (2023). Biopolymers-based microencapsulation technology for sustainable textiles development: A short review. *Case Studies in Chemical and Environmental Engineering*, 100349.
8. Dubey, R. (2009). Microencapsulation technology and applications. *Defence Science Journal, 59*(1), 82.
9. Rani, S., & Goel, A. (2021). Microencapsulation technology in textiles: A review study. *The Pharma Innovation Journal, 10*, 660–663.
10. Abdel-Aziz, E., et al. (2022). Microencapsulation and its application in textile wet processing: A review. *Journal of Textiles, Coloration and Polymer Science, 19*(2), 189–202.
11. Nelson, G. (2002). Application of microencapsulation in textiles. *International journal of pharmaceutics, 242*(1–2), 55–62.
12. Microencapsulation. https://www.reedpacific.com/microencapsulation/
13. Misra, S., et al. (2022). Emerging technologies and coating materials for improved probiotication in food products: A review. *Food and Bioprocess Technology, 15*(5), 998–1039.
14. Šiler-Marinković, S., Bezbradica, D., & Škundrić, P. (2006). Microencapsulation in the textile industry. *Chemical Industry & Chemical Engineering Quarterly, 12*(1), 58–62.
15. Zhao, D., et al. (2016). Preparation of high encapsulation efficiency fragrance microcapsules and their application in textiles. *RSC advances, 6*(84), 80924–80933.
16. Aklujkar, P. S., & Kandasubramanian, B. (2021). A review of microencapsulated thermochromic coatings for sustainable building applications. *Journal of Coatings Technology and Research, 18*, 19–37.
17. Enascuta, C. E., et al. (2018). Microencapsulation of essential oils. *Revista de Chimie, 69*(7), 1612–1615.
18. Iqbal, K., et al. (2019). Phase change materials, their synthesis and application in textiles—a review. *The journal of the Textile Institute, 110*(4), 625–638.
19. Cheng, S. Y., et al. (2009). Cosmetic textiles with biological benefits: Gelatin microcapsules containing Vitamin C. *International Journal of Molecular Medicine, 24*(4), 411–419.
20. Prichystal, P., et al. (2013). Encapsulation of textile dyes and textile auxiliary agents into liposome systems and their use for polyamide dyeing. *Coloration Technology, 129*(1), 64–68.
21. Sarier, N., & Onder, E. (2007). The manufacture of microencapsulated phase change materials suitable for the design of thermally enhanced fabrics. *Thermochimica acta, 452*(2), 149–160.

22. Mytara, A. D., et al. (2021). Synthesis of polyamide-based microcapsules via interfacial polymerization: Effect of key process parameters. *Materials, 14*(19), 5895.
23. Singh, N., & Sheikh, J. (2020). *Microencapsulation and its application in production of functional textiles.* Indian Journal of Fibre & Textile Research, 45, 495–509.
24. Boh Podgornik, B., Šandrić, S., & Kert, M. (2021). Microencapsulation for functional textile coatings with emphasis on biodegradability—a systematic review. *Coatings, 11*(11), 1371.
25. Ghosh, S. K. (2006). Functional coatings and microencapsulation: a general perspective. *Functional Coatings: by Polymer Microencapsulation*, 1–28.
26. Kłosowska, A., Wawrzyńczak, A., & Feliczak-Guzik, A. (2023). Microencapsulation as a route for obtaining encapsulated flavors and fragrances. *Cosmetics, 10*(1), 26.
27. Alias, J., et al. (2021). Self-healing epoxy coating with microencapsulation of linseed oil for the corrosion protection of magnesium (Mg). *Journal of Physics: Conference Series.* IOP Publishing.
28. Topbas, O., et al. (2020). Photochromic microcapsules by coacervation and in situ polymerization methods for product-marking applications. *Iranian Polymer Journal, 29*(2), 117–132.
29. Valle, J. A. B., et al. (2021). Chitosan microcapsules: Methods of the production and use in the textile finishing. *Journal of Applied Polymer Science, 138*(21), 50482.
30. Jyothi, N. V. N., et al. (2010). Microencapsulation techniques, factors influencing encapsulation efficiency. *Journal of Microencapsulation, 27*(3), 187–197.
31. Kumar, A., Sharma, P., & Banik, A. (2011). Microcapsulation as a novel drug delivery system. *International Pharmacutica science*, 1, 46–72.
32. Li, W., et al. (2014). Composition and characterization of thermoregulated fiber containing acrylic-based copolymer microencapsulated phase-change materials (MicroPCMs). *Industrial & Engineering Chemistry Research, 53*(13), 5413–5420.
33. Tylkowski, B., Giamberini, M., & Prieto, S. F. (2020). *Microencapsulation.* Walter de Gruyter GmbH & Co KG.
34. Soo, P. (2002). *Producing method of coated fabric changing its color according to temperature variation and having excellent washing resistance and the coated fabric produced by the method.* Korean Patent KR20020059049A, 12.
35. Ahangaran, F., Navarchian, A. H., & Picchioni, F. (2019). Material encapsulation in poly (methyl methacrylate) shell: A review. *Journal of Applied Polymer Science, 136*(41), 48039.
36. Pasha, A., Utekar, S., & Ghosh, R. (2021). Recent advances in microencapsulation technology and their applications. *The Bombay Technologist*, *68*(1). https://doi.org/10.36664/bt/2021/v68i1/160705
37. Salaün, F., et al. (2010). Thermoregulating response of cotton fabric containing microencapsulated phase change materials. *Thermochimica Acta, 506*(1–2), 82–93.
38. Dridi, D., et al. (2021). Enhanced antibacterial efficiency of cellulosic fibers: Microencapsulation and green grafting strategies. *Coatings, 11*(8), 980.
39. Kazanci, B., Cellat, K., & Paksoy, H. (2020). Preparation, characterization, and thermal properties of novel fire-resistant microencapsulated phase change materials based on paraffin and a polystyrene shell. *RSC Advances, 10*(40), 24134–24144.
40. Geethadevi, R., & Maheshwari, V. (2015). Long-lasting UV protection and mosquito repellent finish on bamboo/tencel blended fabric with microencapsulated essential oil. *Indian Journal of Fibre & Textile Research (IJFTR), 40*(2), 175–179.
41. El-Rafie, H. M., Zahran, M. K., Raoof, G. F. A. (2022). Cotton bandages finished with microcapsules of volatile organic constituents of marine macro-algae for wound healing. *Bioprocess and Biosystems Engineering*, 1–14.
42. Ghayempour, S., & Montazer, M. (2016). Micro/nanoencapsulation of essential oils and fragrances: Focus on perfumed, antimicrobial, mosquito-repellent and medical textiles. *Journal of Microencapsulation, 33*(6), 497–510.
43. Cheng, S., et al. (2008). Development of cosmetic textiles using microencapsulation technology. *Research Journal of Textile and Apparel, 12*(4), 41–51.

Chapter 10
Enzyme-Based Textile Finishing: Leveraging Biological Catalysts in Textile Industry

Abhishek Kasana, Priyadharshini Babu, and Vaishali Verma

Abstract The textile industry constitutes an essential part of the economy and culture of any country. Finishing textile goods is an important step in improving the properties of the fabric. However, using harsh chemicals during textile finishing can lead to generating various types of pollutants. Driven by an increasing demand for sustainable, environment-friendly, and high-quality textiles, a marked shift in the textile finishing processes has been observed. Enzymes function as biochemical catalysts that enhance the rate of chemical reactions while remaining unaltered. They have gained prominence in textile finishing, offering an alternative to traditional chemical methods. Enzyme-based finishing processes are preferred because they are environment-friendly, require less energy, and can be performed under milder processing conditions than conventional chemical methods. Many enzyme-based approaches have been developed and are undergoing additional refinement, focusing on various aspects of textile finishing. This chapter explores the applications of enzymes in the context of textile finishing, ranging from their historical development to the latest advancements, thus providing a comprehensive overview of this environment-friendly approach.

Keywords Textile industry · Sustainable textiles · Environment-friendly finishing · Enzymes

A. Kasana · V. Verma (✉)
Department of Biotechnology, School of Engineering and Applied Sciences, Bennett University, Greater Noida, Uttar Pradesh, India
e-mail: vaishali.verma@bennett.edu.in

A. Kasana
e-mail: E22SOEP0009@bennett.edu.in

P. Babu
Centre for Disaster Mitigation and Management, Vellore Institute of Technology, Vellore, Tamil Nadu, India
e-mail: bpriyadharshini@vit.ac.in

M. Shahid et al. (eds.), *Advancements in Textile Finishing*,
https://doi.org/10.1007/978-981-96-6385-9_10

10.1 Introduction

The textile industry plays an instrumental role in the economic growth and cultural enrichment of any nation. This industry has grown significantly from the earliest handmade fabrics to the advanced textiles available today. Textile finishing methods are essential for improving the properties and aesthetics of textiles [1]. Traditionally, these processes involve chemical-based treatment of fibers, fabrics, or clothing to achieve specific functional outcomes. Properties such as wash-and-wear, flame retardancy, water repellency, and antistatic properties are commonly introduced into fabrics using chemicals to improve their performance [2]. However, such chemical treatments often lead to water, air, and soil pollution [3]. The dyeing and printing processes consume significant quantities of water, resulting in wastewater that contributes to environmental pollution [4]. The high demand for dyed products exacerbates this issue because wastewater often contains thermally and photostable dyes that persist in the environment, leading to long-term pollution [5]. Furthermore, gaseous emissions produced during dyeing and printing processes are also a significant environmental concern [3]. The steam generated during these processes releases a variety of harmful substances and gases such as carbon (C), sulfur (S), carbon monoxide (CO), and carbon dioxide (CO_2) into the air, thereby causing air pollution [6]. In addition, the textile processing units often generate high noise levels, causing noise pollution, and prolonged exposure to high noise increases the risk of hearing damage in workers [7]. The textile industry also relies on the use of raw materials, including flax fiber, cotton, and hemp, all of which require significant amounts of water during different processing steps to yield the finished textile goods [3, 8]. Additionally, for the cultivation of these crops, farmers often use significant amounts of herbicides and pesticides, leading to adverse environmental effects, including soil and water pollution [9, 10].

Consequently, the need of the hour is to explore sustainable alternatives that can mitigate the impact of chemical finishing processes on the environment while retaining their effectiveness. Natural antimicrobial agent-based textile processing [11], laser processing [12], and enzyme-based textile finishing [13] are some of the sustainable approaches being investigated as alternatives to chemical-based textile finishing. Natural antimicrobial agents extracted from plant extracts, essential oils, and microbial by-products can effectively counteract harmful microbes [11]. These agents enhance fabric durability and wash resistance and reduce allergic reactions and skin irritation caused by synthetic antimicrobial agents [14]. Laser technology enables accurate surface design and textile coloration without water and chemicals [12]. Several studies have explored the applications of laser technology in the textile industry [15–18].

Enzymes play important roles in cellular metabolic processes. Typically, the tertiary structure of an enzyme contains an active site where substrate molecules bind and are converted into products [19]. Enzymes are commonly used to produce a variety of food products including cheese, curd, vinegar, and wine [20]. In addition to the food industry, enzyme-based processes are used as viable alternatives to traditional chemicals for reducing the pollution generated by the textile industry [21]. They offer multiple advantages because they are environment-friendly, non-toxic, and highly specific in terms of their action [22]. Further, enzymatic treatments can be conducted at lower temperatures and for shorter durations, contributing to energy conservation and reduced process costs [23]. Amylases, xylanases, cellulases, catalases, and lipases are some of the major enzymes used in the textile industry for different applications [24]. In this chapter, we discuss the role of enzymes in the textile industry, their diverse applications, advantages, opportunities, and associated challenges.

10.2 Classification of Enzymes

Enzymes are biological catalysts that increase the speed of chemical reactions without getting consumed in the process [25]. The classification system established by the Enzyme Commission (EC) categorizes enzymes into six distinct groups that catalyze specific types of reactions [26]. In 2018, a seventh class was introduced as an additional category of enzymes [27].

In the textile industry, hydrolases and oxidoreductases play important roles in various enzymatic applications. Hydrolases are enzymes that play an essential role in breaking down carbon–oxygen, carbon–nitrogen, carbon–carbon, and phosphorus–nitrogen chemical bonds within biomolecules. Hydrolases use water as a hydroxyl group donor, breaking large molecules into smaller ones [28]. Oxidoreductases catalyze the exchange of electrons between an electron donor and an electron acceptor. Oxidoreductases are divided into two nonhydrolytic classes, namely, peroxidases and catalases [23]. Most enzyme applications are specifically related to cotton processing [29]. These include de-sizing, scouring, and bleaching, which helps eliminate impurities from cotton fabrics; bio-finishing; bio-stoning or stone washing; and the cleanup of bleaching agents to eliminate any remaining hydrogen peroxide (H_2O_2) before dyeing the fabric [29–33].

10.3 Application of Enzymes in Textile Finishing

Enzyme-based textile processing began in the mid-nineteenth century. Enzymes were first introduced for de-sizing purposes in 1857 but were successfully implemented in 1912 [34]. In the 1980s, cellulases were used for the de-fuzzing and de-pilling of cellulose-based fabrics [35]. In the 1990s, catalases were introduced in bleaching processes, and pectin-degrading enzymes were used instead of traditional alkaline scouring [25, 35]. The initial use of enzymes was limited, but as the understanding of their catalytic properties grew, so did their applications. Enzymes like cellulases,

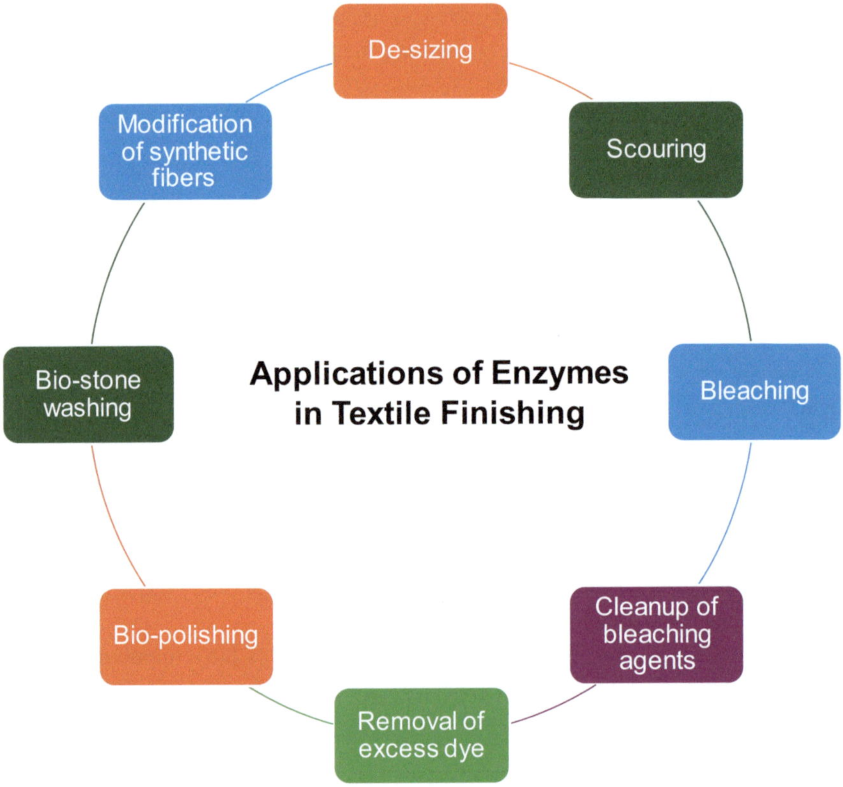

Fig. 10.1 Applications of enzymes in textile finishing

catalases, lipases, and laccases began to be used for various processes such as bio-polishing, bleaching, and dyeing (Fig. 10.1). Enzymes such as proteases, xylanases, amylases, cellulases, and lipases have found applications in the textile, food, paper, and pulp industries since the 1920s [23].

10.3.1 De-sizing

Weaving is an important textile processing step that involves interlacing warp and weft yarns. However, this process exposes warp threads to mechanical strain, leading to breakage [36]. To prevent such breakage, a gelatinous substance known as size is applied to coat the wrap threads, a process referred to as sizing [37]. In cotton fabrics, the commonly used sizing material contains starch along with other polymers such as polyacrylic acid (PAA), polyvinyl alcohol (PVA), or carboxymethyl cellulose (CMC) [38]. Unfortunately, this sizing process affects warp yarns and makes them less absorbent [39]. Consequently, the size must be removed before dyeing or finishing the fabric. However, conventional de-sizing treatments consume substantial amounts of water and involve the use of harmful chemicals, leading to a significant environmental impact [13].

Alternatively, a hydrolytic enzyme, amylase, is used in the de-sizing process [40]. Its primary function is to break down dietary starch into oligosaccharides including short-chain sugars, maltose, and dextrin. This enzymatic action ensures uniform wet processing [41]. The specificity of amylase for starch during the removal process also prevents damage to the supporting fabric [42]. Two types of amylase enzymes are used in textile processing namely, α-amylase and β-amylase, which are used to remove sizing materials from the warp yarn of woven fabrics [23]. De-sizing processes using amylase can be performed at low temperatures (between 30 and 60 °C), with an optimal pH range of 5.5–6.5, leading to energy savings and reduced operational costs [35]. Researchers have also explored the potential of amylase extracted from different microbial strains such as *Aspergillus tubingensis* isolate [43], *Bacillus cereus* AS2 [42], and *Bacillus* sp. KR-8104 [44] for textile de-sizing.

Furthermore, enzyme-based processes can be integrated into continuous processing systems, to improve the overall efficiency of the process. For example, a continuous enzyme-based de-sizing method involves treating textiles in a bath with enzyme mixtures, followed by drying, which simplifies the process and reduces handling time [45]. Enzyme-based de-sizing can also be combined with other processes such as scouring to further enhance efficiency. The use of xylanase enzyme has been described in a combined de-sizing and scouring process, which not only removes sizing but also cleans the fabric of natural impurities [46].

10.3.2 Scouring

Scouring is an important step in textile processing, focused on removing significant impurities from fabrics. These impurities comprise dirt, oils, wax, and gums, as well as non-fibrous impurities such as seed fragments [47]. The scouring process effectively removes natural and synthetic impurities with hydrophobic properties [35]. Typically, in this process, the fabric is subjected to boiling in the presence of an alkali solution in large iron vessels [35]. Scouring enhances fiber wettability, ensuring uniform finishing and dyeing [48]. Traditional hot aqueous alkaline treatments are performed at high pH and temperature [49]. In contrast, bio-scouring using enzymes such as lipases, cellulases, pectinases, and proteases provides good flexibility across a broader pH and temperature range without affecting fabric quality [47]. Marine pectinase has been shown to preserve the fiber structure better than alkaline scouring, preventing deterioration and enhancing dye uptake [50]. Immobilized thermophilic lipase has been used to remove the surface lipid barrier of wool because this barrier enhances the hydrophobicity of wool and affects its wettability during processing [51]. Bacterial lipase from *Bacillus licheniformis* has been used for scouring cotton fabrics and in the reduction of abrasion resistance [52]. Bacterial lipase derived from marine bacteria *Bacillus sonorensis* has been used as a bio-scouring agent to remove wax from the cotton surface [53]. Researchers have also explored the application of microbial cellulase in bioscouring [54]. The bio-scouring of cotton using protease and pectinase was explored, with pectinase performing better than protease. Both enzymes contributed to a weight loss of 10–15% in the cotton fabric at optimal pH and temperature [55]. Badhe et al. reported the bio-scouring of wool using protease enzyme extracted from *Bacillus subtilis*, which was purified through ammonium sulfate precipitation followed by dialysis and then applied for wax removal from wool fibers [56]. Toprak et al. described a one-bath treatment strategy of raw cotton fabrics with amylase, pectinase, and cellulase enzymes as an efficient alternative to traditional methods, significantly reducing processing time and temperature [57]. Enzymes can also be combined with other chemical agents to improve the scouring. A cotton-polyester blend fabric scouring agent that combines 50–88% biological enzymes with 12–50% surfactants has been described to remove impurities while requiring low temperatures, which helps in reducing consumption, enabling faster processing times, and being environment-friendly [58].

10.3.3 Bleaching

Bleaching is a textile process aimed at enhancing the whiteness of materials through the removal of natural colors and various impurities. The objective is to achieve a lasting and uniform white appearance of the fabric, which is required for homogenous dyeing and appropriate printing. Among various bleaching agents, H_2O_2 is the most commonly used industrial bleaching agent over the past few decades. However, H_2O_2 bleaching requires high temperatures and extended processing times, which lead to increased energy consumption and fiber damage that can cause challenges during dyeing [59]. As an alternative, enzyme-based bleaching processes can be employed to achieve the desired levels of whiteness in textiles, often in combination with other agents to enhance the bleaching effect. Eren et al. described a one-bath process that combines de-sizing, bleaching, and dyeing of cotton fabrics using enzymes [60]. In this process, the amyloglucosidase/pullulanase enzyme generates glucose during de-sizing, and glucose oxidase uses it to produce H_2O_2, which is then used for bleaching. Finally, catalase is employed to decompose residual H_2O_2, ensuring fabric safety and environmental compliance. Several studies have also reported the use of laccase enzyme (oxidoreductase) for bleaching [61–64]. Laccase enzymes can bind to the hydroxyl groups of phenolics, which leads to the fading of colored flavonoids [65]. When applied before conventional peroxide bleaching, laccases also improve the final whiteness of the fabric, achieving results comparable to multiple peroxide bleaching cycles. This method is particularly beneficial for reducing the chemical load and energy consumption in textile processing [64]. Unuofin et al. demonstrated successful bio-bleaching of denim using a crude laccase from *Pseudomonas* sp. HRJ16 [62]. Xylanases (hydrolases) also play an important role in bleaching. The use of xylanase isolated from *Bacillus amyloliquefaciens* [66], and *Bacillus pumilus* SV-85S [67] have been described for bleaching. A combination of enzymes xylanase and mannanase (a hydrolase) has also been reported for the enhancement of the bleaching process [68]. A method involving perhydrolase enzymes has also been developed for bleaching synthetic textiles [69]. This process includes a combination of a perhydrolase enzyme, an ester substrate, an H_2O_2 source, and other agents such as surfactants and stabilizers. The method also incorporates fluorescent whitening agents and optional dyes to achieve a measurable whitening effect.

The process of bleaching using H_2O_2 is usually performed after scouring and before dyeing. However, approximately 10–15% of H_2O_2 remains on the fabric, which can lead to cellulose degradation and the formation of pinholes on the surface of the fabric, ultimately leading to a reduction in the strength of the fibers [70]. Traditionally, different reducing agents such as carbon, hydrogen, carbon monoxide, sodium sulfate, etc. are employed to remove the excess H_2O_2, or water is used to rinse out the H_2O_2 bleach [71]. However, in addition to their direct use as bleaching agents, enzymes can also be used to treat the bleaching effluents to reduce their environmental impact. Catalases can be used for removing excess H_2O_2 [71]. This approach eliminates the need for harsh reducing agents and decreases water consumption. Catalase-peroxidase enzymes derived from thermoalkaliphilic *Bacillus* sp. exhibit remarkable stability at elevated pH levels and temperatures, making them ideal for the decomposition of H_2O_2 used in traditional bleaching processes [72, 73]. Additionally, eco-friendly methodologies such as single-bath scouring and bleaching have been described involving the use of pectinase enzymes in conjunction with peracetic acid, which enable simultaneous scouring and bleaching of cotton, enhancing fabric absorbance while preserving fiber integrity [74].

10.3.4 Removal of Excess Dye from Over-Dyed Fabric

The removal of dye from over-dyed fabric is known as “back stripping” or “destructive stripping” [23]. Industries face significant challenges in the removal of coloring dyes from the effluents. Typically, synthetic dyes contain conjugated double bonds and aromatic rings and due to the inherent stability of their structure, they persist for a long time in the environment. For example, azo dyes exhibit a characteristic azo linkage. This specific linkage reduces the susceptibility of azo dyes to oxidative reactions, rendering them resistant to conventional degradation methods [75].

Using biological methods to remove color dyes from textile effluents is often a cost-effective alternative, as it involves minimal processing cost. Fungi can break down textile dye effluents by their intrinsic activity or by producing the enzymes that digest the dye [76]. Enzymes also find applications in decolorizing textile dyes. Enzymes including lignin peroxidases, manganese peroxidases, and laccases are frequently used [77, 78]. Catalases derived from *G. lucidum* have been reported to exhibit the ability to decolorize reactive black B-dyed cotton fabric [79, 80]. Enzymes like laccases, manganese peroxidases, tyrosinases, and lignin peroxidases are also used in textile effluent treatment plants (ETPs) [81].

10.3.5 Bio-polishing/Bio-finishing

Bio-polishing is an important process that contributes to superior textile quality [40]. It involves the enzyme-based removal of micro-hair and fuzzy fibrils from the fabric surfaces [82]. This action also enhances the color brightness, improves the fabric's hand feel, and increases water absorbency [82]. Bio-polishing also reduces the tendency for pill formation [22]. In addition, bio-polishing offers several advantages such as improved breaking strength, increased elongation [83], reduced rugosity [84], reduced water and energy consumption [85, 86], elimination of the need for chemical coating [87], and it is an eco-friendly process [88], generally performed after bleaching and pre-treatment [89]. Cellulase enzyme is majorly used in bio-polishing. Several studies have reported the application of cellulase enzyme in bio-polishing. Ali et al. performed a study on four types of cotton-based fabrics to explore the effect of temperature in bio-polishing using cellulase enzymes. They found that higher temperatures (60 °C) did not decrease the whiteness index, while lower temperatures (40 °C) did. Additionally, a temperature range of 50–60 °C showed a positive effect on reducing surface roughness [90].

Uddin explored the treatment of cotton fibers with acid and neutral cellulase using different parameters and concluded the best outcome of both kinds of cellulase in the biopolishing of cotton fibers. Abrasion resistance was improved in the case of acid cellulase and pilling resistance was improved in both cases [91]. Sankarraj et al. compared free and immobilized cellulase for bio-polishing of cotton fabrics and concluded that immobilized cellulase performs better than free cellulase on quality parameters like tensile strength, whiteness index, and crystallinity index of bio-polished cotton fabric [92]. Ikbal et al. described an eco-friendly method of bio-polishing cotton fabrics through enzymes derived from wasted sugarcane bagasse [32].

10.3.6 Bio-stone Washing of Denim

Denims are very popular due to their unique fading effect and the aging process. Washing denim is a cosmetic process that improves the fabric's flexibility and comfort [93]. Traditionally, the pumic stone washing technique was used for the desired appearance and to improve the softness and flexibility of denim garments [93]. However, this process has several disadvantages. Pumice stones can lead to significant back-staining, cause wear and tear on machines, and make it challenging to remove residual pumice from processed garments. Additionally, pumice dust can clog machine drainage passages and sewer lines [22]. Due to these limitations associated with stone-based washing, enzymes were introduced as an alternative method [94, 95]. Cellulase enzymes are majorly used in bio-stoning, specifically targeting the cellulose component, primarily on the fabric surface. Several studies have reported the applications of cellulase enzymes in bio-stoning [84, 96, 97].

10.3.7 Modification and Functionalization of Synthetic Fibers

Various enzymes have been used for the modification of synthetic fibers. Synthetic fibers exhibit hydrophobic characteristics, and treatment with suitable enzymes can enhance their hydrophilicity and properties like wearing comfort, antipiling, dyeability, and antistatic charge development [23]. Various types of enzymes, including cutinases, lipases, and esterases, have been used for modifying polyester. Various source organisms of cutinases have been reported to hydrolyze polyester [98–100]. Enzymatic modification of Polyamide 6.6 (PA6.6) fabric using a protease derived from *Bacillus licheniformis* (specifically, Subtilisin A) has also been reported [101]. This modification aimed to improve the fabric's compatibility with water-based nanocoatings, which are used to impart electrical conductivity [101]. The enzymatic degradation of three types of polyester using cutinase revealed that the biological degradability depends on the spacing of ester groups and the crystallinity. Also, the balance in the ratio of hydrophilicity and hydrophobicity on the polyester surface plays an important role in determining its enzymatic degradability [102]. Analysis of the effect of lipase treatment on weight loss and surface modification of polyester fabrics has revealed that the enzymatic process results in reduced surface damage and weight loss, enhanced dye and moisture uptake, and low shear rigidity [103].

10.4 Latest Advancements in the Use of Enzymes for Textile Finishing: The Use of Immobilized Enzymes and Nanoparticles

To keep up with the demand for products, improved technologies involving immobilized enzymes and nanoparticles have emerged for textile finishing. New technologies have been developed to enhance the immobilization of enzymes while maintaining their stability to reduce the overall cost. These include microwave-assisted immobilization, the use of mesoporous supports, crosslinked enzyme aggregates (CLEAs), and single-enzyme nanoparticles [41]. For example, amylase can be immobilized through chemical cross-linking with functional magnetite nanoparticles. This process not only increases thermal stability but is also helpful in long-term storage [104]. Covalent immobilization of commercial laccase on green coconut fiber enhances thermal and operational stabilities while maintaining significant activity for dye degradation [105]. Immobilization of proteases results in increased thermal stability and finds application in the production of shrink-free finished woolen garments [106]. Similarly, the immobilization of *Fusarium solani pisi* cutinase in the form of CLEAs has also been explored for the efficient synthesis of nylons [107]. Further, since nanoparticles provide a high surface area for enzyme immobilization, their use can also significantly improve the properties of enzymes. Enzymes immobilized on nanostructured materials like carbon nanotubes and metal–organic frameworks exhibit enhanced stability due to the favorable micro-environments created by these supports [108]. Immobilized enzymes on nanoparticles also exhibit increased catalytic activity, making them highly preferable for increasing production rates [109]. Using nanoparticles, such as silver nanoparticles (AgNPs), in combination with enzymes has also been explored to develop textiles with rapid antibacterial properties. Silk fabrics treated with enzymatically grafted modified polyethyleneimine and AgNPs showed excellent antibacterial activity, maintaining effectiveness even after multiple washes [110]. The use of nanozyme composites, such as cotton textiles modified with iron oxide nanoparticles, demonstrates peroxidase-like activity. These composites can rapidly decolorize dyes, offering a low-cost and efficient solution for textile finishing [111]. Nanoparticles coated with Manganese peroxidase enzyme have also been explored to decolorize textile dyes [112].

10.5 Applications of Enzymes in Reducing the Environmental Issues Related to Textile Finishing

The environmental issues concerning the textile industry have gained attention worldwide due to the production of significant amounts of effluents containing toxic agents and chemicals [3]. Enzyme-based methods are an effective alternative to traditional methods and can replace or minimize the use of these hazardous chemicals. Enzymes including amylases, catalases, proteases, laccases, cellulases, and lipases have been reported to have widespread applications in textile manufacturing processes. The use of enzymes offers several advantages, such as being eco-friendly, producing high-quality product outcomes, and reducing energy and water consumption [113]. Enzymes function effectively at lower temperatures and neutral pH levels, leading to substantial savings in energy and water. Enzymes are also biodegradable and non-toxic, making them safer for the environment [113]. Enzymes such as laccases, protease, nylonase, cellulase, polyesterase also find applications in the composting of textile waste [41]. Jankowska et al. have also reported the removal of dye from model seawater by using immobilized horseradish peroxidase enzyme [114].

10.6 Current Patent Landscape in Enzyme-Based Textile Finishing

The current landscape of patents describing enzyme-based textile finishing strategies is an example of the growing emphasis on adapting to sustainability and efficiency. Enzymes are increasingly being used to replace traditional chemical processes in textile finishing, offering environment-friendly alternatives that reduce energy

Table 10.1 Patent trends in enzyme-based textile finishing

S. No.	Patent number	Year	Enzyme used	References
1.	CN116377703A	2023	Amylase	[115]
2.	CN112981949A	2021	Tyrosinase	[116]
3.	CN109457468A	2019	Laccase	[117]
4.	CN108642854A	2018	Zytase, urase, horseradish peroxidase, and laccase	[118]
5.	CN108251319A	2018	Cutinase	[119]
6.	CN108978232A	2018	Amylase	[120]
7.	CN107287895A	2017	Cellulase	[121]
8.	EP2064385B1	2016	Pectate lyase	[122]
9.	EP3088502B1	2016	Nuclease	[123]
10.	ES2547135T3	2015	Perhydrolase	[124]
11.	WO2010030769A1	2010	Perhydrolase	[125]

consumption and pollution. Table 10.1 provides a summary of patented advancements in enzyme-based textile treatment and finishing processes. The patents highlight the application of diverse enzymes to improve textile manufacturing, including bleaching, dyeing, wrinkle-proofing, and flame-retardancy.

10.7 Challenges and Future Trends

Enzyme-based textile finishing presents a promising alternative to traditional chemical processes, offering environmental and economic benefits. However, the transition to enzymatic processes is not without challenges. Enzymes are proteins by nature and often require specific conditions to function optimally, such as moderate temperatures and pH levels. However, textile processing environments can be harsh, with high temperatures and extreme pH, which can denature enzymes and reduce their effectiveness [22]. This makes it challenging to integrate enzymes into existing industrial processes without significant modifications [126, 127]. Enzymes can also be sensitive to storage conditions and may lose activity over time. Immobilization techniques have been developed to increase the stability and reusability of enzymes, but these methods can be costly and complex to implement on a large scale [126, 128]. Further, the production and purification of enzymes can be expensive, and scaling up enzyme-based processes to meet industrial demands can present a significant hurdle. While enzymes can reduce energy and water consumption, the initial investment and operational costs can be prohibitive for some manufacturers [35, 128]. Incorporating use of enzymes into existing textile processing lines can also prove to be challenging. Enzymes should be compatible with other chemicals and processes used in textile manufacturing, which may require adjustments to current practices.

Ongoing research in the field of enzyme-based finishing is focused on developing more robust enzymes that can withstand harsh industrial conditions. Advancements in genetic engineering provide possibilities for producing modified or new enzymes with improved properties to alleviate existing challenges [129, 130]. The identification of novel enzymes from archaea bacteria that are found in extreme environments and possess unique biochemical properties, which allow them to thrive under conditions such as high temperature, extreme pH, and high salinity, is being explored [131]. Life cycle assessment studies demonstrate that enzymes can effectively reduce pollution, save resources, and reduce overall costs [128].

Beyond traditional applications, enzymes are being explored for novel uses in textile finishing, such as imparting functional properties like antibacterial activity, UV protection, anti-oxidant activity, and deodorization [132–134]. Overall, the future of enzyme-based textile processing is very promising. Research in this area could lead to developing newer enzyme-based treatment strategies that are more efficient and environment-friendly to support the growing demand for sustainable textile processing methods.

10.8 Conclusion

The textile industry represents a significant part of any nation's economic framework and cultural heritage. This chapter has provided a comprehensive overview of the important role of enzymes in textile finishing processes, highlighting diverse applications and the advantages they offer over traditional chemical-based finishing methods. Enzyme-based treatment not only reduces the environmental concerns associated with chemical pollution but also enhances the overall quality and performance of textiles. The ease of including enzymes such as amylases, cellulases, and laccases in various stages of textile finishing demonstrates their versatility and effectiveness. Furthermore, advancements such as enzyme immobilization and the use of nanoparticle-based strategies present interesting avenues for improving enzyme stability and catalytic efficiency, thereby addressing some of the challenges faced in industrial applications of enzymes. Despite the challenges related to cost, scalability, and operational compatibility, ongoing research in this field holds great promise for the future of sustainable textile finishing. As the textile industry undergoes continual growth, implementing enzyme-based strategies is expected to play an important role in mitigating the overall environmental impact while meeting the increasing demand for sustainable textiles.

Acknowledgements The authors would like to acknowledge Bennett University (The Times Group), India for providing the institutional seed grant to VV and Ph.D. fellowship to AK. Authors would also like to acknowledge Department of Science and Technology—Science and Engineering Research Board (DST-SERB) for providing a Start-up Research Grant (SRG/2022/000486) to VV.

Funding This work was supported by an institutional seed grant by Bennett University and Start-up Research Grant by the Department of Science and Technology—Science and Engineering Research Board (DST-SERB) (SRG/2022/000486) to Vaishali Verma.

Declaration The authors declare that they have no known competing financial interests or personal relationships that could have appeared to influence the work reported in this chapter. Generative AI and AI-assisted technologies have been used in the writing process to improve the readability and language of the work only. After using this tool, the authors carefully reviewed and revised the content to ensure accuracy and quality. The authors take full responsibility for the final version of this publication.

References

1. Sfameni, S., Hadhri, M., Rando, G., Drommi, D., Rosace, G., Trovato, V., & Plutino, M. R. (2023). Inorganic finishing for textile fabrics: Recent advances in wear-resistant, UV protection and antimicrobial treatments. *Inorganics, 11*, 19. https://doi.org/10.3390/inorganics11010019
2. Kiron, M. I. (2012). Textile finishing process | Mechanical and chemical finishes in textiles. In Text. Learn. Retrieved May 6, 2024, from https://textilelearner.net/textile-finishing-process/
3. Azanaw, A., Birlie, B., Teshome, B., & Jemberie, M. (2022). Textile effluent treatment methods and eco-friendly resolution of textile wastewater. *Case Studies in Chemical and Environmental Engineering, 6*, Article 100230. https://doi.org/10.1016/j.cscee.2022.100230
4. Lellis, B., Fávaro-Polonio, C. Z., Pamphile, J. A., & Polonio, J. C. (2019). Effects of textile dyes on health and the environment and bioremediation potential of living organisms. *Biotechnology Research and Innovation, 3*, 275–290. https://doi.org/10.1016/j.biori.2019.09.001
5. Sakamoto, M., Ahmed, T., Begum, S., & Huq, H. (2019). Water pollution and the textile industry in Bangladesh: Flawed corporate practices or restrictive opportunities? *Sustainability, 11*, 1951. https://doi.org/10.3390/su11071951
6. Mia, R., Selim, M., Shamim, A. M., Chowdhury, M., Sultana, S., Armin, M., Hossain, M., Akter, R., Dey, S., & Naznin, H. (2019). Review on various types of pollution problem in textile dyeing & printing industries of Bangladesh and recommendation for mitigation. *Journal of Textile Engineering & Fashion Technology, 5*. https://doi.org/10.15406/jteft.2019.05.00205
7. Fuente, A., & Hickson, L. (2011). Noise-induced hearing loss in Asia. *International Journal of Audiology, 50*, S3–S10. https://doi.org/10.3109/14992027.2010.540584
8. Zimniewska, M. (2022). Hemp fibre properties and processing target textile: A review. *Materials, 15*, 1901. https://doi.org/10.3390/ma15051901
9. Alanezi, A. M. (2018). Impact of pollution generated by the textile industry on health and environment. Journal of University Studies for inclusive Research, *2(3)*, 160–176.
10. Hung, Y.-T., Wang, L. K., & Shammas, N. K. (2012). *Handbook of environment and waste management: Air and water pollution control*. World Scientific. https://doi.org/10.1142/7971
11. Hossain, M. M., Islam, T., Jalil, M. A., Rakibuzzaman, S. M., Surid, S. M., Zabed, M. R. I., Talukder, A., & Hossain, S. (2024). Advancements of eco-friendly natural antimicrobial agents and their transformative role in sustainable textiles. *SPE Polymers*, pls2.10135. https://doi.org/10.1002/pls2.10135
12. Kane, F., Shen, J., Morgan, L., Prajapati, C., Tyrer, J., & Smith, E. (2020). Innovative technologies for sustainable textile coloration, patterning, and surface effects. In S. S. Muthu & M. A. Gardetti (Eds.), *Sustainability in the textile and apparel industries* (pp. 99–127). Springer International Publishing.

13. Shrimali, K., & Dedhia, E. (2016). Enzymatic finishing of textiles. *International Journal of Science and Research, 5*(5), 674–677.
14. Ibrahim, A., Laquerre, J.-É., Forcier, P., Deregnaucourt, V., Decaens, J., & Vermeersch, O. (2021). Antimicrobial agents for textiles: Types, mechanisms and analysis standards. In: B. Kumar (Ed.), *Textiles for functional applications*. IntechOpen.
15. Montazer, M., Chizarifard, G., & Harifi, T. (2014). CO_2 laser irradiation of raw and bleached cotton fabrics, with focus on water and dye absorbency. *Coloration Technology, 130*, 13–20. https://doi.org/10.1111/cote.12057
16. Bahtiyari, M. İ. (2011). Laser modification of polyamide fabrics. *Optics & Laser Technology, 43*, 114–118. https://doi.org/10.1016/j.optlastec.2010.05.014
17. Akiwowo, K., Kane, F., Tyrer, J., Weaver, G., & Filarowski, A. (2014). Digital laser-dyeing for polyester fabrics. *Journal of Textile Design Research and Practice, 2*, 133–151. https://doi.org/10.2752/205117814X14228978833457
18. Ondogan, Z., Pamuk, O., Ondogan, E. N., & Ozguney, A. (2005). Improving the appearance of all textile products from clothing to home textile using laser technology. *Optics & Laser Technology, 37*, 631–637. https://doi.org/10.1016/j.optlastec.2004.10.001
19. Robinson, P. K. (2015). Enzymes: Principles and biotechnological applications. *Essays in Biochemistry, 59*, 1–41. https://doi.org/10.1042/bse0590001
20. Raveendran, S., Parameswaran, B., Ummalyma, S. B., Abraham, A., Mathew, A. K., Madhavan, A., Rebello, S., & Pandey, A. (2018). Applications of microbial enzymes in food industry. *Food Technology and Biotechnology, 56*, 16–30. https://doi.org/10.17113/ftb.56.01.18.5491
21. Aydin, D. (2023). Enzymes in textile industry: Towards to sustainable textile processes. *8*. https://doi.org/10.31031/TTEFT.2023.08.000697
22. Madhu, A., & Chakraborty, J. N. (2017). Developments in application of enzymes for textile processing. *Journal of Cleaner Production, 145*, 114–133. https://doi.org/10.1016/j.jclepro.2017.01.013
23. Md. Mamun Kabir, S., & Koh, J. (2022). Sustainable textile processing by enzyme applications. In: K. Ferreira Mendes, R. Nogueira De Sousa, & K. Cabral Mielke (Eds.), *Biodegradation technology of organic and inorganic pollutants*. IntechOpen.
24. Singh, R., Kumar, M., Mittal, A., & Mehta, P. K. (2016). Microbial enzymes: Industrial progress in 21st century. *3 Biotech, 6*, 174. https://doi.org/10.1007/s13205-016-0485-8
25. Cavaco-Paulo, A., & Gübitz, G. M., Textile Institute (Manchester, England). (2003). *Textile processing with enzymes*. CRC; Woodhead.
26. McDonald, A. G., & Tipton, K. F. (2023). Enzyme nomenclature and classification: The state of the art. *FEBS Journal, 290*, 2214–2231. https://doi.org/10.1111/febs.16274
27. Porto De Souza Vandenberghe, L., Karp, S. G., Binder Pagnoncelli, M. G., Von Linsingen Tavares, M., Libardi Junior, N., Valladares Diestra, K., Viesser, J. A., & Soccol, C. R. (2020). Classification of enzymes and catalytic properties. In *Biomass, biofuels, biochemicals* (pp. 11–30). Elsevier.
28. Shukla, E., Bendre, A. D., & Gaikwad, S. M. (2022). Hydrolases: The most diverse class of enzymes. In: S. Haider, A. Haider, & A. Catalá (Eds.), *Biochemistry*. IntechOpen.
29. Guo, C., Li, T., Wang, C., Wang, Y., & Zhang, Y. (2017). Enzymatic treatment on cotton fibers: Degradation kinetics of pectin and influence of shape change on adsorption. *Fibers and Polymers, 18*, 1882–1890. https://doi.org/10.1007/s12221-017-7045-7
30. Lenting, H. B. M., & Warmoeskerken, M. M. C. G. (2004). A fast, continuous enzyme-based pretreatment process concept for cotton containing textiles. *Biocatal Biotransformation, 22*, 361–368. https://doi.org/10.1080/10242420400024557
31. Saravanan, D., Sivasaravanan, S., Sudharshan Prabhu, M., Vasanthi, N. S., Senthil Raja, K., Das, A., & Ramachandran, T. (2012). One-step process for desizing and bleaching of cotton fabrics using the combination of amylase and glucose oxidase enzymes. *Journal of Applied Polymer Science, 123*, 2445–2450. https://doi.org/10.1002/app.34838
32. Ikbal, M. S., Tisha, F. A., Asheque, A. I., Hasnat, E., & Uddin, M. A. (2024). Eco-friendly biopolishing of cotton fabric through wasted sugarcane bagasse-derived enzymes. *Heliyon, 10*, Article e26346. https://doi.org/10.1016/j.heliyon.2024.e26346

33. Shao-Wei, D., & Da-Nian, L. (2008). Kinetics of the thermal inactivation of *Bacillus subtilis* α-amylase and its application on the desizing of cotton fabrics. *Journal of Applied Polymer Science, 109*, 3733–3738. https://doi.org/10.1002/app.28612
34. Aly, A. S., Moustafa, A. B., & Hebeish, A. (2004). Bio-technological treatment of cellulosic textiles. *Journal of Cleaner Production, 12*, 697–705. https://doi.org/10.1016/S0959-6526(03)00074-X
35. Chatha, S. A. S., Asgher, M., & Iqbal, H. M. N. (2017). Enzyme-based solutions for textile processing and dye contaminant biodegradation—A review. *Environmental Science and Pollution Research, 24*, 14005–14018. https://doi.org/10.1007/s11356-017-8998-1
36. Patti, A., & Acierno, D. (2023). Materials, weaving parameters, and tensile responses of woven textiles. *Macromol, 3*, 665–680. https://doi.org/10.3390/macromol3030037
37. Panda, S. K. B. C., Sen, K., & Mukhopadhyay, S. (2023). Sustainable photocatalytic desizing process for the starch-based size. *ACS Omega, 8*, 18726–18734. https://doi.org/10.1021/acsomega.3c00713
38. El-Sayed Mohamed, Z., Amr, A., Knittel, D., & Schollmeyer, E. (2010). Synthesis and application of new sizing and finishing additives based on carboxymethyl cellulose. *Carbohydrate Polymers, 81*, 769–774. https://doi.org/10.1016/j.carbpol.2010.03.013
39. Rahmouni, A., Maatoug, S., & Ladhari, N. (2024). The effect of size concentration, lubricant and plasticizer agents contents on cold sizing process performance of cotton warp yarns: modelling and optimization studies. *The Journal of the Textile Institute*, 1–16.https://doi.org/10.1080/00405000.2024.2406604
40. Montazer, M., & Harifi, T. (2018). Introduction: Textile finishing. In *Nanofinishing of textile materials* (pp 1–17). Elsevier.
41. Roy Choudhury, A. K. (2014). Sustainable textile wet processing: applications of enzymes. In: S. S. Muthu (Ed.), *Roadmap to sustainable textiles and clothing* (pp 203–238). Springer Singapore.
42. Rehman, A., Saeed, A., Asad, W., Khan, I., Hayat, A., Rehman, M. U., Shah, T. A., Sitotaw, B., Dawoud, T. M., & Bourhia, M. (2023). Eco-friendly textile desizing with indigenously produced amylase from Bacillus cereus AS2. *Science and Reports, 13*, 11991. https://doi.org/10.1038/s41598-023-38956-3
43. Yahya, S., Sohail, M., & Khan, S. A. (2022). Characterization, thermal stabilization and desizing potential of amylase from *A. tubingensis* SY 1. *Journal of the Textile Institute, 113*, 993–1000. https://doi.org/10.1080/00405000.2021.1914393
44. Chand, N., Sajedi, R. H., Nateri, A. S., Khajeh, K., & Rassa, M. (2014). Fermentative desizing of cotton fabric using an α-amylase-producing Bacillus strain: Optimization of simultaneous enzyme production and desizing. *Process Biochemistry, 49*, 1884–1888. https://doi.org/10.1016/j.procbio.2014.07.007
45. Wersch, K. V. (2004). Continuous enzymatic desizing method. https://patents.google.com/patent/US20050170484A1/en
46. Rajkumar, G. (2019). Combined de sizing and scouring of cotton fabric using xylanase enzyme. *Journal of Textile Engineering & Fashion Technology, 5*. https://doi.org/10.15406/jteft.2019.05.00214
47. Shahid, M., & Adivarekar, R. (2020). *Advances in functional finishing of textiles*. Springer Singapore.
48. Carmen, Z., & Daniel, S. (2012). Textile organic dyes—Characteristics, polluting effects and separation/elimination procedures from industrial effluents—A critical overview. In: T. Puzyn (Ed.), *Organic pollutants ten years after the stockholm convention—Environmental and analytical update*. InTech.
49. Agrawal, P. B., Nierstrasz, V. A., Klug-Santner, B. G., Gübitz, G. M., Lenting, H. B. M., & Warmoeskerken, M. M. C. G. (2007). Wax removal for accelerated cotton scouring with alkaline pectinase. *Biotechnology Journal, 2*, 306–315. https://doi.org/10.1002/biot.200600169
50. Joshi, M., Nerurkar, M., Badhe, P., & Adivarekar, R. (2013). Scouring of cotton using marine pectinase. *Journal of Molecular Catalysis B: Enzymatic, 98*, 106–113. https://doi.org/10.1016/j.molcatb.2013.10.010

51. Taleb, M. A., Gomaa, S. K., Wahba, M. I., Zaki, R. A., El-Fiky, A. F., El-Refai, H. A., & El-Sayed, H. (2022). Bioscouring of wool fibres using immobilized thermophilic lipase. *International Journal of Biological Macromolecules, 194*, 800–810. https://doi.org/10.1016/j.ijbiomac.2021.11.128
52. Sreelakshmi, S. N., Senthil Raja, K., Vasanthi, N. S., & Saravanan, D. (2013). Low temperature, acidic, bacterial lipase for scouring of raw cotton fabrics. *Journal of the Textile Institute, 104*, 1345–1352. https://doi.org/10.1080/00405000.2013.805874
53. Nerurkar, M., Joshi, M., & Adivarekar, R. (2015). *Applied Biochemistry and Biotechnology, 175*, 253–265. https://doi.org/10.1007/s12010-014-1259-6
54. Jayasekara, S., & Ratnayake, R. (2019). Microbial cellulases: An overview and applications. In: A. Rodríguez Pascual & E. M. Eugenio Martín (Eds.), *Cellulose*. IntechOpen. https://www.intechopen.com/chapters/66517
55. Michael Okpara University of Agriculture, Anab-Atulomah C., & Nwachukwu, E. (2021). Bio-scouring of cotton using protease and pectinase from Bacillus subtilis isolated from market waste. *Path of Science, 7*, 3001–3006. https://doi.org/10.22178/pos.72-3
56. Badhe, P., Damale, M., & Adivarekar, R. (2017). Bioscouring of wool using protease from Bacillus subtilis isolated from abattoir waste. *The Journal of Microbiology, Biotechnology and Food Sciences, 6*, 1012–1018. https://doi.org/10.15414/jmbfs.2017.6.4.1012-1018
57. Toprak, T., & Anis, P. (2017). Combined one-bath desizing–scouring–depilling enzymatic process and effect of some process parameters. *Cellulose, 24*, 383–394. https://doi.org/10.1007/s10570-016-1095-7
58. 倪新峰. (2019). A kind of cotton-polyester blend fabric scouring agent and its method for refining.
59. Mojsov, K. (2019). Enzymatic desizing, bioscouring and enzymatic bleaching of cotton fabric with glucose oxidase. *Journal of the Textile Institute, 110*, 1032–1041. https://doi.org/10.1080/00405000.2018.1535240
60. Eren, H. A., Anis, P., & Davulcu, A. (2009). Enzymatic one-bath desizing—Bleaching—Dyeing process for cotton fabrics. *Textile Research Journal, 79*, 1091–1098. https://doi.org/10.1177/0040517508099388
61. Panwar, V., Sheikh, J. N., & Dutta, T. (2020). Sustainable denim bleaching by a novel thermostable bacterial laccase. *Applied Biochemistry and Biotechnology, 192*, 1238–1254. https://doi.org/10.1007/s12010-020-03390-y
62. Unuofin, J. O., Moloantoa, K. M., & Khetsha, Z. P. (2022). The biobleaching potential of laccase produced from mandarin peelings: Impetus for a circular bio-based economy in textile biofinishing. *Arabian Journal of Chemistry, 15*, Article 104305. https://doi.org/10.1016/j.arabjc.2022.104305
63. Saleem, R., Khurshid, M., & Ahmed, S. (2018). Laccases, manganese peroxidases and xylanases used for the bio-bleaching of paper pulp: An environmental friendly approach. *Protein and Peptide Letters, 25*, 180–186. https://doi.org/10.2174/0929866525666180122100133
64. Tzanov, T., Basto, C., Gübitz, G. M., & Cavaco-Paulo, A. (2003). Laccases to improve the whiteness in a conventional bleaching of cotton. *Macromolecular Materials and Engineering, 288*, 807–810. https://doi.org/10.1002/mame.200300100
65. Jemli, S., Ayadi-Zouari, D., Hlima, H. B., & Bejar, S. (2016). Biocatalysts: Application and engineering for industrial purposes. *Critical Reviews in Biotechnology, 36*, 246–258. https://doi.org/10.3109/07388551.2014.950550
66. Kumar, S., Haq, I., Prakash, J., Singh, S. K., Mishra, S., & Raj, A. (2017). Purification, characterization and thermostability improvement of xylanase from Bacillus amyloliquefaciens and its application in pre-bleaching of kraft pulp. *3 Biotech, 7*, 20. https://doi.org/10.1007/s13205-017-0615-y
67. Nagar, S., Jain, R. K., Thakur, V. V., & Gupta, V. K. (2013). Biobleaching application of cellulase poor and alkali stable xylanase from Bacillus pumilus SV-85S. *3 Biotech, 3*, 277–285. https://doi.org/10.1007/s13205-012-0096-y

68. Angural, S., Bala, I., Kumar, A., Kumar, D., Jassal, S., & Gupta, N. (2021). Bleach enhancement of mixed wood pulp by mixture of thermo-alkali-stable xylanase and mannanase derived through co-culturing of Alkalophilic Bacillus sp. NG-27 and Bacillus nealsonii PN-11. *Heliyon, 7*, e05673. https://doi.org/10.1016/j.heliyon.2020.e05673
69. Donze, J.-J., Redling, E., Pflueger, A., Gysin, B., & Schmidlin, A. (2012). Enzymatic textile bleach-whitening methods. https://patents.google.com/patent/EP2403990A2/en
70. Lin, J.-H., Chang, Y.-H., & Hsu, Y.-H. (2009). Degradation of cotton cellulose treated with hydrochloric acid either in water or in ethanol. *Food Hydrocolloids, 23*, 1548–1553. https://doi.org/10.1016/j.foodhyd.2008.10.005
71. Amorim, A. M., Gasques, M. D. G., Andreaus, J., & Scharf, M. (2002). The application of catalase for the elimination of hydrogen peroxide residues after bleaching of cotton fabrics. *Anais da Academia Brasileira de Ciências, 74*, 433–436. https://doi.org/10.1590/S0001-37652002000300006
72. Gudelj, M., Fruhwirth, G., Paar, A., Lottspeich, F., Robra, K.-H., Cavaco-Paulo, A., & Gübitz, G. (2001). A catalase-peroxidase from a newly isolated thermoalkaliphilic Bacillus sp. with potential for the treatment of textile bleaching effluents. *Extremophiles, 5*, 423–429. https://doi.org/10.1007/s007920100218
73. Fruhwirth, G., Paar, A., Gudelj, M., Cavaco-Paulo, A., Robra, K. H., & Gübitz, G. (2002). An immobilised catalase peroxidase from the alkalothermophilic Bacillus SF for the treatment of textile-bleaching effluents. *Applied Microbiology and Biotechnology, 60*, 313–319. https://doi.org/10.1007/s00253-002-1127-0
74. Preša, P., & Tavčer, P. F. (2008). Bioscouring and bleaching of cotton with pectinase enzyme and peracetic acid in one bath. *Coloration Technology, 124*, 36–42. https://doi.org/10.1111/j.1478-4408.2007.00118.x
75. Maddhinni, V. L., Vurimindi, H. B., & Yerramilli, A. (2006). Degradation of azo dye with horse radish peroxidase (HRP). *Journal of the Indian Institute of Science*, 86(5), 507–514.
76. Rajhans, G., Barik, A., Sen, S., & Raut, S. (2021). Degradation of dyes by fungi: An insight into mycoremediation. *Biotechnologia, 102*, 445–455. https://doi.org/10.5114/bta.2021.111109
77. Tian, Y., Wu, K., Lin, S., Shi, M., Liu, Y., Su, X., & Islam, R. (2024). Biodegradation and decolorization of crystal violet dye by cocultivation with fungi and bacteria. *ACS Omega, 9*, 7668–7678. https://doi.org/10.1021/acsomega.3c06978
78. Zeng, X., Cai, Y., Liao, X., Zeng, X., Li, W., & Zhang, D. (2011). Decolorization of synthetic dyes by crude laccase from a newly isolated Trametes trogii strain cultivated on solid agro-industrial residue. *Journal of Hazardous Materials, 187*, 517–525. https://doi.org/10.1016/j.jhazmat.2011.01.068
79. Chatha, S. A. S., Asgher, M., Ali, S., & Hussain, A. I. (2012). Biological color stripping: A novel technology for removal of dye from cellulose fibers. *Carbohydrate Polymers, 87*, 1476–1481. https://doi.org/10.1016/j.carbpol.2011.09.041
80. Chatha, S., Mallhi, A., Hussain, A., Asgher, M., & Nigam, P. (2014). A biological approach for color-stripping of cotton fabric dyed with C.I. Reactive Black 5 using fungal enzymes from solid state fermentation. *Current Biotechnology, 3*, 166–173. https://doi.org/10.2174/22115501026661402l8003757
81. Le Roes-Hill, M., & Prins, A. (2016). Biotechnological potential of oxidative enzymes from actinobacteria. In: D. Dhanasekaran & Y. Jiang (Eds.), *Actinobacteria—Basics and biotechnological applications*. InTech. https://www.intechopen.com/chapters/49191
82. Kumar, M., Shukla, S. R., Arputharaj, A., Saxena, S., Patil, S., Patil, P. G., Varghese, E., & Amarowicz, R. (2021). Biopolishing of cellulosic fabrics: A study on low-stress mechanical properties, microstructure, and dye uptake. *Fibers and Polymers, 22*, 2803–2814. https://doi.org/10.1007/s12221-021-0356-8
83. Kan, C., Yam, L., & Ng, S. (2013). Effect of stretching on ultraviolet protection of cotton and cotton/coolmax blended weft knitted fabric in a wet state. *Materials, 7*, 58–74. https://doi.org/10.3390/ma7010058
84. Kuhad, R. C., Gupta, R., & Singh, A. (2011). Microbial cellulases and their industrial applications. *Enzyme Research, 2011*, 1–10. https://doi.org/10.4061/2011/280696

85. Yu, Y., Yuan, J., Wang, Q., Fan, X., Ni, X., Wang, P., & Cui, L. (2013). Cellulase immobilization onto the reversibly soluble methacrylate copolymer for denim washing. *Carbohydrate Polymers, 95*, 675–680. https://doi.org/10.1016/j.carbpol.2013.03.043
86. Yu, Y., Yuan, J., Wang, Q., Fan, X., Wang, P., & Cui, L. (2014). A study of surface morphology and structure of cotton fibres with soluble immobilized-cellulase treatment. *Fibers and Polymers, 15*, 1609–1615. https://doi.org/10.1007/s12221-014-1609-6
87. Kan, Z., Yang, M., Yang, W., Liu, Z., & Xie, B. (2015). Investigation on the reactive processing of textile-ramie fiber reinforced anionic polyamide-6 composites. *Composites Science and Technology, 110*, 188–195. https://doi.org/10.1016/j.compscitech.2015.01.023
88. Saravanan, D., Vasanthi, N. S., & Ramachandran, T. (2009). A review on influential behaviour of biopolishing on dyeability and certain physico-mechanical properties of cotton fabrics. *Carbohydrate Polymers, 76*, 1–7. https://doi.org/10.1016/j.carbpol.2008.10.019
89. Kan, C. W. (2015). Washing techniques for denim jeans. In *Denim* (pp. 313–356). Elsevier.
90. Ali, H., Hashem, M., Shaker, N., Ramadan, M., El-Sadek, B., & Hady, M. A. (2012). Cellulase enzyme in bio-finishing of cotton-based fabrics: Effects of process parameters. *Research Journal of Textile and Apparel, 16*, 57–65. https://doi.org/10.1108/RJTA-16-03-2012-B006
91. Uddin, M. G. (2015). Effects of biopolishing on the quality of cotton fabrics using acid and neutral cellulases. *Textiles and Clothing Sustainability, 1*, 9. https://doi.org/10.1186/s40689-015-0009-7
92. Sankarraj, N., & Nallathambi, G. (2018). Enzymatic biopolishing of cotton fabric with free/immobilized cellulase. *Carbohydrate Polymers, 191*, 95–102. https://doi.org/10.1016/j.carbpol.2018.02.067
93. Agrawal, B. J. (2017). Bio-stoning of denim—An environmental-friendly approach. *Current Trends in Biomedical Engineering & Biosciences, 3*. https://doi.org/10.19080/CTBEB.2017.03.555612
94. Shafai, M. M., & Imran, M. (2014). Revolution of biotechnology in finishing sector of textile. *Chemistry and Materials Research,* 6(2), 92–103.
95. Parthiban, M., Srikrishnan, M. R., & Kandhavadivu, P. (2018). *Sustainability in fashion and apparels.* WPI Publishing.
96. Korsa, G., Konwarh, R., Masi, C., Ayele, A., & Haile, S. (2023). Microbial cellulase production and its potential application for textile industries. *Annales de Microbiologie, 73*, 13. https://doi.org/10.1186/s13213-023-01715-w
97. Madhu, & Singh, N. (2023) *A new approaches in community science* (1st ed., Vol. 1). AkiNik Publications.
98. Gübitz, G. M., & Paulo, A. C. (2003). New substrates for reliable enzymes: Enzymatic modification of polymers. *Current Opinion in Biotechnology, 14*, 577–582. https://doi.org/10.1016/j.copbio.2003.09.010
99. Kanelli, M., Vasilakos, S., Nikolaivits, E., Ladas, S., Christakopoulos, P., & Topakas, E. (2015). Surface modification of poly(ethylene terephthalate) (PET) fibers by a cutinase from Fusarium oxysporum. *Process Biochemistry, 50*, 1885–1892. https://doi.org/10.1016/j.procbio.2015.08.013
100. Silva, C., & Cavaco-Paulo, A. (2008). Biotransformations in synthetic fibres. *Biocatal Biotransformation, 26*, 350–356. https://doi.org/10.1080/10242420802357845
101. Jordanov, I., Stevens, D. L., Tarbuk, A., Magovac, E., Bischof, S., & Grunlan, J. C. (2019). Enzymatic modification of polyamide for improving the conductivity of water-based multilayer nanocoatings. *ACS Omega, 4*, 12028–12035. https://doi.org/10.1021/acsomega.9b01052
102. Bai, Z., Liu, Y., Su, T., & Wang, Z. (2018). Effect of hydroxyl monomers on the enzymatic degradation of poly(ethylene succinate), poly(butylene succinate), and poly(hexylene succinate). *Polymers, 10*, 90. https://doi.org/10.3390/polym10010090
103. Kumar, J. A., & Senthil Kumar, M. (2020). A study on improving dyeability of polyester fabric using lipase enzyme. *Autex Research Journal, 20*, 243–249. https://doi.org/10.2478/aut-2019-0030

104. Talekar, S., Ghodake, V., Ghotage, T., Rathod, P., Deshmukh, P., Nadar, S., Mulla, M., & Ladole, M. (2012). Novel magnetic cross-linked enzyme aggregates (magnetic CLEAs) of alpha amylase. *Bioresource Technology, 123*, 542–547. https://doi.org/10.1016/j.biortech.2012.07.044
105. Cristóvão, R. O., Silvério, S. C., Tavares, A. P. M., Brígida, A. I. S., Loureiro, J. M., Boaventura, R. A. R., Macedo, E. A., & Coelho, M. A. Z. (2012). Green coconut fiber: A novel carrier for the immobilization of commercial laccase by covalent attachment for textile dyes decolourization. *World Journal of Microbiology & Biotechnology, 28*, 2827–2838. https://doi.org/10.1007/s11274-012-1092-4
106. Silva, C. J. S. M., Zhang, Q., Shen, J., & Cavaco-Paulo, A. (2006). Immobilization of proteases with a water soluble–insoluble reversible polymer for treatment of wool. *Enyzme and Microbial Technology, 39*, 634–640. https://doi.org/10.1016/j.enzmictec.2005.11.016
107. Stavila, E., Arsyi, R. Z., Petrovic, D. M., & Loos, K. (2013). Fusarium solani pisi cutinase-catalyzed synthesis of polyamides. *European Polymer Journal, 49*, 834–842. https://doi.org/10.1016/j.eurpolymj.2012.12.010
108. Razzaghi, M., Homaei, A., Vianello, F., Azad, T., Sharma, T., Nadda, A. K., Stevanato, R., Bilal, M., & Iqbal, H. M. N. (2022). Industrial applications of immobilized nano-biocatalysts. *Bioprocess and Biosystems Engineering, 45*, 237–256. https://doi.org/10.1007/s00449-021-02647-y
109. Yi, S., Dai, F., Zhao, C., & Si, Y. (2017). A reverse micelle strategy for fabricating magnetic lipase-immobilized nanoparticles with robust enzymatic activity. *Science and Reports, 7*, 9806. https://doi.org/10.1038/s41598-017-10453-4
110. Wang, F., Yan, B., Li, Z., Wang, P., Zhou, M., Yu, Y., Yuan, J., Cui, L., & Wang, Q. (2021). Rapid antibacterial effects of silk fabric constructed through enzymatic grafting of modified PEI and AgNP deposition. *ACS Applied Materials & Interfaces, 13*, 33505–33515. https://doi.org/10.1021/acsami.1c08119
111. Safarik, I., Prochazkova, J., Schroer, M. A., Garamus, V. M., Kopcansky, P., Timko, M., Rajnak, M., Karpets, M., Ivankov, O. I., Avdeev, M. V., Petrenko, V. I., Bulavin, L., & Pospiskova, K. (2021). Cotton textile/iron oxide nanozyme composites with peroxidase-like activity: Preparation, characterization, and application. *ACS Applied Materials & Interfaces, 13*, 23627–23637. https://doi.org/10.1021/acsami.1c02154
112. Chang, Y., Yang, D., Li, R., Wang, T., & Zhu, Y. (2021). Textile dye biodecolorization by manganese peroxidase: A review. *Molecules, 26*, 4403. https://doi.org/10.3390/molecules26154403
113. Kumar, D., Bhardwaj, R., Jassal, S., Goyal, T., Khullar, A., & Gupta, N. (2023). Application of enzymes for an eco-friendly approach to textile processing. *Environmental Science and Pollution Research International, 30*, 71838–71848. https://doi.org/10.1007/s11356-021-16764-4
114. Jankowska, K., Zdarta, J., Grzywaczyk, A., Degórska, O., Kijeńska-Gawrońska, E., Pinelo, M., & Jesionowski, T. (2021). Horseradish peroxidase immobilised onto electrospun fibres and its application in decolourisation of dyes from model sea water. *Process Biochemistry, 102*, 10–21. https://doi.org/10.1016/j.procbio.2020.11.015
115. 朱洪际. (2023). Dyeing and finishing method of fireproof polyester fabric and fireproof polyester fabric obtained by dyeing and finishing method.
116. 周春晓, 李铁军, 陈浩文, 殷勇华. (2021). Flame-retardant method for modified real silk based on biological enzyme.
117. 董爱学, 王维明, 胡玲玲, 钱红飞, 刘越. (2019). A kind of antibiotic finishing method of cellulosic fabrics.
118. 孙祎, 许玉, 张烨. (2018). A kind of pure cotton fabric low temperature oxygen bleaching activator.
119. 吴敬, 洪若宇, 宿玲恰, 孙益荣. (2018). A kind of fermentation of cutinase and cotton fabric enzyme refinery practice.
120. 朱荣芳. (2018). A kind of no added textile finishing process.

121. 郝龙云, 王蕊, 蔡玉青, 房宽峻. (2017). A kind of method for carrying out wrinkle proofing to bafta using biopolishing waste liquid.
122. Poulose, A. J., Schmidt, B., Gastel, F. J. C. V., & Yoon, M.-Y. (2016). Enzymatic treatment of textiles using a pectate lyase from bacillus subtilis.
123. Lant, N. J., Wood, R. L., & Gummel, J. R. M. (2018). Method of treating a fabric.
124. Barnett, C. C., & Sala, R. F. (2015). Oxidative decolorization method of dyes with enzymatically generated peracid.
125. Auterinen, A.-L., Prozzo, B., Redling, E., Vermeersch, L., & Yoon, M.-Y. (2010). Enzymatic textile bleaching compositions and methods of use thereof.
126. Kundu, D., Thakur, M. S., & Patra, S. (2021). Textile fabric processing and their sustainable effluent treatment using enzymes—Insights and challenges. In: A. Tripathi & J. S. Melo (Eds.), *Immobilization strategies* (pp. 645–666). Springer Singapore.
127. Tochetto, G. A., Aragão, A. M. I., De Oliveira, D., & Immich, A. P. S. (2022). Can enzymatic processes transform textile processes? A critical analysis of the industrial application. *Process Biochemistry, 123*, 27–35. https://doi.org/10.1016/j.procbio.2022.10.030
128. Madhu, A. (2022). Immobilization as sustainable solutions to textiles chemical processing. In: S. S. Muthu (Ed.), *Sustainable approaches in textiles and fashion* (pp. 21–67). Springer Singapore.
129. Srivastava, A., & Bandhu, S. (2022). Biotechnological advancements and challenges in textile effluents management for a sustainable bioeconomy: Indian case studies. *Case Studies in Chemical and Environmental Engineering, 5*, Article 100186. https://doi.org/10.1016/j.cscee.2022.100186
130. Rafeeq, H., Afsheen, N., Rafique, S., Arshad, A., Intisar, M., Hussain, A., Bilal, M., & Iqbal, H. M. N. (2023). Genetically engineered microorganisms for environmental remediation. *Chemosphere, 310*, Article 136751. https://doi.org/10.1016/j.chemosphere.2022.136751
131. MaÁ, C., & Blamey, J. M. (2018). Biotechnological applications of archaeal enzymes from extreme environments. *Biological Research, 51*, 37. https://doi.org/10.1186/s40659-018-0186-3
132. Sun, S.-S., Xing, T., & Tang, R.-C. (2015). Enzymatic dyeing and functional finishing of textile fibres with ferulic acid. *Indian Journal of Fibre & Textile Research,* 40(1), 62–69.
133. Vajpayee, M., Dave, H., Singh, M., & Ledwani, L. (2022). Cellulase enzyme based wet-pretreatment of lotus fabric to improve antimicrobial finishing with *A. indica* extract and enhance natural dyeing: Sustainable approach for textile finishing. *ChemistrySelect, 7*, e202200382. https://doi.org/10.1002/slct.202200382
134. Gulati, R., Sharma, S., & Sharma, R. K. (2022). Antimicrobial textile: Recent developments and functional perspective. *Polymer Bulletin, 79*, 5747–5771. https://doi.org/10.1007/s00289-021-03826-3

Chapter 11
Biotechnology for Surface Modification of Polyester Fibers

Irina Singer, Sabrina Kolbe, Isabel Hofmann, Britta Hilgenberg, Amelie Achten, Aleksandr Lobanov, Uta Bergstedt, Michaela Wagner, and Maike Rabe

Abstract Polyester dominates the world fiber market and is used in almost all applications due to its versatile properties and low price. Nevertheless, it has low wettability and high electrostatic charge, which are overcome by partly environmentally harmful hydrophilic/antistatic finishing agents or caustic treatments. Additionally, polyester irreversibly accumulates in the environment as the processing, and the use of textiles made of synthetic polymers is a significant source of fibrous microplastic. As polyesterases degrade the ester bonds of polyesters, leading to free carboxyl and hydroxyl groups, enzyme technology offers extensive potential in industrial polyester finishing processes. Likewise, no vast quantities of chemicals are needed,

I. Singer · S. Kolbe · I. Hofmann · B. Hilgenberg · M. Rabe (✉)
Research Institute for Textile and Clothing, Niederrhein University of Applied Sciences, Mönchengladbach, Germany
e-mail: maike.rabe@hs-niederrhein.de

I. Singer
e-mail: irina.singer@hs-niederrhein.de

S. Kolbe
e-mail: sabrina.kolbe@hs-niederrhein.de

I. Hofmann
e-mail: isabel.hofmann@sergeferrari.com

B. Hilgenberg
e-mail: britta.hilgenberg@hs-niederrhein.de

A. Achten · A. Lobanov · U. Bergstedt · M. Wagner
Institute for Coatings and Surface Chemistry, Niederrhein University of Applied Sciences, Krefeld, Germany
e-mail: amelie.achten@hs-niederrhein.de

A. Lobanov
e-mail: aleksandr.lobanov@hs-niederrhein.de

U. Bergstedt
e-mail: uta.bergstedt@hs-niederrhein.de

M. Wagner
e-mail: michaela.wagner@hs-niederrhein.de

M. Shahid et al. (eds.), *Advancements in Textile Finishing*,
https://doi.org/10.1007/978-981-96-6385-9_11

resulting in a relatively low environmental impact. However, this fundamental understanding has not yet been fully implemented in the industry. Research in the field of polyester-degrading enzymes has found several new microorganisms and enzymes that are capable of attacking the ester bonds of synthetic polyester. Still, these "1st generation" enzymes are not effective enough for industrial needs. A solution is the development of next-generation polyesterases with adapted properties so that biotechnical surface modification, as well as degradation of polyester to recover valuable building blocks, can become economically feasible. The enzymes available on the market were investigated and applied in laboratory tests on both yarns and textile surfaces. The commercially available enzymes tested were not found to be effective enough to functionalize the materials with any measurable effect. However, the 2nd generation enzymes did: the tests showed an improved surface functionality. The best enzymatic treatments were achieved already after a few hours at an optimum use of neutral pH value and moderate temperature. The functional groups were primarily detected via the methylene blue test. The Ninhydrin test confirmed the successful removal of enzymes by washing the textiles after the enzymatic treatment. PET yarn with a low degree of crystallinity was completely degraded within only several days, while stretched filaments showed a low level of functionalization. All in all, 2nd generation polyesterases can be integrated into an optimized enzymatic treatment process of (amorphous) polyester, regardless of origin, to generate functional groups on the fiber surface for an innovative approach of functionalization, which may be able to replace chemical-intensive processes in the textile industry in the future.

Keywords Textile finishing · Sustainable finishing · Enzymatic treatment · Biodegradation · Protein engineering · Hydrophilization

11.1 Introduction

The textile market currently produces 60 million tonnes of polyester fibers per year [1]. The fiber is used in clothing, home textiles, and technical applications and has become an integral part of the textile market. The demand for polyester fibers is not leveling off but continues to rise. Although textile polyester (PET; polyethylene terephthalate) has many advantages, such as good washing properties, high color fastness, high mechanical strength, and cost-efficient production, it also has some disadvantages. For example, it must be hydrophilized to ensure good wearing comfort. Furthermore, chemical finishes, e.g., for antistatic properties, are applied using resource-intensive thermosol processes.

The biotechnological application of enzymes offers sustainable and eco-friendly alternatives to conventional chemical processes across various industries. Enzymes, as highly specific biological catalysts, enable reactions under mild conditions, reducing the need for harsh chemicals and excessive energy. This makes them valuable for the textile industry. For instance, in the textile industry, enzymatic treatments enhance fabric properties and process efficiency while minimizing environmental impact. Despite their potential, industrial enzyme applications face challenges such as optimizing stability and activity. Advances in protein engineering and molecular biology are crucial to overcoming these hurdles, paving the way for broader and more efficient enzyme use in biotechnology. This chapter explores the diverse roles of enzymes in biotechnology for surface modification of polyester fibers, emphasizing their benefits, challenges, and future prospects.

11.2 Enzymes

11.2.1 Theoretical Background

Enzymes play a crucial role in facilitating biochemical reactions within living organisms. These biological catalysts accelerate reactions by lowering the activation energy required, enabling complex processes to occur efficiently and under mild conditions [2]. Enzymes are proteins [3] made of amino acid chains. There are 20 proteinogenic amino acids, each with a central carbon atom (α-C) connected to an amino group ($-NH_2$), a carboxyl group (–COOH), and a variable side chain. These side chains differ in size, polarity, and charge, contributing to the diversity of amino acids. In proteins, amino acids link via peptide bonds, forming chains. A peptide bond forms when the –COOH group of one amino acid bonds with the $-NH_2$ group of another by releasing water. Peptides can range from dipeptides to long polypeptides, with proteins consisting of about 50–2000 amino acids.

11.2.2 Types of Enzymes Used for Surface Modification of Polyester

Textile and clothing manufacturers increasingly prioritize environmentally friendly products and processes to avoid harmful chemicals. Enzymatic treatments have gained attention due to their specificity, non-toxic nature, environmental friendliness, and energy-saving attributes. Enzymes are used across all stages of textile processing, such as amylases for desizing, cellulases for bio-polishing and denim washing, and proteases for wool modifications [4, 5].

Enzymatic surface modification of textiles involves altering fibers' physical and chemical properties. For example, enzymes remove cotton impurities, improve appearance, and cause a used look. Specifically, lipases are used for the surface modification of synthetic fibers, bioscouring, and desizing cotton. Enzymatic treatments can be refreshed during laundry with enzyme-containing detergents, unlike harsh chemical treatments. Figure 11.1 gives an overview of the enzyme classes used in the textile industry.

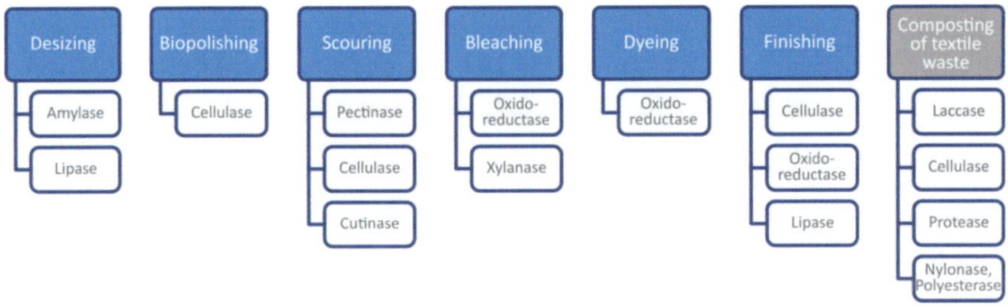

Fig. 11.1 Enzymes used in textile processes, reprinted from [6]

For achieving further positive properties of polyester surface modification, researchers are constantly looking for new enzymes and optimizing processes.

11.3 Enzymatic Modification of PET

11.3.1 Types of Polyester

Polyesters are categorized into two types: aliphatic and aromatic. Aromatic polyesters, like polyethylene terephthalate (PET), have higher melting points due to their structure [7]. Polyesters are characterized by the ester group (COO-R), represented as–(COOR)x [8]. PET, the most common textile type, is made from terephthalic acid and ethylene glycol via polycondensation. Other broadly used polyesters are polytrimethylene terephthalate (PTT), polybutylene terephthalate (PBT), polyethylene naphthalate (PEN), polytrimethylene naphthalate (PTN) and polybutylene naphthalate (PBN) [9].

11.3.2 Crystallinity of PET

The crystallinity of PET determines its mechanical, thermal, and chemical properties and, thus, the performance of fibers used in the textile industry. A higher degree of crystallinity improves the tensile strength and rigidity of PET, making it more durable. The crystalline materials have higher resistance to chemicals and solvents [7, 10–12]. The crystallinity of PET is favored by chemical structure, including the short CH_2 sequences, lack of asymmetric functional groups, and presence of benzene rings or polar ester groups [10]. Extrinsic factors like thermal treatment under mechanical stress increase fiber crystallinity and stability [12] (Fig. 11.2).

Although high crystallinity is beneficial for the strength, thermal stability, and durability of PET fibers in industrial applications [13], it significantly reduces the biodegradability of the material. This persistence in the environment contributes

Fig. 11.2 The chemical structure of polyethylene terephthalate (PET) consists of repeating units of ethylene glycol and terephthalic acid

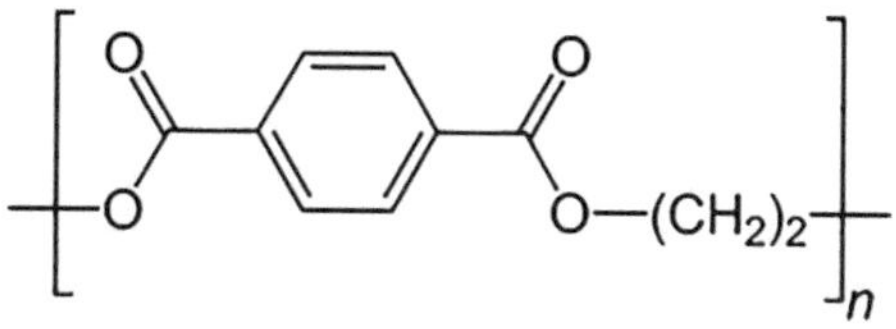

substantially to global pollution, as PET textiles do not break down easily and accumulate over time.

Another important drawback of high crystallinity is their decreased dyeability. Unstretched polyester yarns and fabrics made thereof can be dyed more easily with disperse dyes than crystalline fibers, as they have a higher number of amorphous areas. Therefore, the most effective and homogenous dyeing can be achieved with disperse dyes at dyeing temperatures which open the free volume (130 °C for exhaust dyeing and 200 °C for thermosol dyeing) after a pre-treatment by thermosetting. If these high temperatures are not possible, carriers can be used to accelerate the dyeing process [7]. By using enzymes, the dyeing of polyester could be made more efficient, as the result is a more hydrophilic fiber.

11.3.3 Ecological Issues and Recycling of PET

Polyester fibers face significant challenges in textile applications, primarily due to plastic waste, poor biodegradability, and high crystallinity.

Polyester contributes significantly to plastic waste, including microplastics (MP). Microplastics are plastic particles smaller than 5 mm [14], including fiber fragments released during textile production, processing, and use. Estimates of polyester microplastic release vary. The Eunomia study postulates an input of 190,000 tonnes of fibrous microplastic (FMP) per year [15] into the marine system. In countries with a three-step sewage treatment plant system, up to 84% (European Union) [16] of the microplastic fibers released into the environment during washing are retained. In Germany, a retention rate of up to 95% is even postulated [17].

One possible way to solve the problem of the increasing amount of plastic waste is to use biodegradable materials. Biodegradable polyesters such as PLA (polylactic acid) have similar properties like density or moisture absorption [18] and in principle could replace conventional PET or recycled PET (rPET), but has not yet been established on the market due to other difficulties with regard to textile applications and long term stability. However, most biodegradable plastics are only biodegradable under industrial composting conditions, which can be tested according to EN 13,432 [19], and not under real-life circumstances. The conditions to biodegrade PLA are high temperatures > 60 °C and a high humidity > 80% RH. Under these conditions, the primary degradation phase of PLA starts with chemical hydrolysis, which does not involve microorganisms. After that, the microorganisms in the soil start to digest the lower molecular weight oligomers and lactic acid, resulting in a production of carbon dioxide and water [20]. Furthermore, the biodegradability of PLA depends on the composition of the molecule, meaning the proportions of L-lactide and DL-lactide copolymers. For comparison: pure poly-L-lactide is approximately 37% crystalline and degrades much slower than the amorphous form. Consequently, it can take up to two years to disintegrate completely [13].

Another way to solve the problem of plastic pollution and microplastics is the application of polyester degrading enzymes, which are referred to as polyesterases (PETase). This group of enzymes comprises of e.g. hydrolases and oxidoreductases like cutinases, lipases, and esterases. Until today, around 80 known PET-active enzymes from bacteria and fungi have been established. More than 35 wild-type enzymes capable of catalyzing the partial degradation of PET into oligomers or even monomers have been identified [21, 22]. Furthermore, an archaeal PETase has been functionally verified [22]. Most PETases show increased catalytic activity around 65 °C but also at lower environmental temperatures [21]. Many of them show promising results regarding PET-degradation or surface modification.

11.3.4 Biotechnological Approach for Processing of Textiles as an Eco-Friendly Alternative

11.3.4.1 Enzymatic Mechanism for Degradation of PET

Enzymatic treatment of polyester involves the use of specific enzymes, such as polyesterases, which catalyse the breakdown of polyester molecules into smaller fragments. These enzymes target the ester bonds present in the polyester polymer chains, leading to their hydrolysis as shown in Fig. 11.3.

The biodegradation of polymers, particularly polyesters like PET, occurs through a two-stage enzymatic chain cleavage process involving oxidation and hydrolysis. This process begins with depolymerization, where polymer chains are cleaved outside the organism. Following this, mineralization occurs inside the cell, breaking down smaller oligomers into water, carbon dioxide, minerals, and biomass. The degradation mechanism can be classified into two types based on the enzyme's point of attack on the polymer chain [23]:

- **Endomechanism**: Enzymes break down monomers or short-chain oligomers without a favoured point of attack, statistically distributed along the chain
- **Exomechanism**: Enzymes cleave monomers or short-chain oligomers from the end of the chain.

$$\text{R-COOR'} \xrightarrow{\text{enzyme}} \text{R-COOH} + \text{R' -OH}$$

Fig. 11.3 Functionality of an enzymatic surface treatment of polyester

During the initial stages of biotechnological treatment with isolated enzymes, partial degradation of the polyester substrate occurs, producing both long polymer chains and shorter fragments. At this stage, the pH of the solution may drop slightly due to the release of acidic by-products from the hydrolysis reaction. However, the weight loss of the substrate is usually minimal. As the enzymatic treatment progresses, the polyester chains continue to be hydrolysed, leading to the accumulation of shorter fragments and the eventual depletion of long polymer chains. In this later stage, both the pH reduction and the weight loss become more pronounced, indicating a more advanced stage of polyester degradation.

As explained before, polyester fibers have significant crystalline regions, making enzymatic degradation challenging since enzymes, being large biomolecules, cannot penetrate these areas, as shown in Fig. 11.4 [24]. Commercial enzymes tested by the authors lack affinity for crystalline polymer chains and can only act on the fiber surface with difficulty. Controlling process kinetics allows regulation of polymer degradation, aiding biochemical recycling concepts, although the reactions are currently too slow for practical use [25–28].

If the enzymatic treatment is performed long enough, biodegradation of the polyester fibers can also be achieved. The degradation rate of polymers depends on several factors. A correlation can be established between the rate of degradation and the polymer melting temperature; as the rate decreases, the higher the melting temperature of the polymer gets [29].

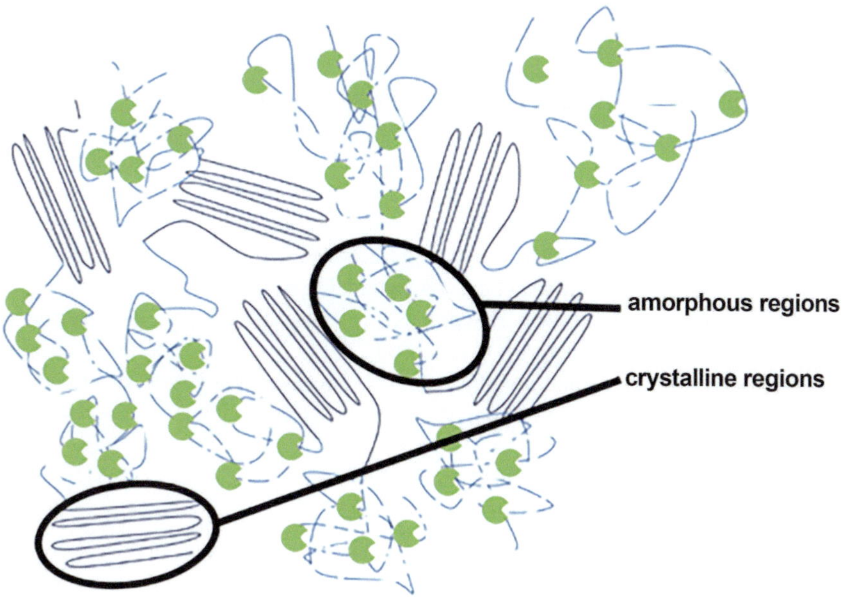

Fig. 11.4 Schematic representation of amorphous and crystalline areas of polyester and the attacking enzymes (shown here in green)

Understanding the enzymatic mechanisms behind the degradation of PET is crucial for the rational design and optimization of effective enzymes. Identified polyesterases, which play a significant role in PET degradation, share common structural features. These enzymes belong to the α/β hydrolase fold family, characterized by a conserved catalytic triad composed of serine, histidine, and aspartate residues. This catalytic triad is essential for the hydrolytic activity of the enzymes, facilitating the cleavage of ester bonds in PET [30, 31].

11.3.5 Comparison of Finishing Techniques for Polyester

In many textile applications, fibers with great hydrophilicity are needed. This property ensures, especially in fields of clothing, good wearing comfort, anti-static behavior, and moisture management. Besides that, it provides good dyeability. With the help of chemical modification of polyester fiber, a higher hydrophilicity performance can be achieved, and poor wettability can be overcome. The aim is to create a material that becomes more similar to cotton [32]. On a molecular basis, more functional groups, especially hydroxyl groups, must be created on the polyester surface. Hydroxyl groups can interact with other polar substances, especially with water molecules. Therefore, the existing aliphatic polymer-carbon chain needs to be disrupted and modified. To create this change in molecular structure in a conventional way, alkaline treatment dominates the hydrophilization of polyester. However, this can lead to weight losses and material damage [5]. Other options are using hydrophilic or antistatic finishing agents and plasma treatment. The latter involves high-energy consumption with a medium–high effect level and short durability of the formed functional groups [2].

As the existing processes have various disadvantages, new, gentler ways of hydrophilizing polyester surfaces are constantly being sought and are in the focus of application research. A sustainable method for surface modification is the use of enzymes like polyesterases. During enzyme treatment, hydrolysis at the ester bonds creates new hydroxyl and carboxyl groups, enhancing properties such as wettability, adhesion, and dyeability [33, 34]. This process prevents pilling and greying, similar to using cellulases in laundry detergents for cotton [25] [35]. It also improves dyeability with less surface damage and weight loss [36]. However, the results of enzymatic treatment regarding surface modification do not yet come close to the results of conventional methods.

The comparison of alkaline treatment with sodium hydroxide (NaOH), plasma, and enzymatic treatment are summarized in Table 11.1.

Table 11.1 The comparison of alkaline treatment with sodium hydroxide (NaOH), plasma and enzymatic treatment

NaOH	Plasma	Enzymatic treatment
Advantages		
• Effective	• The actual reaction is quick	• Is not consumed
• Cheap		• Reduction of water, energy and raw material consumption
		• Reduction of waste
		• Low temperatures
		• Specific reaction
		• Uniform surface effects
Disadvantages		
• High amount of chemicals	• Cost-intensive	• Current reaction times are too long
• Pitting corrosion	• Apparatus effort very high	• High costs • Effect depends on the degree of crystallinity
• Consumed after a single reaction	• Reaction times, including the evacuation and aeration of the used vacuum reactors, are too high for use in textile industry	

11.3.6 Analytical Methods for Monitoring PET Surface Modification and Degradation

The development of effective surface modification methods for polyester fibers necessitates comprehensive analytical techniques capable of characterizing and quantifying the modifications at the molecular, microstructural, and macroscopic levels. This chapter explores various analytical methods employed to assess surface modifications of polyester fibers, encompassing gravimetric, microscopic, spectroscopic, and chromatographic techniques. By elucidating the principles, applications, and advantages of these analytical methods, this article aims to provide insights into the diverse approaches for evaluating surface modifications of polyester fibers, thereby facilitating the advancement of tailored polyester materials with enhanced functionalities and performance attributes.

11.3.6.1 Gravimetric Method

During enzymatic treatment, textile fibers lose weight as the enzymes break down surface fibrils and other components. Gravimetric studies can assess the scale of weight loss by comparing the weight of textiles before and after enzymatic treatment. The degree of weight loss is an important indicator of the effectiveness of the enzymatic process and its impact on the fabrics‘ integrity and performance [37]. The gravimetric method is a simple, easy-to-handle technique that requires minimal apparatus.

11.3.6.2 Microscopy

Microscopy is the most suitable technique for analyzing the surface of textiles after enzymatic treatments with enzymes such as chitinases, lipases, and esterases. These enzymatic treatments can modify the fiber surface, making it smoother and less prone to pilling by gently etching the fiber surface and reducing the tendency for fibers to break and form pills. Depending on their resolution, microscopic techniques can visualize structural changes, such as porosity, at different scales.

Optical microscopy shows the overall surface topography and changes in fiber morphology after enzymatic treatment, providing a general overview of the fibers' appearance. Scanning electron microscopy (SEM) offers higher resolution images, revealing detailed surface features, such as fibril structure and roughness, and allowing for quantitative measurements of fiber dimensions and surface roughness [38–43]. Transmission electron microscopy (TEM) goes even further, providing insights into the internal structure and nanoscale alterations of the fibers [44]. Atomic force microscopy (AFM) can map the surface at the nanometer scale, detecting fine surface irregularities and changes in nanomechanical properties [42, 45–47]. Laser microscopy, including laser scanning confocal microscopy (LSCM), adds another dimension by providing high-resolution, three-dimensional images of the textile surface. This technique can penetrate deeper into the sample, allowing for the visualization of subsurface features and providing detailed information on the fiber's structural integrity and any enzymatic modifications that have occurred [48].

11.3.6.3 Spectroscopic Methods in Combination with Coloration Tests:

The microscopic techniques are limited when it comes to an understanding the molecular-level changes on the fiber surface. Therefore, spectroscopic techniques can provide a better understanding of the presence and contribution of specific functional groups. The introduction of groups like hydroxyl (-OH), carboxyl (-COOH), and amino (-NH_2) by enzymatic treatments increases the hydrophilicity of fibers. Enzymatic treatment can also improve the dyeability of fibers by creating new sites for dye attachment, enhancing dye uptake and colour fastness.

By employing spectroscopic methods such as Fourier Transform Infrared Spectroscopy (FTIR) [43, 49–51], X-ray Photoelectron Spectroscopy (XPS) [40, 47, 52], ultraviolet–visible spectroscopy (UV–Vis) [43, 53], and fluorescence spectroscopy [52, 54], researchers can gain detailed insights into the chemical modifications occurring on the fiber surface. These techniques allow for the identification and quantification of functional groups, elucidating the mechanisms through which enzymatic treatments alter fiber properties.

Combining UV–Vis spectroscopy with colouring methods provides important information about the enzymatic modification of polyester fibers. A methylene blue test is carried out to provide quantitative evidence of successful hydrophilization of the polyester yarn using enzymes. The treated yarn is coloured with the dye methylene blue. Essentially, this is a cationic dye that can bind to negatively charged molecules due to its positive charge. Partially negative charges are present on hydroxyl and real negative charges on carboxyl groups, which are to be formed by enzymatic treatment of the polyester. The higher the concentration of functional groups on the surface, the more dye is bound, which results in a decrease in the dye concentration in the reaction solution. The decrease can be determined photometrically [55, 56].

Another widely used test is the ninhydrin test. The yarns were washed and dried several times after enzymatic treatment. A ninhydrin test can be carried out to rule out the possibility of enzyme residues remaining on the yarns that could falsify the result of methylene blue test. The ninhydrin test can be used to detect proteins and therefore also enzymes. If proteins are present, they are visualised by a colour change, as the ninhydrin reacts with amino acids, polypeptides and proteins. This reaction can take up to three days. The result is a dark purple colour [57, 58].

11.3.6.4 Mass Spectrometry

Another powerful technique for identifying functional groups on the textile surface is mass spectrometry (MS). MS can analyse the molecular composition and structural changes of textile fibers after enzymatic treatment. The detection of functional groups on fiber surfaces can be used to assess the surface hydrophilicity. Also, the detection of remaining enzyme residues is possible.

Matrix-Assisted Laser Desorption/Ionization Mass Spectrometry (MALDI-MS) is useful for detecting large biomolecules like proteins and polymers on the fiber surface after treatment. This technique ionizes the sample with a laser, allowing for the analysis of enzymatic degradation products and newly introduced functional groups based on their mass-to-charge ratio [59, 60].

Electrospray Ionization Mass Spectrometry (ESI–MS) is ideal for smaller molecules, detecting enzymatic reaction products on the fiber surface. It ionizes the sample by applying high voltage to create an aerosol, providing detailed information about the molecular weight and structure of modification products [59].

Tandem Mass Spectrometry (MS/MS) techniques provide detailed insights into enzymatic modifications of textile fibers. Collision-induced dissociation (CID) or electron transfer dissociation (ETD) fragmentation of precursor ions generates characteristic product ions, elucidating specific molecular structures and chemical bonds. MS/MS analysis facilitates the identification of modification sites, cleavage patterns, and functional group changes induced by enzymatic treatment, offering comprehensive insights into surface chemistry and molecular transformations [60, 61].

11.3.6.5 Chromatographic Methods (HPLC)

High-Performance Liquid Chromatography (HPLC) is an analytical technique that separates molecular mixtures based on their interactions with a stationary phase under high pressure. It can identify various molecules formed, cleaved or being attached to the textile surface during enzymatic processes [62–64]. Detection can be enhanced using UV–Vis, fluorescence, or mass spectrometric detectors. Monitoring enzyme activity and stability with HPLC during enzymatic treatments provides critical information for optimizing enzymatic processes [65].

11.3.6.6 Research Examples

In order to investigate the effect of polyesterases on the modification of polyester and polylactic acid surfaces, various tests were carried out with both commercially available enzymes and recombinantly produced polyesterases (further designated as 2nd generation polyesterases). Initially, the tests were carried out on yarns and subsequently on textile surfaces at varied temperatures (20, 40, and 60 °C) and pH levels (7, 8, 9, and 11) for each enzyme type. The pH was adjusted with a buffer solution before the treatment and measured again afterwards.

It turned out that the commercially available enzymes were not effective enough to functionalize the materials on the surface with a measurable effect. Therefore, 2nd generation polyesterases were used [32, 66]. Here, the recombinant enzyme *Thermobifida fusca* cutinase 2 (TfCut2) was used (developed by the department of biotechnology HSNR) and also a cutinase from *Humicola insolens* (HiC) which was provided by BOKU. These enzymes were also tested in the same way as the enzymes available on the market and proved to be more effective. It could be shown that the amorphous areas of PET and PLA are attacked, recognizable by a rough surface as shown in Fig. 11.5.

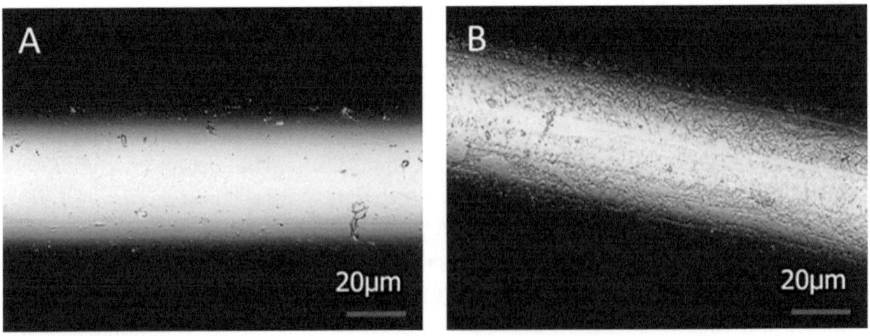

Fig. 11.5 Laser microscope image (20 μm) of an undrawn polyester yarn, untreated (**a**) and treated with TfCut2 (**b**), 24 h treatment time, pH 8 and 60 °C

The laboratory tests resulted in an optimum use of the enzymes at a pH value of 8 and a treatment time of 4 h. A temperature optimum of 60 °C was found for the HiC, and for the TfCut2 an optimum of 40 °C. The intensity of the surface modification was primarily determined using the methylene blue test and confirmed the results of the microscopic examination for undrawn yarn. By extending the incubation period, an undrawn PET yarn was completely degraded within 7 days. The surfaces of stretched yarns (PET and PLA) were also functionalized by enzyme treatment with HiC, which could be proved by absorption of methylene blue (Fig. 11.6). Although the results were significantly lower in comparison to the undrawn yarn which is due to the higher crystallinity.

In the next step, fabrics made of polyester were enzymatically treated (Fig. 11.7).

TfCut2 inserted new functional groups on the textiles. The temperature and the pH value are important factors for optimal enzymatic treatment and must be adjusted depending on the enzyme. This is important to note, as otherwise, the enzymes will not function properly or could even denature if the temperature is set too high. Furthermore, the treatment must be carried out under agitation to guide the enzymes to the yarn accordingly. The treatment of the textile surfaces shows slightly poorer results, as the yarns offer more surface area for the enzymes to attack.

11.3.6.7 Properties of Enzymatically Treated Polyester

Enzymatic treatment of polyester creates and improves certain properties. The most important achievement to mention is the improvement of wettability due to higher hydrophilicity as a result of new functional groups. The crystallinity is also increased with extended treatment time (30 and 120 min were tested). The reason for this are the amorphous areas, which are attacked and modified by the enzymes, resulting in statistically more crystalline areas. Infrared absorption spectroscopy (ATR-technique) also shows that there is an increase in functional groups (hydroxyl and carboxyl groups). At a wave number of 793 cm^{-1}, the presence of ester groups results in increased absorption. If correspondingly fewer ester groups are present because they have been cleaved, the absorption at this wave number is lower. This is offset by increased absorption at a wave number of 1406 cm^{-1} if hydrophilic regions (in this case OH and COOH groups) are present. It can be seen that increased absorption can be measured with longer treatment times. In addition, an increase in the polar components could be achieved, which was verified by a wettability test with a polar liquid. Good wettability is a particularly important factor in finishing processes, as it means that additional auxiliaries can be reduced [34].

To summarise these research examples, it can be stated that the 2nd generation enzymes already show significantly better results than the commercial available polyesterases. Nevertheless, a significant increase in function still needs to be achieved so that the surface functionalization process can be used in the textile industry on larger scale. Thus, the enzymes must be modified via biotechnological methods to achieve higher activity.

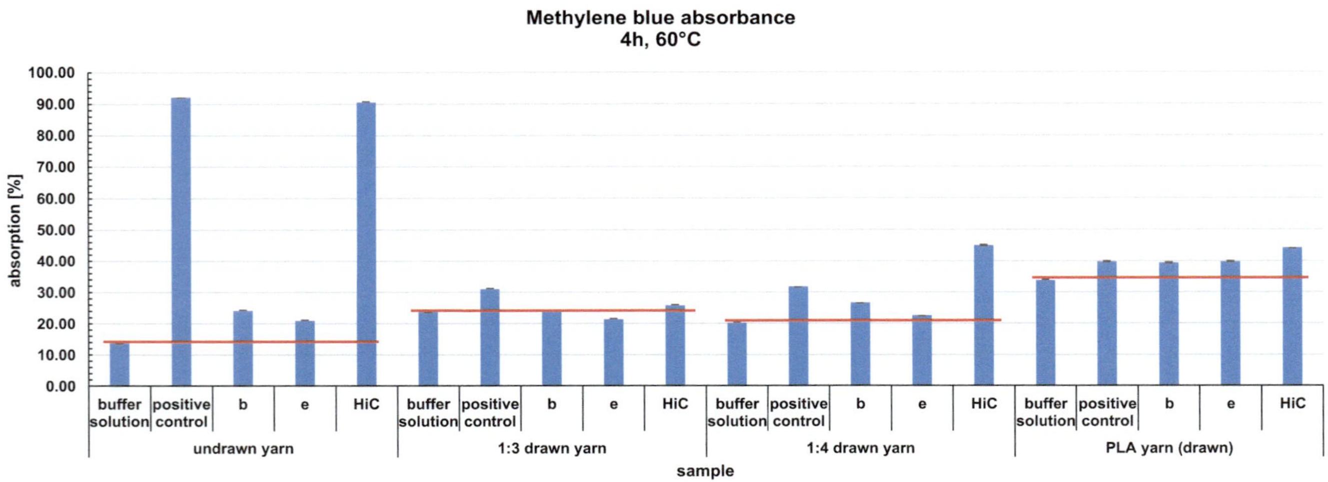

Fig. 11.6 Determination of the carboxyl group content of the fiber surfaces of enzymatically modified polyester yarns using the methylene blue test, 4 h 60 °C

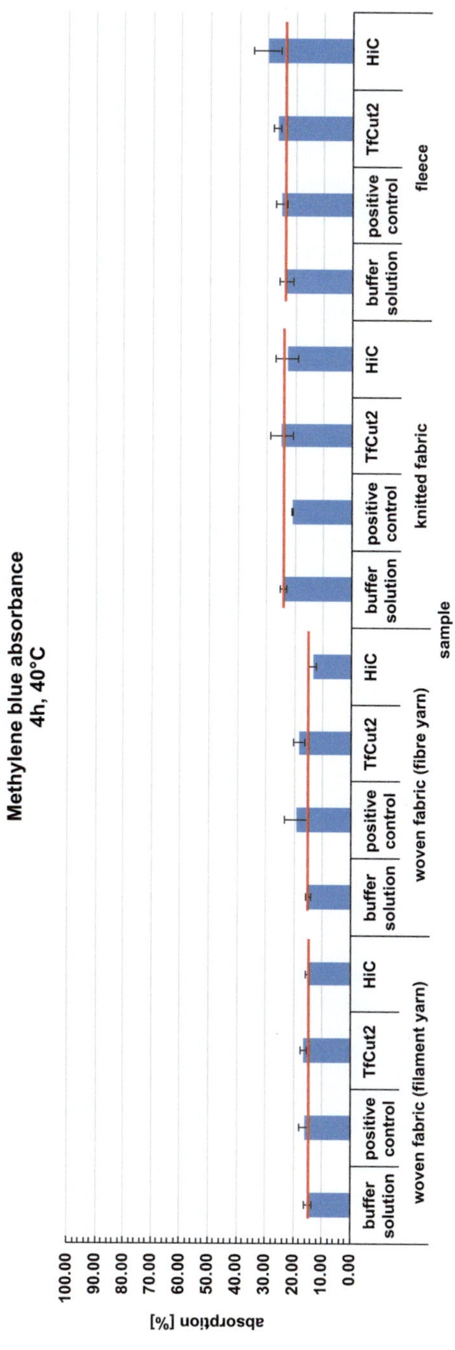

Fig. 11.7 Determination of the carboxyl group content of the fiber surfaces of enzymatically modified polyester fabrics using the methylene blue test, 4 h 40 °C

11.4 Biotechnological Process Development and Optimization of Enzymes

Development of processes for enzymatic treatment of fibers is challenging but fascinating at the same time. In the first step researchers need to find the enzymes that perform the desired modification. These enzymes can be sourced from industrial suppliers or obtained through recombinant expression in suitable microbial systems. Following enzyme acquisition, the development and optimization of the process can start. Every enzyme has preferences, which includes temperature, pH, osmolarity, presence of certain co-factors. In order to develop an efficient process, the optimal parameters have to be identified and maintained during the whole process. The concentration of the enzyme is also important because it is the main cost factor. Therefore, in order to decrease the cost, highly active proteins have to be found or developed. Another important factor is the duration of the process. The activity of enzymes can be improved due to protein engineering strategies which are shown in Fig. 11.8.

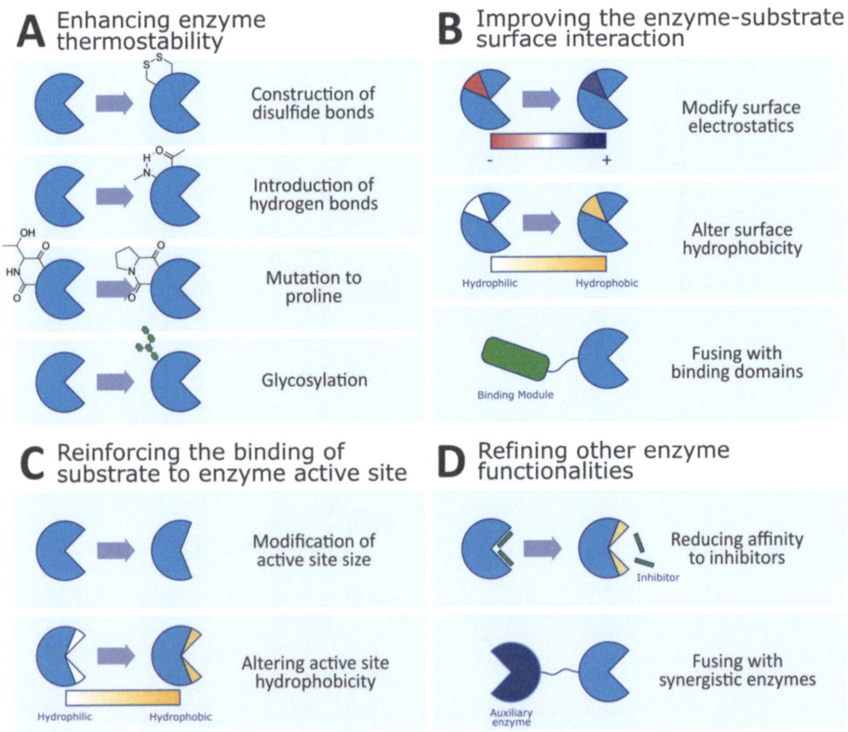

Fig. 11.8 Protein engineering strategies to improve plastic-degrading enzymes. **a**–**d**, reprinted from [67]

11.4.1 Latest Development

In nature, enzymes that degrade PET usually have the task of splitting natural polymer substrates with ester bonds, such as lipids. The properties of these enzymes are designed for the catalytic reaction of the natural substrates under the optimum temperature conditions of the natural enzyme producers and therefore often show lower activity compared to the unnatural substrate PET. Various strategies for enzyme optimization are being pursued to improve catalytic activity against PET. One approach is to improve thermostability. PET has a glass transition temperature (T_g) of approx. 65–80 °C. Above this temperature, the polymer chains exhibit greater flexibility and mobility, making them more susceptible to enzymatic reactions [68]. Increased thermostability of enzymes can be achieved by the introduction of disulphide bridges, salt bridges, hydrogen bonds, the insertion of proline residues or glycosylation (Fig. 11.8a) [69]. When modifying enzymes, however, care must be taken to ensure that the activity is not impaired by the stiffer enzyme structure.

There are some options to increase the interaction between the enzyme and the substrate (Fig. 11.8b). PET itself is hydrophobic. Increasing the hydrophobicity or changing the electrostatic properties of the active site or the enzyme surface can lead to higher substrate binding [72], [27]. It should be noted that the introduction of too many hydrophobic residues can lead to protein aggregation or changes in the protein structure, which in turn can lead to lower catalytic activity. An approach to increase catalytic activity is to improve enzyme-substrate binding (Fig. 11.8c) [69]. This can be achieved, for example, by opening up the active centre, which increases the accessibility of polymer substrates to the catalytic centre [70]. However, increasing the size of the active centre can also be accompanied by a reduction in substrate affinity and thus reduce the reaction rate [71]. The modification of the active centre should therefore be substrate-dependent and should be checked by in silico computer simulations as well as in vitro enzyme activity tests. Substrate binding can also be increased by attaching special binding domains. Some carbohydrate-binding modules (CBMs), which occur in natural carbohydrate-active enzymes, also bind to PET. The fusion of CBMs to enzymes that do not naturally possess CBMs, or the replacement of natural CBMs with CBMs with a higher affinity for PET, can ultimately also increase catalytic activity due to the proximity to the substrate. Also refining a lot of other enzyme functionalities is possible (Fig. 11.8d). To summarize, the approaches to activity-improving enzyme modifications are diverse and are often carried out in combination. Examples described in the literature are discussed in more detail in the following sections.

11.4.2 Limitation of Enzymatic Treatment

The research on enzymatic surface functionalization and microplastic degradation is hindered by the lack of commercially available enzyme systems. Many enzymes that are available on the market are limited to the amorphous regions of PET. They are hindered by dense structure of crystalline areas. Therefore, the processes for enzymatic modification and degradation needs to be carried out at glass transition temperature of PET, which varies depending on the used buffer and additives. The natural plastic-degrading enzymes are currently unsuitable for industrial applications involving synthetic polymers due to weak enzyme–substrate interactions, low catalytic activity, and poor thermostability [69]. Conventional methods like random mutagenesis for optimizing these enzymes have largely failed [73]. Thus, novel protein engineering strategies utilizing innovative experimental and computational approaches are necessary.

Enzymes also require careful storage and handling, their reuse is challenging, and their economic viability is often need improvisation. Despite these challenges, enzyme-based biotechnology offers an environmentally friendly and sustainable solution for the textile industry, but significant development is still required.

11.4.3 Latest Research and Developments in the Field of Enzymes

In recent years, inspiring results have been produced in searching, characterizing and engineering enzymes that can decompose PET. For the application of enzymes on an industrial scale, the enzymes must be able to withstand the often harsh conditions in the respective industry, and must be highly active in order to be used cost-effectively. Various techniques are used to adapt the enzymes to the high requirements and achieve ever better results by increasing protein expression, stability, and function. Recent research into enzymes for industrial applications has pursued a number of methods to drive innovation. Traditional techniques, including the exploration for novel microorganisms, are complemented by contemporary, computer-assisted strategies in protein engineering.

11.4.3.1 Utilization of Known Enzymes and Search for New, Suitable Enzymes E.G. Metagenome Analyses, Extremophilic / Thermophilic Organisms

Traditionally, enzymes are sourced from a variety of natural organisms with most natural PETase resources belonging to the group of bacteria and fungi, but PET-degrading enzymes have also been discovered in microalgae and archaea [22, 74–76]. These wild-type microorganisms that degrade PET have been found in various habitats such as soils, landfills and aquatic habitats [76] Diverse microbial communities of heterotrophs, autotrophs, predators and symbionts have been discovered on marine plastic particles, referred to as the ‘plastisphere’ [75, 77]. Naturally PET-degrading microorganisms and enzymes as well as optimized enzymes are listed systematically in several databases like the PAZy database; The plastics-active enzymes database (https://www.pazy.eu.), [21], PMBD: Plastics Microbial Biodegradation Database (http://pmbd.genome-mining.cn/home/), [78] or PlasticDB: a database of microorganisms and proteins linked to plastic biodegradation (https://plasticdb.org/), [79].

The direct screening of PETase-producing microorganisms in the laboratory can only utilize a small proportion of the natural potential. According to environmental microbiologists, only about 2% of naturally occurring microorganisms can be cultivated in research laboratories, leaving a large proportion of uncultivable fungi, bacteria and other microorganisms undiscovered. The development of computer tools for analysing large data sets and next-generation sequencing techniques enables metagenome analyses to be carried out [80]. Beyond traditional microbial sources, metagenomic analyses have proven invaluable in the search for novel enzymes [31, 81]. It enables the parallel analysis of many biological samples and the processing of millions of DNA/RNA fragments from one habitat, thus opening up the expansion of the natural potential for analysing unculturable microorganisms. Another way to find new PETases is to search for non-redundant protein homologs in metagenome databases like open-access MGnify metagenomic database (MGnify - EBI), [82].

The best-known example of the successful discovery of a highly active PETase via metagenome assays is the leaf compost cutinase (LCC) from a leaf compost metagenome [83]. It has attracted considerable attention due to its superior activity in the hydrolysis of polyethylene terephthalate (PET) compared to other known enzymes [84]. However, the authors report that the LCC reaction stopped when approaching the glass transition temperature with a conversion efficiency of only 31%, indicating that the thermostability of the LCC was a limiting factor [84]. As shown in this example, the stability and hydrolytic activity of naturally occurring enzymes usually do not meet the requirements of industrial applications.

Therefore, another approach focuses on extremophilic organisms—those thriving under extreme environmental conditions such as high temperatures or acidic conditions—which host enzymes (extremozymes) with exceptional stability and activity profiles. Thermophilic organisms, for example, grow successfully in high-temperature environments and produce thermostable enzymes that are ideal for processes requiring high heat. Thermophilic enzymes are especially advantageous for polyester depolymerization as higher temperatures increase flexibility and accessibility of polyester chains for enzymatic hydrolysis [85, 86].

11.4.3.2 Optimization and Design of Known Enzymes by Modification: Protein Engineering by Means of Rational Design or Directed Evolution

Recent advancements in computational biology and data science, including machine learning and deep neural networks, have revolutionized our capacity to analyse and interpret vast datasets. These tools are instrumental in identifying potential modifications that can lead to the improvement of enzyme properties. Using state-of-the-art methods known enzymes can be optimized by protein engineering, using both rational design and directed evolution. Although these approaches differ fundamentally in their design, they converge in their goals:

- Enhancing the enzyme thermostability
- Improvement of hydrolytic activity
- Modifying the substrate-binding site to increase binding
- Modifying enzyme surface to increase enzyme-PET contact

Rational design involves modifying the target protein by leveraging an understanding of its structural and mechanistic properties, alongside computational simulation and modelling. The prevalent application of rational design in the engineering of plastic-degrading enzymes is largely attributed to the availability of extensive structural and mechanistic data for these enzymes. Conversely, the absence of effective high-throughput screening methods constitutes a significant obstacle to the application of directed evolution in optimizing plastic-degrading enzymes [69].

Rational Design

Protein engineering through rational design involves deliberate modifications to a protein's structure to enhance specific characteristics such as activity, stability, or substrate specificity. The optimization focuses not only on enhancing the fundamental activity and stability of enzymes but also on tailoring them to meet specific industrial requirements effectively. This method utilizes a deep understanding of the protein's molecular structure to make educated predictions about the effects of specific alterations. Computational tools, including molecular simulation, deep learning models, and computational docking experiments, are pivotal in predicting how changes in amino acid sequences could impact the enzyme's function.

A cornerstone of rational design is the ability to predict and model enzyme–substrate interactions, which allows researchers to modify the enzyme in a way that optimizes these interactions. For instance, altering the substrate-binding pose can significantly increase the enzyme's efficiency and specificity. By predicting the substrate-binding molecular simulation and docking helped to identify candidate residues for engineering [87]. To improve LCCs activity and thermostability, protein modelling and molecular docking were employed to identify 11 amino acid positions for mutagenesis. [84]. Recent studies, such as those on the polyester-degrading enzyme LCC, have shown that strategic mutations can result in enhanced thermal stability and catalytic activity, crucial for industrial applications. Additionally, the study by [84] demonstrates that terephthalic acid, recovered from enzyme-mediated hydrolytic reactions, can be utilized to resynthesize virgin PET, thereby confirming the feasibility of enzyme-mediated PET recycling.

Directed Evolution

Directed evolution mimics natural evolutionary processes in the laboratory to evolve proteins with enhanced or novel functions. This technique does not require prior knowledge of the protein structure; instead, it generates a wide array of variants and selects those with desirable traits. The process involves creating a library of mutations, expressing these variants, and screening them to identify those that exhibit improved characteristics. The power of directed evolution lies in its ability to explore a vast genetic landscape, where even unforeseen beneficial mutations can be discovered [88]. Combining random mutagenesis with high-throughput screening methodologies allows for the efficient evaluation of thousands of enzyme variants. Directed evolution is labour-intensive and relies heavily on efficient high-throughput screening to assess numerous variants. This necessity poses challenges for PET-degrading enzymes, where standard efficacy tests involve HPLC analysis; however, [89] mitigated this issue by developing a novel PET analog that allows quick enzyme activity assessment through the measurement of fluorescence from the hydrolyzed product. Moreover, machine learning has become an invaluable tool in directed evolution, helping to predict the outcomes of mutations more accurately and suggesting potential candidates for further testing. Machine learning is a subcategory of artificial intelligence where algorithms recognize patterns within a data set and make predictions. Machine learning methods support directed evolution by predicting how sequences correlate with function and using characterized variant properties to select sequences with likely improved properties without needing a detailed model of the underlying physics or biology [90]. In the future, enhanced screens and selections that facilitate higher throughput or enable more continuous rounds of evolution can expand the exploration of the fitness landscape. Simultaneously, innovative mutagenesis strategies can improve the efficiency of the search process.

Both rational design and directed evolution serve crucial roles in the field of protein engineering, each with its strengths and limitations. Rational design is suited for targeted modifications guided by detailed structural knowledge, while directed evolution offers a broader approach to discovering novel mutations that confer desired traits. Together, these methodologies contribute significantly to the optimization of enzymes for various industrial processes, including the degradation of synthetic polymers such as PET, thereby supporting environmental sustainability and recycling efforts.

11.5 Outlook

Enzymatic surface modification and biodegradation of polyester, particularly PET, is a rapidly advancing research area with significant potential for the textile industry. Implementing enzymes in textile processing can drastically reduce the use of hazardous chemicals, energy consumption and mitigate waste pollution including the release of microplastics in the environment. Future biotechnology holds promise for addressing PET waste accumulation through the development of circular recycling processes, where PET can be enzymatically degraded into monomers and subsequently used for the enzymatic polymerization of novel PET. During textile processing, small particles from textiles accumulate in the machines, necessitating frequent cleaning. This results in extended downtimes and additional costs, as well as reduced machine efficiency. Enzymatic cleaning can alleviate these issues by effectively removing the particles, thereby reducing downtime and increasing overall production efficiency.

By leveraging enzymatic degradation, a closed-loop system can be established, efficiently converting polyester waste back into raw materials. This approach reduces reliance on virgin resources and minimizes environmental impact. Furthermore, cascading the utilization of residual materials for producing biochar or green hydrogen can improve the CO_2 footprint, contributing to the overall sustainability of the system [91].

The high degree of crystallinity of PET presents a significant challenge for enzymatic degradation, as it hinders enzyme access to the polymer backbone. Reducing the crystallinity of PET can be achieved by processing the fibers at the glass transition temperature of around 70 °C. However, not many enzymes maintain stability at this elevated temperature.

While this chapter highlights recent advancements, it also emphasizes the vast potential for discovering novel enzymes to address the global issue of plastic waste. Continued progress in enzymology and molecular biology will be essential for adapting enzymes to meet industrial needs, paving the way for more effective and sustainable solutions.

Figure 11.9 shows the development of enzymatic processing in the textile industry. Firstly, there is a focus on developing economically viable enzyme processes suitable for large-scale industrial implementation. This highlights the industry's drive towards sustainable practices while ensuring cost-effectiveness. Secondly, supplementary enzyme assistance is emphasized, indicating a trend towards leveraging multiple enzymes to enhance the efficiency and effectiveness of textile enzymatic treatments. This approach reflects ongoing efforts to optimize enzymatic processes for improved outcomes.

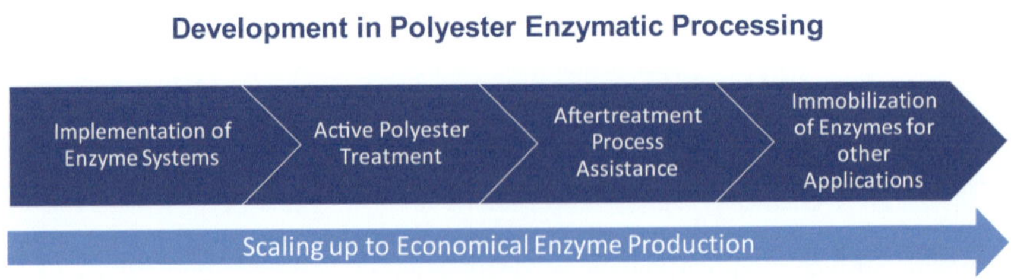

Fig. 11.9 Development in application of enzymes for textile processing

The concept of combined enzymatic pre-treatment, which integrates different enzymatic processes to achieve synergy effects in textile processing, is also significant. This integrated approach underlines the industry's quest for comprehensive and efficient enzymatic treatment methods.

Enzyme immobilization is a significant strategy in enzymatic processing, stabilizing enzymes, extending their lifespan, and enhancing their efficiency. This technique involves attaching enzymes to support materials, reducing their mobility while retaining catalytic activity, enabling their repeated use. Recent advances in protein engineering have led to more efficient industrial catalysts, with enzyme immobilization playing a key role in enhancing their properties. This approach contributes to the economic viability, sustainability, and efficiency of enzymatic processes, particularly in industries like textiles.

References

1. CIRFS. (2024). CIRFS European Man-made Fibres Association. https://www.cirfs.org/about-cirfs-1/activities. Accessed Mär 2024.
2. Gamerith, C., Gajda, M., Ortner, A., Acero, E. H., Guebitz, G., & Ulbricht, M. (2017). Enzymatic hydrolysis of poly(ethylenterephthalate) used for and analysed by pore modification of track-etched membranes. *New Biotechnology, 39*, 42–50.
3. Galuschka, A. (2015). *Biochemie für Ahnungslose.* Stuttgart.
4. Madhu, A., & Chkraborty, J. N. (2017). Developments in application of enzymes for textile processing. *Journal of Cleaner Production, 145*(1), 114–133.
5. Shen, J., & Smith, E. (2015). Enzymatic treatments for sustainable textile processing. In R. Blackburn (Ed.), *Sustainable Apparel: Production, Processing and Recycling*, Cambridge.
6. Korkmaz, G., Kılınç, M., Kılınç, N., & Kut, D. (2024). The role of surface modification methods for sustainable textiles. In *Roadmap to Sustainable Textiles.* IntechOpen. https://doi.org/10.5772/intechopen.107551.
7. Tetzlaff, G., Wulfhorst, B., Koch & Albrecht, W. (1993). Faserstoff-Tabellen. Polyesterfasern nach P.-A. Koch. In *Faserstoff-Tabellen.* Frankfurt am Main, Dt. Fachverlag.
8. Zhang, C. (2014). Biodegradable polyesters: synthesis, properties, applications. In *Biodegradable Polyesters.* Auckland.
9. Loy, W. (2008). *Chemiefasern für technische Textilprodukte.* Frankfurt am Main.
10. Loy, W. (1978). Chemiefaserstoffe, Berlin.
11. Cook, J. G. (2001). *Handbook of Textile Fibres: Man-made Fibres*, vol. II. Cambridge: Woodhead Publishing
12. Mather, R., & Wardman, R. (2010). Chemistry of textile fibres. Cambridge.
13. East, J. (2009). The structure of polyester fibers. In S. J. Eichhorn (Eds), *Handbook of Textile Fibre Structure: Vol 1. Fundamentals and Manufactured Polymer Fibres*, Oxford.
14. Boucher, J., Friot, D. (2017). *Primary microplastics in the oceans: a global evaluation of sources.* IUCN.
15. Eunomia. (2016). *Plastics in the marine environment*. Eunomia, Bristol.
16. Hann, S., Sherrington, C., Jamieson, O., Hickman, M., Kershaw, P., Bapasola A., & Cole, G. (2018). *Investigation options for reducing releases in the aquatic environment of microplastics emitted by (but not intentionally added in) products.* Eunomia.
17. Bertling, J., Bertling R., & Hamann, L. (2018). Kunststoffe in der Umwelt: Mikro- und Makroplastik: Ursachen, Mengen, Umweltschicksale, Wirkungen, Lösungsansätze, Empfehlungen.
18. Sabale A. G, Rane, V. M & Toke, M. (2019). PLA (Polylactid Acid): New Era of Textiles. www.fibre2fashion.com

19. DIN EN 13432 Berichtigung 2:2007–10 Packaging - *Requirements for packaging recoverable through composting and biodegradation - Test scheme and evaluation criteria for the final acceptance of packaging,* Berlin: Beuth Verlag, 2007.
20. Mochizuki, M. (2009). Synthesis, properties and structure of polylactic acid fibres. In *Handbook of Textile Fibre Structure: Vol 1. Fundamentals and Manufactured Polymer Fibres.*
21. Buchholz, P. C. F., Feuerriegel, G., Zhang, H., Perez-Garcia, P., Nover, P. P., Chow, J., Streit, W. R., & Pleiss, J. (2022). Plastics degradation by hydrolytic enzymes: The plastics-active enzymes database-PAZy. *Proteins,* 1443–1456.
22. Perez-Garcia, P., Chow, J., Costanzi, E., Gurschke, M., Dittrich, J., Dierkes, R. F., Molitor, R., Applegate, V., Feuerriegel, G., Tete, P., Danso, D., Thies, S., Schumacher, J., Pfleger, C., Jaeger, K.-E., Gohlke, H., Smits, S. H., Schmitz R. A., & Streit W. R. (2023) An archaeal lid-containing feruloyl esterase degrades polyethylene terephthalate. *Communications Chemistry,* 1–13.
23. Kaplan, D. Fundamentals of Biodegradable Polymers. In C. Ching (Ed.), *Biodegradable Polymers and Packaging.* Basel.
24. Xu, R., Xie, J., & Lei, C. (2017). Influence of melt-draw ratio on the crystalline behaviour of a Polylactic Acid Casr film with a chi structure. *RSC Advances, 7*(63), 39914–39921.
25. Perz, V., Zumstein, M., Sander, M., Zitzenbacher, S., Ribitsch, D., & Guebitz, G. (2015). Biomimetic approach to enhance enzymatic hydrolysis of the synthetic polyester poly(1,4-butylene adipate): fusing binding modules to esterases. *Biomacromolecules, 16*(12), 3889–3896.
26. Ribitsch, D., Yebra, A., Zitzenbacher, S., Wu, J., Nowitsch, S., Steinkeller, G., Greimel, K., Doliska, A., Oberdorfer, G., Gruber, C., Gruber, K., Schwab, H., Stana-Kleinschek, K., Acero, E., & Guebitz, G. (2013). Fusion of binding domains to Thermobifida cellulosilytica cutinase to tune sorption characteristics and enhancing PET hydrolysis. *Biomacromolecules, 14*(6), 1769–1776.
27. Acero, E., Ribitsch, D., Dellacher, A., Zitzenbacher, S., Marold, A., Steinkellner, G., Gruber, K., Schwab, H., & Guebitz, G. (2013). Thermobifida cellulosilytica for improved polyester hydrolysis. *Biotechnology and Bioengineering, 110*(10), 2581–2590.
28. Espino-Rammer, L., Ribitsch, D., Przylucka, A., Marold, A., Greimel, K., Acero, E., Guebitz, G., Kubicek, C., & Druzhinina, I. (2013). Two novel class ii hydrophobins from trichoderma spp. stimulate enzymatic hydrolysis of poly(ethylene terephthalate) when expressed as fusion proteins. *Applied and Environmental Microbiology, 79*(14), 4230–4238.
29. Tokiwa, Y., & Suzuki, T. (1981). Hydrolysis of copolyesters containing aromatic and aliphatic ester blocks by lipase. *Journal of Applied Polymer Science, 26*, 441–448.
30. Richter, P. K., Blázquez-Sánchez, P., Zhao, Z., Engelberger, F., Wiebeler, C., Künze, G., & Sonnendecker, C. (2023). Structure and function of the metagenomic plastic-degrading polyester hydrolase PHL7 bound to its product. *Nature Communications.*
31. Danso, D., Schmeisser, C., Chow, J., Zimmermann, W., Wei, R., Leggewie, C., & Streit, W. R. (2018). New insights into the function and global distribution of polyethylene terephthalate (PET)-degrading bacteria and enzymes in marine and terrestrial metagenomes. *Applied and Environmental Microbiology*, 1–13.
32. Hofmann, I., Singer, I., Kolbe, S., & Rabe, M. (2023). Increased surface functionality through enzymatic treatments of bio-based and conventional polyesters. In *Aachen-Dresden-Denkendorf International Textile Conference*, Dresden.
33. Kumar, J. A., Kumar, M. S. (2019). A study on improving Dyeability of polyester fabric using lipase enzyme. *AUTEX Research Journal,* 1–7.
34. Kardas, I., Lipp-Symonowicz, B., & Sztajnowski, S. (2011). The influence of enzymatic treatment on the surface modification of PET Fibers. *Journal of Applied Polymer Science, 119*, 3117–3126.
35. Lee, S. H., & Song, W. S. (2011). Enzymatic Hydrolysis of Polylactid Acid Fiber. *Applied Biochemistry and Biotechnology, 164*(1), 89–102.
36. Kumar, D., Bhardwaj, R., Jassal, S., Goyal, T., Khullar, A., & Gupta, N. (2021). Application of enzymes for an eco-friendly approach to textile processing. *Environmental Science and Pollution Research.*

37. Esfandiari, A., Firouzi-Pouyaei, E., & Aghaei-Meibodi, P. (2014). Effect of enzymatic and mechanical treatment on combined desizing and bio-polishing of cotton fabrics. *The Journal of the Textile Institute*, 1193–1202.
38. Mahltig, B., & Grethe, T. (2022). High-performance and functional fiber materials—a review of properties, scanning electron microscopy SEM and electron dispersive spectroscopy EDS. *Textiles*, 209–251.
39. Chippindale, P. (2008). Replication of textile fibres in electron microscopy. *Journal of the Textile Institute Transactions,* T392-T397.
40. George, M., Mussone, P. G., & Bressler, D. C. (2014). Surface and thermal characterization of natural fibres treated with enzymes. *Industrial Crops and Products*, 365–373.
41. Qi, H., Chen, H., Mao, K., Qiu, Z., Zhang, L., & Zhou, J. (2019). Investigation of the structure of ramie fibers by enzymatic peeling. *Cellulose*, 2955–2968.
42. Nguyen, H. L., Bechtold, T., Fabbri, F., Pellis, A., Guebitz, G. M., & Pham, T. (2022). Characterisation of enzyme catalysed hydrolysation stage of poly (lactic acid) fibre surface by nanoscale thermal analysis: New mechanistic insight. *Materials & Design*, 1–11.
43. Zwane, P. E., Ndlovu, T., Mkhonta, T. T., Masarirambi, M. T., & Thwala, J. M. (2019). Effects of enzymatic treatment of sisal fibres on tensile strength and morphology. *Scientific African*, 1–7.
44. Radulescu, I. R., Surdu, L., Scarlat, R., Constantin, C., Mitu, B., Morari, C., & Costea, M. (2021). Modelling the woven structures with inserted conductive yarns coated with magnetron plasma and testing their shielding effectiveness. *Textiles*, 4–20.
45. George, M., Mussone, P. G., Abboud, Z., & Bressler, D. C. (2014). Characterization of chemically and enzymatically treated hemp fibres using atomic force microscopy and spectroscopy. *Applied Surface Science*, 1019–1025.
46. Lee, J. M., Heitmann, J. A., & Pawlak, J. J. (2007). Local morphological and dimensional changes of enzyme-degraded cellulose materials measured by atomic force microscopy. *Cellulose*, 643–653.
47. Karaca, B., & Özdoğan, E. (2013). A study on surface characterization of enzyme treated polyethylene terephthalate fibers by Xps and AFM. *TEKSTİL VE KONFEKSİYON,* 16–22.
48. Song, Y., Srinivasarao, M., Tonelli, A., Balik, C. M., & McGregor, R. (2000). Laser scanning confocal microscopy study of dye diffusion in fibers. *Macromolecules, 33*(12), 4478–4485.
49. Jayamani, E., Loong, T. G., & Bakri, M. K. B. (2020). Comparative study of Fourier transform infrared spectroscopy (FTIR) analysis of natural fibres treated with chemical, physical and biological methods. *Polymer Bulletin*, 1605–1629.
50. Wojciechowska, E., Rom, M., Włochowicz, A., Wysocki, M., & Wesełucha-Birczyńska, A. (2004). The use of Fourier transform-infrared (FTIR) and Raman spectroscopy (FTR) for the investigation of structural changes in wool fibre keratin after enzymatic treatment. *Journal of Molecular Structure*, 315–321.
51. Garside, P., & Wyeth, P. (2003). Identification of cellulosic fibres by FTIR spectroscopy-thread and single fibre analysis by attenuated total reflectance. *Studies in conservation*, 269–275.
52. Ramesh, D., Ayre, B. G., Webber, C. L., & D'Souza, N. A. (2015). Dynamic mechanical analysis, surface chemistry and morphology of alkali and enzymatic retted kenaf fibers. *Textile Research Journal*, 2059–2070.
53. Jus, S., Guebitz, G. M., & Kokol, V. (2010). Enzyme-catalysed coupling of functional antioxidants onto protein fibres. *Medical and Healthcare Textiles,* 126–134.
54. Peciulyte, A., Kiskis, J., Larsson, P. T., Olsson, L., & Enejder, A. (2016). Visualization of structural changes in cellulosic substrates during enzymatic hydrolysis using multimodal nonlinear microscopy. *Cellulose,* 1521–1536.
55. Bekka, E. (2022). *Zur Toxizität der chromodiagnostischen Farbstoffe Methylenblau.* Toluidinblau und Indigokarmin in vitro.
56. Rouette, H. K. (2001). *Encyclopedia of Textile Finishing*. Aachen.
57. Bader, H. J & Rothweil, M. (2009). *Forensische Chemie-Mit Chemie auf Verbrecherjagd. Chemitry and Industry for Teachers in European Schools*

58. Haeger, G., Bongaerts, J., & Siegert, P. (2022). A convenient ninhydrin assay in 96-well format for amino acid-releasing enzymes using an air-stable reagent. *Analytical Biochemistry, 654.*
59. Rizzarelli, P., & Carroccio, S. (2014). Modern mass spectrometry in the characterization and degradation of biodegradable polymers. *Analytica Chimica Acta*, 18–43.
60. Chiliveri, S. R., & Linga, V. R. (2014). A novel thermostable, alkaline pectate lyase from Bacillus tequilensis SV11 with potential in textile industry. *Carbohydrate polymers*, 264–272.
61. Feng, C., Sultana, N., Sui, X., Chen, Y., Brooks, E., & Vinueza, N. R. (2022). High-resolution mass spectrometry analysis of reactive dye derivatives removed from biodegraded dyed cotton by chemical and enzymatic methods. *AATCC Journal of Research.*
62. Pinelli, P., Ieri, F., Vignolini, P., Bacci, L., Baronti, S., & Romani, A. (2008). Extraction and HPLC analysis of phenolic compounds in leaves, stalks, and textile fibers of Urtica dioica L. *Journal of agricultural and food chemistry*, 9127–9132.
63. Asgher, M., Yasmeen, Q., & Iqbal, H. M. N. (2014). Development of novel enzymatic bioremediation process for textile industry effluents through response surface methodology. *Ecological Engineering*, 1–11.
64. Phugare, S. S., Kalyani, D. C., Surwase, S. N., & Jadhav, J. P. (2011). Ecofriendly degradation, decolorization and detoxification of textile effluent by a developed bacterial consortium. *Ecotoxicology and Environmental Safety*, 1288–1296.
65. Michiniewicz, A., Ledakowicz, S., Ulrich R., Hofrichter, M. (2008). Stanisław. *Dyes and Pigments,* 295–302.
66. Singer, I. (2022). *Untersuchung von Prozessparametern für eine enzymatische Oberflächenbehandlung von Polyester.* Mönchengladbach.
67. Choi, J., Kim, H., Ahn, Y. R., Kim, M., Yu, S., Kim, N., & Kim, H. O. (2024). Recent advances in microbial and enzymatic engineering for the biodegradation of micro-and nanoplastics. *RSC Advances*, 9943–9966.
68. Wei, R., & Zimmermann, W. (2017). Microbial enzymes for the recycling of recalcitrant petroleum-based plastics: How far are we? *Microbial Biotechnology, 10*(11), 1308–1322.
69. Zhu, B., Wang, D., & Wei, N. (2022). Enzyme discovery and engineering for sustainable plastic recycling. *Trends in Biotechnology*, 22–37.
70. Araújo, R., Casal, M., & Cavaco-Paulo, A. (2008). Application of enzymes for textile fibres processing. *Biocatalysis and Biotransformation, 26*(5), 332–349.
71. Kawabata, T., Oda, M., & Kawai, F. (2017). Mutational analysis of cutinase-like enzyme, Cut190, based on the 3D docking structure with model compounds of polyethylene terephthalate. *Journal of Bioscience and Bioengineering*, 28–35.
72. Ma, Y., Yao, M., Li, B., Ding, M., He, B., Chen, S., & Yuan, Y. (2018). Enhanced poly (ethylene terephthalate) hydrolase activity by protein engineering. *Engineering*, 888–893.
73. Tan, L.-T., Hiraishi, T., Sudesh, K., & Maeda, M. (2013). Directed evolution of poly(R)-3-hydroxybutyrate depolymerase using cell surface display system: Functional importance of asparagine at position 285. *Applied Microbiology and Biotechnology, 97*, 4859–4871.
74. Srikanthe, M., Sandeep, T., Sucharitha, K., & Godi, S. (2022). Biodegradation of plastic polymers by fungi: A brief review. *Bioresources and Bioprocessing*, 42–52.
75. Zeghal, E., Vaksmaa, A., Vielfaure, H., Boekhout, T., & Niemann, H. (2021). The potential role of marine fungi in plastic degradation—A review. *Sec Aquatic Microbiology*, 1–17.
76. Benavides Fernández, C. D., Guzmán Castillo, M. P., Quijano Pérez, S. A., & Carvajal Rodríguez, L. V. (2022). Microbial degradation of polyethylene terephthalate: A systematic review. *SN Applied Sciences.*
77. Zettler, E. R., Mincer, T. J., & Amaral-Zettler, L. A. (2013). Life in the "plastisphere": microbial communities on plastic marine debris. *Environmental Science & Technology*, 7137–7146.
78. Gan, Z., & Zhang, H. (2019). PMBD: a comprehensive plastics microbial biodegradation database. *Database*, 1–11.
79. Gambarini, V., Pantos, O., Kingsbury, J. M., Weaver, L., Handley, K. M., & Lear, G. (2022). PlasticDB: A database of microorganisms and proteins linked to plastic biodegradation. *Database*, 1–12.

80. Reddy, C. N., Kallem, P., Mounika, K., Muqeet, A., Raj, J. C. J., Aishwarya, C. V. S., & Banat, F. (2023). Review of microplastic degradation: Understanding metagenomic approaches for microplastic degrading organisms. *Polymer Testing*, 1–13.
81. Eiamthong, B., Meesawat, P., Wongsatit, T., Jitdee, J., Sangsri, R., Patchsung, M., & Uttamapinant, C. (2022). Discovery and genetic code expansion of a polyethylene. *Bioremediation*, 1–11.
82. White, M. F., & Wallace, S. (2023). A new PETase from the human saliva metagenome and its functional modification via genetic code expansion in bacteria. *Angewandte Chemie International Edition*, 1–3.
83. Sulaiman, S., Yamato, S., Kanaya, E., Kim, J. J., Koga, Y., Takano, K., & Kanaya, S. (2012). Isolation of a novel cutinase homolog with polyethylene terephthalate-degrading activity from leaf-branch compost by using a metagenomic approach. *Applied and Environmental Microbiology*, 1556–1562.
84. Tournier, V., Topham, C. M., Gilles, A., David, B., Folgoas, C., Moya-Leclair, E., Desrousseaux, M. L Texier, H., Gavalda, S., Cot, M., Guémard, E., Dalibey, M., Nomme, J., Cioci, G., Barbe, S., Chateau, M., André, I., Duquesne, S., & Marty, A. (2020). An engineered PET depolymerase to break down and recycle plastic bottles. *Nature*, 216–219
85. Atanasova, N. (2021). Plastic degradation by extremophilic bacteria. *International Journal of Molecular Sciences*, 1–19.
86. Williams, G. B., Ma, H., Khusnutdinova, A. N., Yakunin, A. F., & Golyshin, P. N. (2023). Harnessing extremophilic carboxylesterases for applications in polyester depolymerisation and plastic waste recycling. *Essays in Biochemistry*, 715–729.
87. Guo, R. T., Li, X., Yang, Y., Huang, J. W., Shen, P., Liew, R. K., & Chen, C. C. (2024). Natural and engineered enzymes for polyester degradation: A review. *Environmental Chemistry Letters*, 1275–1296.
88. Packer, M. S., & Liu, D. R. (2015). Methods for the directed evolution of proteins. *Nature Reviews Genetics*, 379–394.
89. Shi, L., Liu, P., Tan, Z., Zhao, W., Gao, J., Gu, Q., & Zhu, L. (2023). Complete depolymerization of PET wastes by an evolved PET hydrolase from directed evolution. *Angewandte Chemie International Edition*, 1–11.
90. Yang, K. K., Wu, Z., & Arnold, F. H. (2019). Machine-learning-guided directed evolution for protein engineering. *Nature Methods*, 687–694.
91. Sonnendecker, C., Oeser, J., Richter, P. K., Hille, P., Zhao, Z., Fischer, C., & Zimmermann, W. (2022). Low carbon footprint recycling of post-consumer PET plastic with a metagenomic polyester hydrolase. ChemSusChem *15*.

Chapter 12
Advancements in Plasma Technology for Textile Surface Modification

Priyanka Gupta and Ankur Shukla

Abstract Plasma treatment in textile finishing is an innovative, eco-friendly approach that transforms traditional fabric modification methods. This chapter examines plasma technology, its advancements, and its applications in textile finishing. It provides an overview of plasma and its properties, detailing different types used in textile processing, including atmospheric and low-pressure plasma. The mechanisms behind plasma treatment are explored, particularly its role in modifying surface properties such as wettability, dyeability, and adhesion. The environmental benefits of plasma treatment, including reduced chemical and water usage, are also discussed. The chapter highlights opportunities for customization and innovation, paving the way for future developments in textile manufacturing. It concludes by addressing current challenges and future prospects, offering a comprehensive understanding of plasma treatment in textile finishing.

Keywords Functional fabric · Plasma treatment · Surface modification · Sustainability · Textile finishing

12.1 Introduction

Molecules, atoms, ions, radicals, and other reactive species make up plasma, the fourth state of matter. The term plasma was coined by Irving Langmuir in 1928 [1]. Most of the universe's apparent matter, including interstellar matter and the interiors of stars, is actually in a plasma state. Most people associate plasma with high-energy applications such as nuclear fusion. The application of plasma has recently grown in various fields, such as plasma television sets, fluorescent lighting tubes, and microelectronic circuits and microprocessors manufacturing [2–4].

P. Gupta
Dr. B.R. Ambedkar National Institute of Technology, Jalandhar, Punjab, India
e-mail: guptap@nitj.ac.in

A. Shukla (✉)
Venus Safety & Health Pvt. Ltd., Navi Mumbai, Maharashtra, India
e-mail: ankurshukla1893@gmail.com

M. Shahid et al. (eds.), *Advancements in Textile Finishing*,
https://doi.org/10.1007/978-981-96-6385-9_12

Plasma has numerous applications in various fields. Plasma technology is widely used worldwide for energy production from trash, biomass, and coal because of its exceptional and distinctive characteristics. Plasma-assisted combustion, gasification, and pyrolysis are distinct techniques employed for energy generation [5]. Plasma Medicine is an emerging discipline within the medical sector that significantly contributes to various critical aspects of patient care. It integrates the principles of plasma physics, biology, and clinical medicine [6]. In the past, plasmas were utilized in the field of medicine for electrosurgery purposes, specifically for cauterization and non-contact hemostasis. Nonthermal plasmas are widely utilized in medicine because of their efficacy and compatibility with biological systems.

When a gas that is neither positively nor negatively charged is excited, plasma is created. Multiple methods exist to provide the requisite energy for plasma formation to a neutral gas. First method is providing heat energy, which leads to the creation of charge carriers. Another method of providing energy to the gas reservoir is using energetic beams that decrease intensity within a gas volume. For technological and technical purposes, the predominant technique for creating and maintaining a low-temperature plasma involves the introduction of an electric field to a gas that is initially neutral [7]. Plasma generation necessitates using a plasma discharge apparatus called a plasma cathode. Plasma cathodes can generate higher levels of current density compared to hot cathodes. These devices possess the capability to release electrons in a pulsed manner and are not affected by poor vacuum conditions. Initially, plasma cathodes were seen as a potential substitute for hot cathodes in the early stages of research. However, their applicability has since been restricted to a rather narrow field. This technique employs a discharge mechanism featuring a negative potential drop near the anode. Hence, the emitted electrons will reach the anode solely if they surpass a potential barrier. Consequently, this leads to the emission of electrons, also known as plasma [8].

The current chapter shows various methods for generating plasma and the latest advancements in applying plasma treatment on textiles. Plasma treatment in textiles is used for surface modification, various chemical treatments, and to make the process sustainable.

12.2 Fundamentals of Plasma Technology

High energy must be applied to a gas in order to create plasma. This gas can preferably be neutral in nature, or it can be constituted of charged particles depending on the application. These charged particles are energised using different techniques. The collision of these high-energy particles with other atoms, molecules or electrodes creates new charged particles, leading to plasma generation. The methods generally used for plasma generation are discussed below.

12.2.1 DC Discharge

Direct current can be applied across electrodes to energize the gas or air molecules between the electrodes. This creates a plasma discharge, which depends on the voltage applied and the discharge current across the electrodes. DC discharge is primarily divided into three categories: dark, glow, and arc discharge [7]. Each category has been divided into different sub-categories based on the voltage applied.

As the current is increased, the plasma discharge transitions from townsend (self-sustaining) discharge to glow discharge (non-equilibrium) and arc (thermal) discharge. The voltage across the electrodes changes with different categories of discharge [9]. Townsend discharge is where the ionisation takes place, and the free electrons are accelerated by using an electric field, leading to free additional electrons. Townsend discharge could be defined as the sustaining voltage that is independent of the current and by insufficient space charge. A breakdown voltage is required to cause the discharge of electrons. In Townsend discharge, the breakdown voltage and the discharge voltage are equal. The Townsend discharge covers the whole area of the electrode, but this decreases as the current increases [10]. As the current is increased, glow discharge takes place. Glow discharge is comparable to sputtering, where the atoms are expelled by high-energy gas atoms from the analyte surface. Electron ejection could be achieved either by using a DC or an RF AC. The analyte is charged negatively, which collides with the gas atoms. The positively charged gas atoms/ions move towards the sample surface. A glow is produced when the high-energy atoms or ions deexcite and release photons [11]. The glow discharge method ionises the gas partially. Gases such as argon should contain many neutral atoms [12]. The arc discharge method generates an arc between the electrodes placed at a very small distance apart. The arc is produced due to the transfer of electrical current through the inert gas surrounding the electrodes. This occurs due to the breakdown of inert gas at higher voltage and sub-atmospheric pressure [13].

12.2.2 Pulsed DC Discharges

Compared to continuous current, pulsed DC discharge can be advantageous. It could operate at a higher power and provide more control over plasma formation. The rapid gas exchange between the pulses maintains the gas composition and supports homogeneous film formation [7]. Pulsed DC discharge has higher efficiency, lower cost and considerable control than other sources like RF discharge. Within the pressure range of 2 to 7 Torr and the applied voltage in pulse duration of 10 to 4000 μsec, the DC discharge current is considerably less than the discharge current during the early phases of plasma formation. Other than charged particles, DC discharge also produces metastable atoms, radicals and molecules, which might be possessing less ionisation potential than the dominant gas molecules. Hossain et al. developed a ring-shaped plasma source using a single pole magnet setup for pulsed DC discharge plasma, which was highly efficient compared to other setups. Multiple magnets were arranged circularly around a centre magnet, forming 1, 2 and 3 circles. The setup required high gas pressure and higher frequency to generate plasma discharge [14].

12.2.3 RF and Microwave Discharges

RF and microwave discharges are enthusiastic and persistent using electromagnetic waves at extremely high frequencies. This frequency is usually in the microwave range since the electron density available in the microwave range is greater than in radio frequency. The electron density is generally evaluated using the electron energy distribution function (EEDF). The majority of commercial plasma reactors run at 13.56 MHz and 2.45 GHz, authorised by ISM frequencies. When the frequency is increased, the charged particle production is also increased. Higher frequency also helps in polymerisation when a monomer is introduced in gas form to the reaction chamber [15]. In RF discharges, power coupling can be achieved inductively or capacitively. The electrodes of the capacitively coupled plasma are often circular discs spaced a few centimetres apart. The electrodes are separated by a dielectric, where a gas can be filled, with pressures ranging from $1\text{–}10^3$ Pa. During operation, the space between the electrodes gets filled with bulk plasma. In capacitively coupled plasma, the electron density is $10^9\text{–}10^{10}$ cm^{-3}, which might further be increased at higher frequencies [7, 16]. In the plasma reactor with inductive coupling, a transformer from an RF current in the conductor is used. The current-carrying coil is wound helically or spirally either outside or within the treatment section. The electron densities in inductively coupled plasma (ICP) can be as high as 10^{12} cm^{-3} at low ion energies [17, 18]. Different variations of inductively coupled plasma have occurred, such as neutral loop discharge (NLD) and helicon discharge. NLD was developed for applications in microelectronics. It consists of multiple coils acting together to form a neutral magnetic loop. Compared to a traditional ICP source, an NLD source's etching rate can be up to three times higher. Also, the NLD source can work well at low processing pressures (~0.1 Pa) [19]. A helicon discharge plasma is produced in a cylindrical vacuum vessel contained in a longitudinal homogenous magnetic field over 100 G or higher. This process requires specially shaped antennas for plasma generation [20].

12.2.4 Dielectric Barrier Discharge (DBD)

DBD consists of dielectric barrier between the electrodes generating plasma. Usually, this gap of a few millimetres is filled with gas. The charges generated in this gap can be sustained using a 1 to 100 kV voltage and 50 Hz to 1 MHz. The plasma formation looks like streams of charge within the gap of the electrodes. The lifetime of such streams is only 1 to 10 ns. These discharges can also be used to develop plasma displays [21, 22].

12.2.5 *Plasma Production Using Beams*

The plasma can also be produced using electron beams or laser beams. However, sustaining the produced plasma requires using a gaseous medium where the interaction of the electron or laser beam takes place leading to high turbulence in plasma oscillations. The beam could be heated to sustain the discharge plasma [23]. The plasma produced using electron beams is highly applicable for processing the materials having higher surface area [24].

12.3 Advances in Plasma Treatment Techniques for Textiles

Conventionally, textile processing, such as dyeing, coating, and finishing, requires a lot of water treatment. This leads to water pollution due to waste chemicals. Due to environmental concerns, the consumption of water needs to be reduced. For this purpose, different processes are used, such as supercritical CO_2 dyeing for different textile fibers, such as nylon, acrylic, and polyester [25]. Each textile process, such as designing, dyeing, mercerisation, and bleaching, requires separate energy consumption, which increases the price of the product.

In recent advancements, this conventional method of textile treatment can now be done by using cold plasma for surface modification. Plasma treatment in the presence of O_2 showed etching behaviour, which helped remove polyvinyl alcohol (PVA) and polyacrylic acid (PAA) based size. Thus, plasma treatment increased the efficiency of size removal from the cotton fabric. The presence of oxygen gas also increased the fabric hydrophilicity [26]. For fibers and polymers, the cold plasma technique can etch the surface, change its chemical and physical characteristics, activate the surface by rupturing surface bonds, and remove impurities from the surface [27]. Plasma treatment results in weight loss of the cotton fabric, which increases with plasma treatment time. Pretreatment of cotton fabric with plasma has been shown to reduce enzyme consumption in desizing, along with lower fabric weight loss after scouring and bleaching [26]. Primc et al. discussed the mechanism of plasma for treatment over textiles. The treatments with gas at low pressure require high frequency between 10 kHz to 10 GHz. The presence of a gas environment ensures a uniform treatment and prevents arcing [28]. Upon inserting a textile sample into the high-frequency plasma, there is a difference in the potential of the textile and plasma. This creates a negative potential of about 10 V over the textile surface [29]. Therefore, when using microwave discharge, no extremely energetic gaseous species will come into contact with the fabric. In the case of RF plasma, as the frequency of the power source and the density of the charged particles decrease, the depth of plasma penetration

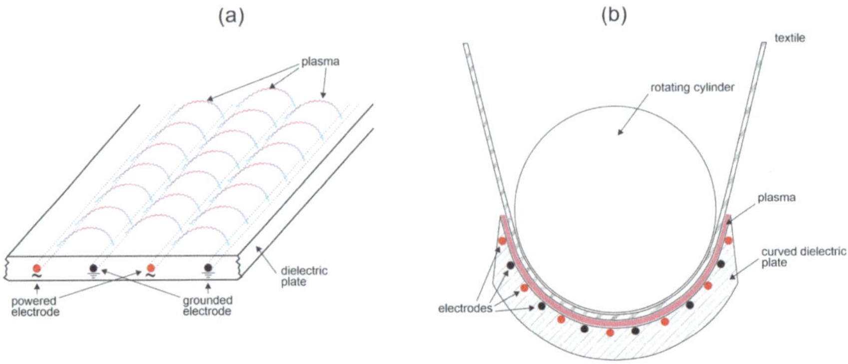

Fig. 12.1 **a** Schematic of surface charges sustained in DBD mode, and **b** treatment of textile with plasma in continuous mode (Reprinted with permission from [28])

increases [30]. The textiles are often treated with plasma at atmospheric pressure or atmospheric pressure plasma (APP) in automatic mode running at speeds up to 1 m/ s. However, continuous plasma treatment was not possible earlier, but recently, due to technological advancements, continuous plasma treatment can be done using a special electrode setup design, as shown in Fig. 12.1.

Since the breaking voltage of air is lower at atmospheric pressure, streamers are generated, which are streams of electrical charges. Highly resistant coatings of ceramic or glass are placed over the electrodes to protect the conducting wires from such streamers. The streamers are dense, which shows glowing plasma between the electrodes. However, the distance of the textile from the electrode is of great importance since the plasma depends upon the distance between the electrodes [31]. Dielectric coplanar surface barrier discharge (DCSBD) and volume dielectric barrier discharge (VDBD) methods have shown similar results for the treatment of textile fabrics. The DCSBD setup was relatively gentle compared to the VDBD setup since it has a more destructive effect over the substrate surface [32].

The surface modification of textiles requires a large surface area to be exposed to plasma. For this, the energy required is also higher. Textile materials are porous; in such cases, the treated area and the area projected to plasma are not the same but higher. However, a tightly woven fabric may behave like a film but with more surface area. This higher surface area requires about 57% more energy for plasma treatment over textile fabric. For an idealised fabric with n parallel circular filaments stacked per unit length (or width, W_F) in a tight array, the surface area available for treatment can be calculated using Eq. 12.1.

$$R_s = \frac{\sum(n\pi d_f)}{2} \tag{12.1}$$

where Rs is the surface area available for plasma treatment, d_f is the fiber diameter, and n is the number of filaments present in the textile structure.

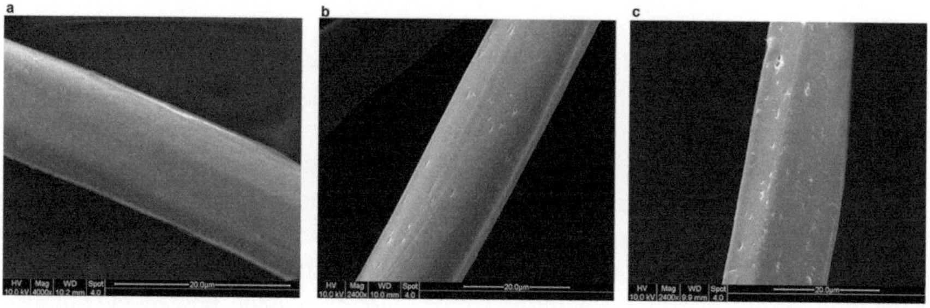

Fig. 12.2 SEM images of **a** Untreated, **b** Air plasma treated, and **c** Air/Ar plasma treated polyester fiber (Reprinted with permission from [35])

As plasma formation occurs, the concentration of reactive species formed during the process affects the extent of penetration in textiles [33]. The hydrophilisation effect of plasma was studied extensively by Hossain et al. [34] using polyester woven and knitted fabrics. Polyester spun yarn-based knitted and woven fabric was compared with the effect on polyester filament-based fabric with the help of a capillary test. The hydrophilicity was highest in woven fabric, followed by knitted and filament-based fabric. However, the textile geometry also largely influenced hydrophilicity, which could be enhanced by plasma treatment. The effect of plasma over polyester fiber was further analysed by Zhang and Fang [35]. The surface morphology showed increased roughness after the treatment (Fig. 12.2).

The surface composition of polyester fiber was also changed since the oxygen/carbon atoms ratio increased by about 21 %, and some traces of nitrogen were also observed after the plasma treatment. This increased oxygen was present in the form of C = O or COOH groups, which resulted in increased hydrophilicity of the fibers.

Recently, there has been a new term known as plasma textile. A voltage of up to 10 kV is delivered, and the plasma electrodes are embedded into the textiles. This has proved effective for filtration of sub-micrometre particles with 100 % efficiency. Submicrometre particles acquire an electric charge through the triboelectric charging process. To further add up to this charge, an external charge source such as corona discharge can also be used [36]. This pre-charges the particles before filtration and then filters them out using an electrified fiber filter. Using this principle, a filter bag was developed to collect and store the charged particles. The corona discharged wire was axially placed in the fabric, and the gas flow was tangential to the wire [37]. Rasipuram et al. integrated wires in the polypropylene woven fabric and applied high DC voltage to generate corona. An aerosol consisting of NaCl was directed towards the fabric, and filtration efficiency was studied. The particle penetration in the fabric was found to increase with an increase in voltage. Similarly, at initial voltages, the filtration efficiency was 30 %, which increased to 70 % after the corona was ignited in the fabric [38]. This shows the benefits of plasma treatments combined with textile materials.

Taking a step towards sustainability, Zhou et al. [39] developed a new process for pretreatment of fabric in open width where plasma treatment, padding, enzyme washing, etc. can be performed. The plasma treatment helped destroy the oily and waxy layer over the knitted cotton fabric. This increased the product quality since the treatment did not alter the microstructure of the fiber. The water consumption was also reduced. The quantity of water required was 26.4 % of that used in the conventional process. This also resulted in the reduction of pollutants generated during the treatment process. The plasma treatment is independent of the fiber being processed, and the extent of modification can be tuned by using the correct set of parameters. The parameters that affect the surface modification include the gas being used for processing, the working temperature, frequency, power, and duration of exposure to plasma [40].

12.4 Summary

The current chapter shows the most used methods for generating plasma. The production of plasma requires energy that can be transferred in the form of thermal or electrical energy. Textile processing, including dyeing, coating, and finishing, requires extensive water treatment, leading to water pollution and environmental concerns. Various processes, like supercritical CO_2 dyeing, are used to reduce water consumption in different textile fibers. Advancements in plasma techniques introduced cold plasma treatment, which can be used for surface modification and enhance the efficiency of size removal and fabric hydrophilicity. However, plasma treatment results in weight loss of cotton fabric, which increases with treatment time. Advanced plasma treatment requires a high frequency between 10 kHz and 10 GHz to ensure uniform treatment and prevent arcing. The negative potential of the textile and plasma prevents highly energetic gaseous species from interacting with the textile. The penetration depth of plasma increases with decreasing frequency and density of charged particles. Plasma treatment of textiles requires a higher energy and surface area than traditional textiles. The hydrophilisation effect of plasma has been studied extensively, with the highest hydrophilicity found in woven fabric, followed by knitted and filament-based fabric. The surface morphology showed increased roughness after treatment. The surface composition of polyester fiber also changed, with an increase in oxygen/carbon atoms and nitrogen. Plasma textiles, which use electrodes integrated within textiles and voltages up to 10 kV, for effective filtration of submicrometre particles with 100 % efficiency. A new process for treatment of fabric width wise using plasma combined with other methods for reducing the wastage of water is also introduced. The plasma treatment is independent of the fiber to be processed. It can be tuned using the correct parameters like the working temperature, frequency, gas used, power, and plasma exposure time. These advancements indicate a strong future for plasma technology in the textile industry, offering innovative solutions that enhance the performance and sustainability of textile products.

Declaration The authors declare that they have no known competing financial interests or personal relationships that could have appeared to influence the work reported in this chapter. Generative AI tools were used in the writing process to improve the readability and language of the work only. After using this tool, the authors carefully reviewed and revised the content to ensure accuracy and quality. The authors take full responsibility for the final version of this publication.

References

1. Langmuir, I. (1928). Oscillations in ionized gases. *Proceedings of the National Academy of Sciences of the United States of America, 14*, 627–637.
2. d'Agostino, R., Favia, P., Oehr, C., & Wertheimer, M. R. (2005). Low-temperature plasma processing of materials: past, present, and future. *Plasma Processes and Polymers, 2*, 7–15. https://doi.org/10.1002/ppap.200400074
3. Khan, N., & Abas, N. (2011). Comparative study of energy saving light sources. *Renewable and Sustainable Energy Reviews, 15*, 296–309. https://doi.org/10.1016/j.rser.2010.07.072
4. Thejo Kalyani, N., & Dhoble, S. J. (2012). Organic light emitting diodes: Energy saving lighting technology—a review. *Renewable and Sustainable Energy Reviews, 16*, 2696–2723. https://doi.org/10.1016/j.rser.2012.02.021
5. Yazıcıoğlu, Ö., & Katırcıoğlu yaşar, T. (2017). Applications of plasma technology in energy sector. *Kırklareli Üniversitesi Mühendis Ve Fen Bilim Derg, 3*, 18–44.
6. Ryan, T. P., Stalder, K. R., & Woloszko, J. (2013). Overview of plasma technology used in medicine. In *Energy-based Treatment of Tissue and Assessment VII* (pp. 175–195). SPIE.
7. Conrads, H., & Schmidt, M. (2000). Plasma generation and plasma sources. *Plasma Sources Science and Technology, 9*, 441–454. https://doi.org/10.1088/0963-0252/9/4/301
8. Oks, E. M. (1992). Physics and technique of plasma electron sources. *Plasma Sources Science and Technology, 1*, 249. https://doi.org/10.1088/0963-0252/1/4/004
9. Shohet, J. L. (2003). Plasma science and engineering. In *Encyclopedia of Physical Science and Technology* (pp. 401–423). Elsevier.
10. Kolobov, V. I., & Fiala, A. (1994). Transition from a townsend discharge to a normal discharge via two-dimensional modeling. *Physical Review E, 50*, 3018–3032. https://doi.org/10.1103/PhysRevE.50.3018
11. Twyman, R. M. (2005). Atomic emission spectrometry | Principles and Instrumentation. In P. Worsfold, A. Townshend, & C. Poole (Eds.), *Encyclopedia of analytical science* (2nd ed., pp. 190–198). Elsevier.
12. Bogaerts, A. (1999). Glow discharge mass spectrometry, methods. In J. C. Lindon (Ed.), *Encyclopedia of spectroscopy and spectrometry* (pp. 669–676). Elsevier.
13. Park, Y. S., Kodama, S., & Sekiguchi, H. (2021). Preparation of metal nitride particles using arc discharge in liquid nitrogen. *Nanomaterials, 11*, 2214. https://doi.org/10.3390/nano11092214
14. Hossain, Md. A., Patwary, M. A. M., Bhuiyan, M. M. R., et al. (2019). Observation of ring–shaped pulsed DC discharge plasma source using single pole magnet setups for material processing. *Radiation Effects and Defects in Solids, 174*, 380–396. https://doi.org/10.1080/10420150.2019.1596105
15. Moisan, M., & Wertheimer, M. R. (1993). Comparison of microwave and r.f. plasmas: Fundamentals and applications. *Surface & Coatings Technology, 59*, 1–13. https://doi.org/10.1016/0257-8972(93)90047-R
16. Lister, G. G. (1992). Low-pressure gas discharge modelling. *Journal of Physics D: Applied Physics, 25*, 1649–1680. https://doi.org/10.1088/0022-3727/25/12/001
17. Holland, G. (1991). *Applications of plasma source mass spectrometry*. Royal Society of Chemistry.
18. Hopwood, J. (1992). Review of inductively coupled plasmas for plasma processing. *Plasma Sources Science and Technology, 1*, 109–116. https://doi.org/10.1088/0963-0252/1/2/006
19. Tsuboi, H., Itoh, M., Tanabe, M., et al. (1995). Usefulness of magnetic neutral loop discharge plasma in plasma processing. *Japanese Journal of Applied Physics, 34*, 2476. https://doi.org/10.1143/JJAP.34.2476

20. Chen, F. F., Sudit, I. D., & Light, M. (1996). Downstream physics of the helicon discharge. *Plasma Sources Science and Technology, 5*, 173. https://doi.org/10.1088/0963-0252/5/2/009
21. Esena, P., Riccardi, C., Zanini, S., et al. (2005). Surface modification of PET film by a DBD device at atmospheric pressure. *Surface & Coatings Technology, 200*, 664–667. https://doi.org/10.1016/j.surfcoat.2005.02.188
22. De Geyter, N., Morent, R., & Leys, C. (2006). Surface modification of a polyester non-woven with a dielectric barrier discharge in air at medium pressure. *Surface & Coatings Technology, 201*, 2460–2466. https://doi.org/10.1016/j.surfcoat.2006.04.004
23. Schmidt, M., Föste, T., Michel, P., & Šapkin, V. V. (1982). Device for producing and investigating an electron beam plasma discharge. *Beiträge aus der Plasmaphysik, 22*, 435–441. https://doi.org/10.1002/ctpp.19820220506
24. Meger, R. A., Blackwell, D. D., Fernsler, R. F., et al. (2001). Beam-generated plasmas for processing applications. *Physics of Plasmas, 8*, 2558–2564. https://doi.org/10.1063/1.1345506
25. Bai, T., Kobayashi, K., Tamura, K., et al. (2019). Supercritical CO2 dyeing for nylon, acrylic, polyester, and casein buttons and their optimum dyeing conditions by design of experiments. *Journal of CO2 Utilization, 33*, 253–261. https://doi.org/10.1016/j.jcou.2019.05.013
26. Samei, N., Shahidi, S., & Mongkholrattanasit, R. (2022). Dry finishing using plasma treatment for shortening the initial wet finishing of cotton fabrics. *Fibers Polym, 23*, 3442–3451. https://doi.org/10.1007/s12221-022-4150-z
27. Bhat, N., Netravali, A., Gore, A., et al. (2011). Surface modification of cotton fabrics using plasma technology. *Textile Research Journal, 81*, 1014–1026. https://doi.org/10.1177/0040517510397574
28. Primc, G., Zaplotnik, R., Vesel, A., & Mozetič, M. (2022). Mechanisms involved in the modification of textiles by non-equilibrium plasma treatment. *Molecules, 27*, 9064. https://doi.org/10.3390/molecules27249064
29. Apostol, M. (2020). Penetration depth of an electric field in a semi-infinite classical plasma. *Optik, 220*, Article 165009. https://doi.org/10.1016/j.ijleo.2020.165009
30. Otsuka, F., Hada, T., Shinohara, S., & Tanikawa, T. (2015). Penetration of a radio frequency electromagnetic field into a magnetized plasma: One-dimensional PIC simulation studies. *Earth, Planets and Space, 67*, 85. https://doi.org/10.1186/s40623-015-0244-9
31. Homola, T., Kelar, J., Černák, M., & Kováčik, D. (2022). Large-area open air plasma sources for roll-to-roll manufacture. *Vakuum in Forschung und Praxis, 34*, 21–25. https://doi.org/10.1002/vipr.202200785
32. Šrámková, P., Kelar Tučeková, Z., Fleischer, M., et al. (2021). Changes in surface characteristics of BOPP foil after treatment by ambient air plasma generated by coplanar and volume dielectric barrier discharge. *Polymers, 13*, 4173. https://doi.org/10.3390/polym13234173
33. Krentsel, E., Yasuda, H., Miyama, M., & Yasuda, T. (1995). Penetration of plasma surface modification into porous media. III. Multiple samples exposed to CF4 and C2F4 low temperature cascade arc torch. *Journal of Polymer Science Part A: Polymer Chemistry, 33*, 2887–2892. https://doi.org/10.1002/pola.1995.080331705
34. Hossain, M. M., Herrmann, A. S., & Hegemann, D. (2006). Plasma hydrophilization effect on different textile structures. *Plasma Processes and Polymers, 3*, 299–307. https://doi.org/10.1002/ppap.200500152
35. Zhang, C., & Fang, K. (2009). Surface modification of polyester fabrics for inkjet printing with atmospheric-pressure air/Ar plasma. *Surface & Coatings Technology, 203*, 2058–2063. https://doi.org/10.1016/j.surfcoat.2009.02.006
36. Mizuno, A. (2000). Electrostatic precipitation. *IEEE Transactions on Dielectrics and Electrical Insulation, 7*, 615–624. https://doi.org/10.1109/94.879357
37. Plaks, N. (1988). Fabric filtration with integral particle charging and collection in a combined electric and flow field: Part I. Background, experimental work, analysis of data, and approach to the development of a mathematical engineering design model. *Journal of Electrostatics, 20*, 247–266. https://doi.org/10.1016/0304-3886(88)90001-0
38. Rasipuram, S. C., Wu, M., Kuznetsov, I. A., et al. (2013). Submicrometre particle filtration with a dc activated plasma textile. *Journal of Physics D: Applied Physics, 47*, Article 025201. https://doi.org/10.1088/0022-3727/47/2/025201

39. Zhou, L., Bai, Y., Zhou, H., & Guo, S. (2019). Environmentally friendly textile production: Continuous pretreatment of knitted cotton fabric with normal temperature plasma and padding. *Cellulose, 26*, 6943–6958. https://doi.org/10.1007/s10570-019-02508-8
40. Haji, A., & Kan, C. -W. (2021). Chapter 19-plasma treatment for sustainable functionalization of textiles. In N. Ibrahim & C. M. Hussain (Eds.), *Green chemistry for sustainable textiles* (pp. 265–277). Woodhead Publishing.

Chapter 13
Conductive and Smart Textiles: Innovations and Environmental Impact

Mir Abdul Gaffar, Muntaha Mahin, Mst. Sadia Islam, Muntaha Rahman Raha, and Md. Rashedul Islam

Abstract The rapid advancement of technology is reshaping traditional textiles into materials with capabilities once thought impossible. Conductive and smart textiles are at the forefront of this revolution by transforming fabrics into dynamic surfaces that can monitor health, communicate data, and respond to external stimuli. These devices are not just limited to laboratories but are finding real-world applications very quickly. For example, smart textiles have notable applications including tracking vital signs in medical wearables and equipping communication systems in military uniforms. These evolutions are pushing the boundaries of what a piece of fabric can do by turning everyday items into sophisticated devices. However, integrating electronics into textiles comes with great challenges. This raises concerns about the durability, washability, and long-term performance of the devices. There is also the growing issue over the environmental impact of producing smart textiles. Using metals and synthetic materials poses risks related to energy consumption and waste generation. This chapter addresses both the innovations of smart textiles and their environmental impact while also providing insight into the future directions and challenges that could balance innovation with sustainability.

Keywords Conductive textiles · Smart fabrics · Wearable technology · Sustainable materials · Environmental impact · Textile innovations

---

M. A. Gaffar · Mst. S. Islam · M. R. Raha · Md. R. Islam (✉)
Department of Wet Process Engineering, Faculty of Textile Chemical Engineering, Bangladesh University of Textiles (BUTEX), Tejgaon, Dhaka, Bangladesh
e-mail: rashedul.islam@wpe.butex.edu.bd

M. Mahin
Department of Yarn Engineering, Faculty of Textile Engineering, Bangladesh University of Textiles (BUTEX), Tejgaon, Dhaka, Bangladesh

M. Shahid et al. (eds.), *Advancements in Textile Finishing*,
https://doi.org/10.1007/978-981-96-6385-9_13

13.1 Introduction

Smart textiles have advanced characteristics, including electronic functionality ranging from conductive tracks to communications, sensing/actuating, and micro-processing [1]. Efforts are underway to embed technologies like self-healing polymers, conductive fibers and yarns, nanocomposites, and biosensors with textiles, transforming them into dynamic materials capable of responding to external stimuli. Conductive and smart textiles are transforming the convergence of fashion, electronics, material science, and sustainability, focusing on enabling textiles to perform advanced functions like activity and health monitoring, energy harvesting and electricity generation, environmental sensing, etc. [2–7]. Such evolution has created new opportunities across the healthcare, military, lifestyle, and industrial sectors. For example, bio-sensing textiles can monitor blood pressure, pulse, and respiratory rates, and specially designed hydro suits can allow the submarine crew to regulate underwater breathing [8]. Smart textiles can even potentially replace smartphones and other devices [9] with functionalities including power generation or storage, RF functionality, assistive technology, and human interface elements [10]. These functionalities are incorporated into textile structures, exploring diversified processes, including embroidering, weaving, knitting, braiding, spinning, printing, stitching, coating, and chemical treatments .

The potential of wearable smart textiles is significant, offering a comfortable and reliable platform in clothing, enhancing user compliance, and mitigating the stigma of wearable devices. It has facilitated the integration of multiple sensors in garments and typical everyday items, providing real-time feedback [11]. Despite such potential, numerous challenges, such as durability, washability, cost management, and wearability, are still imposed on the commercialization and scalability of smart textiles. There is also a substantial challenge with smart textiles, as it is associated with two alarming global waste streams- textiles and electronics, leading to extended difficulty in reusing or recycling them. At present, the waste coming from smart textiles is very low in quantity. But its sustainability issue is slowly becoming a central debate due to its rapid market expansion [12]. The global market for smart textiles is estimated to reach around $15 billion by 2028 [6]. Therefore, the balance between functionality and sustainability remains crucial to ongoing research.

This chapter delves comprehensively into the advancements in the conductive and smart textiles segment. It explores the recent technological innovations and numerous integration methods of smart textiles, offering an in-depth look at the diversified applications of smart textiles with a focus on emerging technologies, commercialization challenges, and strategies for aligning innovation with environmental responsibility.

13.2 Technological Innovations in Conductive and Smart Textiles

Smart textiles are defined as textiles that interact with their environment and respond according to the applied stimuli. Conductive and smart textiles can be created by incorporating metallic yarn, conductive materials, or compounds onto the fabric substrates [13]. This section briefly elaborates on the type of conductive materials used to create smart textiles, the fabrication and integration methods of conductive materials/devices, and current technological innovations emerging in the conductive and smart textile sectors.

13.2.1 Conductive Materials

Integrating conductive components into textiles enables a change from static fabrics to an active, responsive material that environmental changes can trigger, can create extra comfort for patients or athletes, and can open up new possibilities for application in areas such as healthcare and personal clothing. Conductive materials can be classified according to the raw materials utilized in their fabrication. This primarily comprises metallic fibers and yarns, carbon-based materials, and hybrid composites.

13.2.1.1 Metallic Fibers and Yarns

Metallic fibers are composed of electrically conductive metals like ferrous alloys, nickel, stainless steel, titanium, aluminum, copper, and silver. These materials have a diameter that varies from 1 to 80 μm, which is thinner than human hair (70–100 μm). These highly electrically conductive metal fibers enhance the performance and functionality of textiles utilized in wearable applications [14]. It is manufactured via bundle-drawing or the shaving process and may also be derived from metal or carbon powders [15]. Metallic fibers provide superior conductivity, corrosion resistance (notably silver and copper), and mechanical strength [16]. Among metallic fibers, silver-coated fibers are primarily utilized for their exceptional conductivity. Silver-coated fibers possess a surface resistance below 0.2 Ω/square. Silver's low surface resistivity and intrinsic antibacterial characteristics are especially beneficial for health-monitoring textiles and medical applications. This conductive fabric's capacity for elongation with negligible resistance variation makes it a viable alternative to traditional strain sensors [17]. Copper is a more economical substitute for silver while exhibiting comparable conductivity. Copper is more prone to corrosion, which may restrict its application in specific settings. Stainless steel fibers (SSF) are known for their remarkable durability and moderate conductivity. A study by Shahzad et al. [18], discovered that hybrid spun yarns integrating polyester or viscose with stainless steel fibers exhibited enhanced electrical conductivity, especially with a higher twist multiplier, although showing susceptibility to humidity.

13.2.1.2 Carbon-Based Materials

Carbon materials such as carbon black, carbon nanotubes (CNTs), and graphene have revolutionized the field of smart and conducting textiles. These materials show excellent electrical conductivity, mechanical strength, thermal stability, and flexibility. Carbon black (CB) is powdery and lightweight, possesses a loose consistency, and has an extensive specific surface area. It is mostly used in strain sensors or pressure sensors [19]. Similarly, Carbon nanotubes (CNTs) are allotropes of carbon. It is mainly used for the production of carbon-based conductive fibers [14]. CNTs have exceptional electrical properties, a mobility electron rate of 100,000 $cm^2/V \cdot s$, and a density of current of 10^9 A/cm^2. CNTs have a modulus range of up to 5 TPa, and an elastic strain of 5–12% [19]. Carbon materials are incorporated into textiles using spray coating, immersion and layer-by-layer deposition, and fiber incorporation are achieved through melt or wet spinning [20]. CNT-coated threads have been shown to transform standard cotton into conductive yarns with increased sensitivity and selectivity for biomonitoring applications [21]. When used with graphene/CNT hybrid coatings, the coated fabrics show significant reductions in electrical resistance, with conductivity as high as 9.02×10^{-2} S/m and good thermal management with a temperature of 42.7 °C for 12 V supply voltage, making it suitable for heating applications [22]. Graphene is a monolayer allotrope of carbon exhibiting sp^2 hybridization in a hexagonal lattice and has contributed to wearable electronics for its exceptional electrical, mechanical, and thermal characteristics. Reduced graphene oxide (rGO) is a derivative of graphene that has numerous applications in wearable textiles due to the extensible production of its graphene oxide (GO) precursor, the incorporation of rGO into textiles via traditional printing, coating, and dyeing methods and the development of flexible and washable e-textiles through efficient integration with textile fibers [23]. Graphene-coated textiles have an impressive conductivity as high as 2505.68 S/cm and represent a reliable alternative to conventional electric wires in wearable applications [24].

13.2.1.3 Hybrid Composites

Hybrid composites are materials consisting of two or more distinct types of reinforcements or fillers integrated into a matrix. These materials have great mechanical strength, electrical conductivity, and functional responsiveness. Hybrid composites include conductive polymers, carbon-based nanomaterials, and metallic fillers to make smart textiles capable of sensing, energy storage, or responsive behavior. The common techniques for fabricating these advanced materials include solution mixing, electrospinning, and resin transfer molding [25]. Advanced fillers are added to hybrid composites such as multi-walled carbon nanotubes, silver nanobelts, and titanium carbide or MXene to enhance the electrical conductivity and thermal properties of the composites for making smart textiles [26]. Moreover, there are a few methods of electrospinning to fabricate nanofibers with conductive networks included. Such materials guarantee the improved flexibility and structural stability of composite materials [27]. Post-fabrication treatments, including heat curing and crosslinking, can be used to increase the stability and functional efficiency of composite materials. Poly(vinyl alcohol) (PVA) nanocomposite fibers incorporating silver nanobelts (AgNBs) and multi-walled carbon nanotubes (MWCNTs) reached up to 1600 S/cm, while a gain of 533 times as compared to pristine PVA. These materials are very suitable for applications in e-textiles, such as water leakage detectors and wearable heating devices [28]. Therefore, these fibers prove their capability for application in thermal energy management and heat dissipation in smart textiles [29]. Graphene and MoS_2-infused smart fiber-reinforced polymer (FRP) composites can achieve multifunctionality, including sensing, heating, and energy storage capabilities. It has a real capacitance of 182 $\mu F/cm^2$ and can sustain 90% capacitance retention after 1000 cycles.

13.2.2 Fabrication of Conductive Materials onto Textiles

The development of advanced smart and conductive textiles that are equipped with sensing capabilities, actuating, and adapting to external factors is mainly dependent upon the fabrication of conductive materials into textiles [30]. Some of the important fabrication techniques can be classified into sensors, actuators, printed electronics, coatings, and conventional textile procedures such as weaving, knitting, embroidery, and spinning.

Smart textile integrates conductive materials to develop sensors and actuators that react to environmental triggers such as temperature, pressure, or mechanical strain. Among various classifications of sensors, resistive and capacitive sensors have high sensing capabilities under variable states. Resistive-based sensors, constructed using fiber-based electrodes, detect mechanical strain through resistance changes, whereas capacitive-based sensors, composed of double-fiber electrodes with a dielectric medium, respond to applied voltage through charge accumulation [31]. Actuators, particularly shape actuators, offer durability, adaptability, and light-weight design and are categorized into thermal, optical, humidity, and electrochemical types [32].

Screen printing is an effective fabrication method that employs pressure to transfer ink through a mesh, resulting in patterns on fabric. This process enables layer-by-layer deposition of multifunctional electronics [19, 16]. This method yields constant resistance values under 10 Ω/sq and is extensively utilized in the production of flexible sensors [33]. In contrast, inkjet printing is performed by systematically applying coatings to designated areas, ensuring material efficiency and spatial resolution [32]. This approach allows graphene-based inks to exhibit electrical conductivity beyond 10^4 S/m while maintaining washability [34]. In 3D printing, sequential integration of conductive elements enables the incorporation of complete electronic systems into textile materials (Fig. 13.1).

Coating is a technique that deposits a nanometer to micrometer scale conductive film onto fibers by using an auxiliary instrument [31]. Among the various coating techniques, dip coating, spin coating, and spray coating are mostly utilized in wearable textiles. Dip coating refers to immersing textiles into conductive material solutions for uniform coating. Using the dip-coating method, a highly stretchable composite fiber can be produced with an electrical conductivity of over 10^3 S/cm and

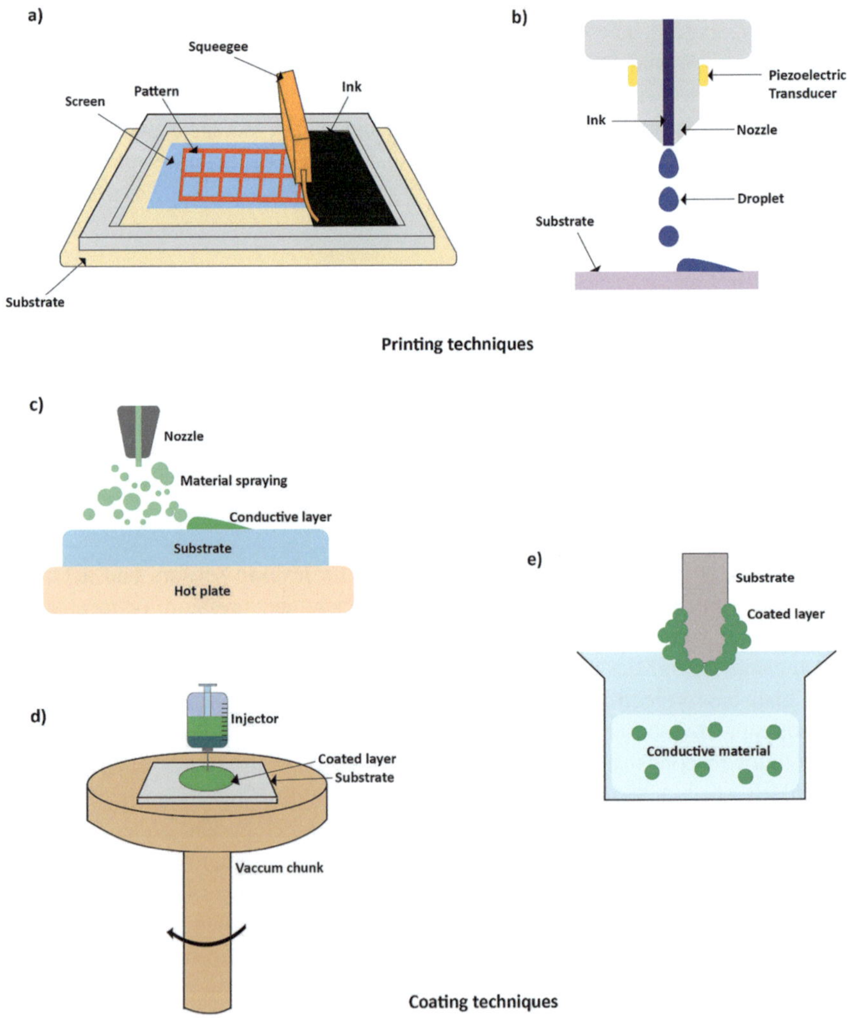

Fig. 13.1 Printing techniques **a** Screen printing, **b** Ink-jet printing and coating techniques, **c** Spray coating, **d** Spin coating, **e** Dip coating [35]

a maximum strain of 500% [3]. Whereas, spray coating involves the precise deposition of conductive materials onto the substrate's surface by using a high-pressure air gun [19]. In spin coating, functional materials are applied to a spinning substrate. The material spreads evenly due to centrifugal force, forming a thin, uniform film.

Conductive yarns can be created by metal wire filaments, polymer yarns coated by metal, blended with metal fibers, wet spinning from conductive materials, or coated by graphene. These conductive yarns can be embedded into textiles via weaving, knitting, and embroidery [12]. Weaving incorporates electronic components into fabrics by merging conductive threads with conventional fibers. Knitting allows for the integration of stretchy conductive yarns. This is appropriate for the production of elastic and flexible fabrics for wearable applications [36]. Embroidery is the placement of conductive yarns or filament fibers onto textiles or other materials using a needle [37].

13.2.3 Integration of Electrodes and Interconnectors

The integration of electronics and interconnectors into smart textiles involves embedding conductive pathways and electronic components within the fabric substrate. This allows the fabric to show multi-functionalities such as sensing, data processing, energy harvesting, and communication. Conventional electronic systems house components using rigid printed circuit boards (PCBs). It is a fundamental requirement of smart textiles to seamlessly incorporate rigid elements (electronics, interconnectors) into flexible substrates. Methods such as weaving or embroidering conductive fibers, conductive coatings, adhesives, or thermo-compression bonding are utilized to attach such components. Despite advancements in making textile-PCB hybrids and encapsulation techniques, there is a continuous concern about factors such as durability, scalability, and performance degradation under mechanical stress and washing [12].

13.2.4 Emerging Technologies

13.2.4.1 Self-Healing Polymers

Self-healing polymer has become a promising technology in the area of smart textiles owing to its enhancing durability and functionality with diversified applications. This breakthrough opens new doors of opportunities in wearable electronics and sensors, adaptive and responsive clothing, sustainable textiles, eco-friendly materials, health monitoring, and biomedical technologies, revolutionizing industries from fashion to healthcare. The successful attempt to create fibers and yarns from bio-based self-healable thermoplastic polyurethane has showcased its potential in producing smart and stretchable textiles, skin-related applications with skin-mimicking properties and

Table 13.1 Overview of self-healable 3D and 4D printing materials [43, 44]

Material	Printing technology	Healing mechanism	Healing efficiency (%)	Potential application	Reference
Photo elastomer	Stereolithography (SLA)	Reversible thiol group	90–100% (Mechanical/ Tensile)	Shoe soles with self-healing characteristics	[45]
PPy and chitosan hydrogel	Direct ink writing (DIW)	Ionic interactions between ferrous ions and carboxylic groups of PAA	Full restoration of mechanical performance and 90% recovery of electrical conductivity	Wearable devices and sensing technologies with human skin mimicking properties	[46]
DMSO-mixed PEDOT: PSS and Triton X-100 (healing agent)	Direct ink writing (DIW)	Hydrogen bonding	> 85% of initial power output	Wearable thermoelectric generators	[47]
PUSA, PUHA, HEA	Digital light processing (DLP)	Disulfide bond	95%	Honeycomb structure	[48]
Wax, Pluronic F127	Direct ink writing (DIW)	Vascular	Up to 100%	Coating protection	[49]

promising biomedical applications with surface nanopores [38]. Self-healing elastomeric polymers from 2-methyl-2-benzyloxycarbonyl propylene carbonate, showcase viable application in surgical sutures, and polyurethane nanohybrid can make the wound-healing duration upto 9 days with its 95.0% shape recovery capacity [39, 40]. Hydrogel dressings incorporated with smart bandages are being explored in terms of several parameters including pressure, temperature, and pH level for wound healing and monitoring purposes [41]. Polymethyl acrylate (PMA) and polynorbornene anhydride (PNA); self-healing, stretchable, and conductive hydrogel fibers have paved new headway toward strain sensing and energy harvesting triboelectric nanogenerator (TENG) textiles [42]. By enhancing functionality to dynamic 3D and 4D printing, self-healing materials are explored for a varied range of applications, as shown in Table 13.1.

13.2.4.2 Bio-Sensing and Energy Harvesting Textiles

Bio-sensing textiles are fabrics that incorporate biosensors to monitor health and the environment in real-time. These can be worn as part of everyday activities, during sports, or for medical management. Biosensors can detect and monitor vital signs, including pulse rate, blood pressure, and respiration rate, and can predict early signs of diseases [50, 8, 51]. One of the most prominent applications is to monitor vital signs for athletes, patients, older adults, firefighters, and even astronauts. Textile-based sensors allow patients to track their health from the comfort of their own home. The application of connected wearable devices for self-monitoring of medical conditions could potentially lower NHS (National Health Service) costs by around 60% [52].

A key challenge with wearable electronics and e-textiles is providing a reliable power source [53]. Most smart textiles rely on traditional rechargeable batteries, but these are often cumbersome and heavy. For example, batteries make up around 20% (9 kg) of the load carried by soldiers on a mission. Additionally, due to their size, inflexibility, and inability to withstand washing, these batteries cannot be seamlessly integrated into the textile structure. As a substitute, energy can be harvested from different sources utilizing textile products as a platform [3]. They are lightweight, breathable, and stretchable. Several types of energy harvesting mechanisms relevant to smart textiles are piezoelectric, photovoltaic, thermoelectric, and triboelectric.

E-textiles capable of capturing bio-signals, combined with flexible energy storage solutions, represent a significant advancement for customized healthcare applications [52, 54, 55]. Graphene-based e-textiles have gained notable interest for wearable electronics in recent years due to graphene's remarkable properties, such as its high specific surface area, excellent mechanical strength, and superior electrical and thermal conductivity. Graphene-based textiles have the potential to offer a versatile platform for creating next-generation innovative e-textile garments that can simultaneously function as energy generators and storage devices, sensors, actuators, and electrodes for bio-signal detection.

13.2.4.3 Nanotechnologies

Nanomaterials are used to add properties such as antimicrobial, UV resistance, electrical conductivity, optical, hydrophobicity, and flame retardancy to fabrics and garments. Smart nanomaterials-based devices are now being incorporated into textiles to enable functions like sensing, energy harvesting and storage, drug release, and optics. Nowadays, green nanomaterials such as nanocellulose, chitosan nanoparticles, silver nanoparticles, graphene, and graphene oxide, etc., are on trend in the field of smart textiles because of their eco-friendly and sustainable approaches. Smart textiles integrated with green nanomaterials are truly a breakthrough in the field of healthcare (Table 13.2).

Table 13.2 A glimpse of recent progress in smart textiles integrated with various nanomaterials and their applications [56, 57]

Textile substrate	Integration method	Nanomaterials	Applications	Ref
Cotton	Coating	Ag nanoparticles	Antibacterial and Water-repellent textiles	[58]
Synthetic fibers	Dip coating, Screen Printing, Electroless, Immersion, Electrospinning	Gold Nanoparticles	Conductive and Antimicrobial textiles for wearable electronics	[59]
Cotton	Sol–gel method (with a green solvent)	Zn Oxide nanoparticles	UV-resistant and antibacterial textiles	[60]
Silicone fibers	Warp and weft weaving	Geniomer 200 (polysiloxane-urea copolymer with a polysiloxane)	Pressure sensor based flexible optical fibers for textiles	[61]
Cotton, Nylon, Polyester	Coating	Graphene oxide	Biomedical applications	[62]
Cotton fabric	Impregnation	NanoTiO_2@DNA	Delivery of drugs based on nanomedicine	[63]
Grooved solid and hollow hydrogel fibers	Weaving, braiding, and embroidering	Polylactic acid/ polydimethylsiloxane(PDMS)	Wearable medical devices, Tissue engineering, soft robotics	[64]
Nanofiber	Electro spraying	Mn@ZnO/CNF	Energy storage on textile (High conductivity)	[65]

13.3 Applications of Smart and Conductive Textiles

Smart and conductive textiles have versatile applications across multiple sectors. In healthcare, they are used for patient monitoring, wearable diagnostics, and therapeutic systems. This enhances real-time data collection and personalized care. Similarly, these textiles enable performance tracking and injury prevention through integrated sensors and adaptive materials in sports and fitness. Advanced camouflage and communication systems are proven to be crucial applications in the military sector, while industrial use spans safety gear and performance enhancement devices. Smart textiles also have numerous applications in environmental monitoring and consumer electronics, with products ranging from pollution sensors to interactive fashion.

Table 13.3 Type of smart textile products [66]

Field of application	Examples of products
Healthcare and medicine	Tracking vital parameters (e.g., blood pressure, heart rate, ECG, EEG, glucose levels), Monitoring wound healing, Patches for drug delivery, Movement analysis
Fitness, sports, and wellness	Activity tracking (e.g., steps, heart rate), Monitoring stress levels, Muscle activation systems, Devices for sleep monitoring
Military and industrial use	Protective gear (PPE), Enhancements for ergonomic performance, Systems for counterfeit prevention, Focus monitoring tools, Powered exoskeletons
Home and building design	Smart displays and controls integrated into structures, Monitoring the structural integrity of buildings, Sensors for movement and fall detection
Fashion and lifestyle	Interactive displays and environmental controls, Wearables with visual and tactile feedback

Table 13.3 shows the type of products used across different sectors. Furthermore, their broad implication among sectors is highlighted in Table 13.4.

13.4 Environmental Impact and Technical Challenges of Conductive and Smart Textiles

13.4.1 Impact of Manufacturing

13.4.1.1 Raw Material Impact

The production of smart textiles comes with significant sustainability issues. Non-renewable resources like petroleum-based polymers or rare earth metals are used as raw materials for smart textiles. The primary components of smart textiles are natural and synthetic fibers. The production of cotton comes with significant environmental concerns. Although cotton is biodegradable and of natural origin, the production of cotton requires large amounts of water and pesticides. Synthetic fibres such as polyester production are responsible for over 706 million tons of greenhouse gas emissions, which is equivalent to the emissions from 185 coal-fired power plants or the energy use of 64 million American homes [90]. Mining rare earth metals and minerals is crucial for components like sensors and batteries. Such electronic components lead to environmental pollution, soil erosion, and water contamination. In addition, the extraction and refinement processes contribute to global warming through greenhouse gas emissions and pollution.

Table 13.4 An overview of applications of smart and conductive textiles [67]

Category	Name of technology	Conductive material used	Sensor used	Parameters monitored/ Applications	References
Healthcare	Seamlessly integrated into textile underwear	Stainless steel and silver-plated conductive yarns	Urine leakage sensors	Prevention of nocturnal enuresis	[68]
	Smart textile based on polymer-encapsulated FBG	Fiber bragg grating (FBG)-based material	Fiber bragg grating (FBG)-based sensor	Cardiac activity	[69]
	Light emitting fabrics	Optical fibers	Fiber optic sensors	Photodynamic therapy	[70]
	Sewn textile	Silver-coated threads, electrodes	Wearable electrocardiogram (ECG) sensors	Cardiac health	[71]
	Wearable sweat loss measuring device	Cotton, copper, PTFE substrate	Antenna-based sensor, RFID sensor	Track sweat loss within range of the human perspiration rate (300–2100 $g/m^2/h$) of healthy individuals	[72]
	Multifunctional graphene-based wearable e-textiles	Graphene	Bio signal monitoring sensors, motion sensors, and supercapacitors	Monitoring EEG bio-signals and energy monitoring	[13]
	Smart T shirt	Piezoresistive textile (hybrid knit fabric metallized with pure silver) (Medtex P130)	Strain sensor	Respiratory rate monitoring	[73]
	Palpreast-stretchy bra (wearable device)	Zebra-patterned conductive fabric	Pressure-sensitive sensor	Breast lumps/tissue thickness	[74]

(continued)

Table 13.4 (continued)

Category	Name of technology	Conductive material used	Sensor used	Parameters monitored/ Applications	References
Sports and fitness	Xiaomi Mijia Cardiogram T-shirt (Sports Smart T-Shirt)	Innovative fiber marked COTECH	Advanced optical measuring sensor	Advanced heart rate monitor (ECG)	[75]
	Highly sensitive and stretchable MXene/CNTs/ TPU Composite	Ti3C2Tx MXene/carbon nanotubes (CNTs)/ thermoplastic polyurethane (TPU)	Strain sensor	Human movement monitoring	[76]
	E-textile with a dual tactile and tension stimulus response	Conductive rGO/CNT PAN fibres, spacer fabric	Tactile tension sensors, capacitive sensor	Monitoring athletic motion during taekwondo	[77]
	Stretchable wearable textile	Gold fibers	Electrochemical biosensor	Lactate monitoring in sweat during exercise	[78]
	Nadi X smart yoga pants	Embedded woven-in conductive fibers	Accelerometers (Tiny Motion Sensors)	Movement and body alignment during yoga, real-time haptic feedback to improve posture and technique	[79]
	Biosensor underwear	Carbon electrode	Amperometric sensor	Mechanical stress of a wearer's daily activity (flexing and stretching)	[80]
Military, industry	Heated gloves	resistive heating elements (wires or the tapes)	Temperature Sensors	Hand protection for military and other extreme conditions by monitoring temperature and moisture	[81]

(continued)

Table 13.4 (continued)

Category	Name of technology	Conductive material used	Sensor used	Parameters monitored/ Applications	References
	Conductive smart clothing	Silver nanowires and MXene coatings	Strain and photothermal sensors	Thermal management and stress	[82]
	Multifunctional conductive cellulose fabric	Ag nanoparticles (Ag NPs),PANI,PFOTS, cellulose fabric	Fire warning sensor	Abnormal temperature and flame proximity	[83]
Consumer lifestyle	E-Textiles for daily monitoring	PEDOT:PSS/PDMS and cotton fabric	EEG monitoring sensors	Brain activity and cognitive responses	[84]
	High-performance wearable supercapacitor	MoS_2 and graphene	Wearable supercapacitors	Energy storage device	[85]
	Ultra-flexible graphene-based wearable e-textile	rGO coated cotton yarn	Pressure and breathing sensors	Human movement detection and bio-signal detection	[86]
	Multifunctional MXene-Based Textiles	MXene with polydopamine and PDMS coatings	Strain and temperature sensors	Human motion and photothermal performance	[87]
Environmental monitoring	Vertically aligned carbon nanotube-coated conductive cotton	Carbon nanotubes (CNTs) and silica sol–gel	Humidity sensors	Environmental moisture levels	[88]
	Wearable temperature sensor integrated in non-woven fabric	Reduced graphene oxide (rGO) and carbon nanotubes (CNTs)	Temperature sensors	Human body and environment temperature	[89]

13.4.1.2 Energy Consumption

The creation of smart textiles with conductive and energy-harvesting properties requires significant energy expenditure and sophisticated manufacturing methods. Such energy-demanding production process presents a significant obstacle to sustainability. Smart textiles and conductive materials, such as carbon fiber, silver, and graphene require high-temperature processing and treatments that greatly increase energy demands. Embedding phase change materials (PCMs) in textiles improves energy efficiency during utilization but escalates energy requirements during production due to the need for supplementary chemical additives and advanced fabrication techniques [91]. Similarly, the production of smart conductive fabrics that incorporate multi-walled carbon nanotubes (MWCNTs) and conductive polymers involves energy-intensive techniques such as electrospinning and dip coating [92]. Furthermore, the manufacturing of yarn-based supercapacitors for incorporation into smart textiles involves significant energy expenditure during the carbonization and material processing stages. These yarns provide a power density of up to 27.5 $\mu W\ cm^{-1}$ [93]. Carbon fibers undergo energy-demanding thermal processes, including stabilization at temperatures ranging from 200 °C to 300 °C, followed by carbonization, which can approach 1000 °C. The energy requirements for these procedures are substantial, with studies indicating that the manufacture of carbon fibers consumes roughly 295 to 740 kWh per kilogram [94].

13.4.1.3 Generation of E-waste

Integrating smart devices onto textiles requires brazing, welding, soldering, or gluing. This type of bonding generally creates a critical weak point which is bound to failure after repetitive usage. To improve the overall usability and washability notable advancements are made to incorporate conductive threads into fabric through techniques like weaving, sewing, stitching, and knitting [95]. Even after such innovations, these small electronic devices can be difficult to recycle, and electronic hardware poses challenges even after disposal. On average, an electronic device often contains up to 40 distinct materials, each demanding specific recycling procedures due to differences in their physical and chemical properties. Although some of these components have some economic value, most of them consist of hazardous substances such as toxic metals and halogenated organic matter [96]. The most common recycling methods include shredding which fails to segregate these materials completely [97]. Moreover, there is a risk of releasing these toxic substances into the environment due to an unsophisticated recycling system. Experiments show that each washing cycle can cause a textile to lose more than 1900 fibers [98]. Research by Wäger et al. [99], highlighted that RFID tags could disrupt conventional recycling processes for non-electronic materials. Government initiatives are also taken to promote responsible recycling and disposal practices among producers [100]. However, the effectiveness of such measures is limited as the fragmented nature of e-waste legislation leads to confusion for both producers and consumers [101]. For this concerning reason, researchers have proposed certain methodologies to assess the end-of-life (EOL) of electronic devices to enhance recycling and reuse [102].

The e-waste (electronic waste) recycling issue is dependent on factors, such as the generation of larger volumes of electronic devices produced, the presence of toxic materials in electronics, the potential of risking human health and the environment by these materials, and the inclusion of costly materials in such devices [103]. These waste-related issues can be drastically mitigated by practicing environmentally conscious design while manufacturing. Implementation of such waste-preventive strategies during the early development stages can prove to be effective, especially when e-textiles are yet to be market-ready.

13.4.2 Challenges in Commercialization

13.4.2.1 Production Barriers

Smart textiles face several production challenges, including:

- **Production Costs**: A key hurdle in the commercialization of smart textiles is that they are costly to produce due to the need for specialized materials, electronics integration, and labor-intensive processes. Additionally, there is a lack of economies of scale, since many smart textile products are still in the initial stages of market development.
- **Scalability**: Scaling up the production of smart textiles from prototypes to mass production is a major challenge. This includes the advancement of standardized production techniques, supply chain issues, and the sourcing of specialized materials.
- **Integration of Electronics with Textiles**: One of the key challenges in smart textile production is the integration of electronic components like sensors, actuators, and conductive materials into traditional textile materials without compromising comfort, wearability, and durability. Textile fibers and fabrics are inherently flexible, but adding electronics without increasing weight or stiffness is complex.
- **Lack of Industry Standards**: The absence of universal standards for smart textiles makes it difficult to ensure consistent quality and interoperability between different smart textile products, which can hinder commercialization.
- **Durability and Washability**: Smart textiles are often sensitive to washing, stretching, or exposure to environmental factors (humidity, UV light, etc.), which can impair the function of the embedded electronics. Ensuring that these textiles retain their smart functionalities over repeated use and washing is a major challenge.

13.4.2.2 Durability and Long-Time Performance Issues

Even though wearable textile accessories are highly associated with a wider portion of the market share, there is an obstacle of low retention and high abandonment rates of such devices in less than one year [104]. Technical issues, such as sensor accuracy, power storage, security, washability, thermal regulations, and comfort are all challenging factors for commercialization [83].

The limited durability due to machine washing is a significant barrier to broader consumer adoption of e-textiles. Figure 13.2 demonstrates the durability issue of e-textiles during machine washing, spanning from basic molecular bonds to structural breakdown and ultimately liberating micro and nanoparticles [5]. For instance, the surface of smart textiles is often integrated with conductive materials through coating or dipping and it shows a tendency to peel off and deformations on frequent washes [105]. Hence, greater effort is necessary to improve the stability and cleaning of smart textiles along with preserving their essential functional properties.

Regular power supply, routine maintenance, and water-repellent characteristics are yet to be overcome before going to bulk production of smart textiles. Executing complex operations of smart textiles implies excessive consumption of energy and resources in most cases, resulting in considerably reduced battery health [106]. Smart textile shows comparatively low performance compared to conventional bulk or thin-film devices due to its smaller diameter and lower conductivity [107]. The ideal balance between its electrical and mechanical properties remains unachieved [108].

13.5 Environmental Management Strategies and Sustainable Production of Conductive Smart Textiles

The global e-waste generation is rising at a significant pace necessitating proper recycling practices. However, improper recycling methods without maintaining proper precautions can pose serious risks to human health. In addition, these activities serve

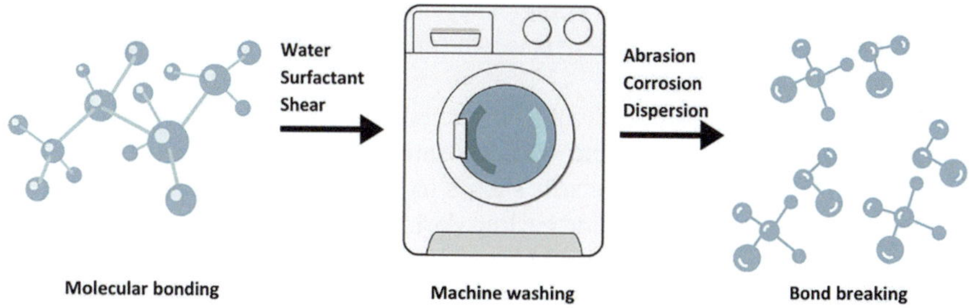

Fig. 13.2 The challenge of smart textile durability during washing [5].

as the main source of income for a large sum of workers mainly in developing countries. Tackling this challenge solely through legislative measures and organized recycling systems may not be sufficient. This is why a holistic approach is needed by raising public awareness about the risks of e-waste, strengthening the current recycling sector, improving product design and manufacturing, and expanding research on the health impacts of e-waste. The following sections explore these aspects in greater detail [109].

13.5.1 Solutions for Sustainable Smart Textile Products

13.5.1.1 Design Principles for Sustainable Smart-Textile Products

Achieving full sustainability of smart textile products is a significant challenge but proper development guidelines for such materials can reduce the overall environmental impact. Such considerations emphasizing simplicity, material efficiency, and environmental friendliness are discussed in this section [110].

- **Simplified Sensor Integration**: Smart textile's accuracy can be improved by integrating multiple sensors rather than only relying on highly intricate ones. This enables data combination from a broader range ensuring collective analysis. This simplified design also reduces manufacturing complexity and costs.
- **Scalable Designs**: Unnecessary material usage can be avoided if sensor systems can be scaled up or down based on application.
- **User-Centric Features**: Incorporating only essential functionalities based on user requirements can avoid adding redundant components.
- **Maximum Utilization of Textile Surface**: Smart textiles usually have a larger surface area where bigger electrical components can be embedded seamlessly within the fabric. This reduces the usage of extremely small sensors or electrodes.
- **Functional Layers**: Usage of multi-functional layers can make way to simultaneously integrate sensors, conductors, and other components over large areas.
- **Aesthetic and Functional Balance**: The smart textiles should be designed to maintain their visual appeal to customers as well as maintain proper functionality.
- **Reliability in Smart Textiles**: Redundancy in data pathways and parallel circuitry can improve dependability. This ensures the product functionality even when some components fail.
- **Durability**: Smart textiles should be subjected to durability testing, such as washing, mechanical stress, and heat exposure.
- **Fail-Safe Mechanisms**: The textile should include fail-safe systems to notify the users of malfunctions or provide backup during failure.
- **Material Selection and Minimal Usage**: Using single material designs, such as 100% polyester or organic fibers like wool, simplifies the recycling processes and enhances sustainability.

- **Biodegradable Alternative**: Biodegradable or compostable materials should be incorporated wherever possible to reduce post-consumer waste.
- **Lightweight Design**: Focus should be given to lightweight materials to reduce material consumption.
- **Solar Integration**: In outdoor applications, flexible photovoltaic panels should be embedded into textiles for energy harvesting.
- **Ensuring Recycling and End-of-Life Management**: To prevent e-waste from entering general waste streams, designs should prioritize easy disassembly and separation of electronic elements from textile bases.
- **Modular Components**: Usage of detachable or modular electronic components that can be easily removed for recycling should be encouraged.
- **Life Cycle Assessments (LCA)**: LCA should be conducted at the design phase to identify and mitigate environmental impact at every stage of production.
- **Multi-Use Application**: Developing multi-functional products can reduce the need for separate items & devices.

13.5.1.2 Sustainable Raw Materials

Conventional functional materials, such as conductive polymers, carbon-based nanomaterials, and metal-based compounds are widely utilized in smart textiles to achieve intended properties. Major environmental impacts associated with the manufacturing of smart textiles are related to the consumption of non-renewable resources, energy-intensive production, and non-readily recyclable materials commonly applied to manufacture smart textiles. Several studies have indicated that switching to biodegradable and recyclable materials could lessen the negative environmental impacts [111].

Biodegradable and Renewable Fibers

Petroleum based synthetic fibers, such as polyester and nylon are not biodegradable. Because of such origin, their production process has very high carbon emissions. The energy consumption of per kg polyester production is 125 MJ and emits about 14.2 kg CO_2/kg in the process, thereby contributing to greenhouse gases [96]. However, cellulose-based fibers, such as cotton and linen, along with regenerated cellulose, provide renewable and biodegradable options. These fibers exhibit compatibility with conductive materials, enabling their application in diverse functionalized textile treatments and coatings. Plant-based CNCs (cellulose nanocrystals) are suitable for flexible electronic textiles due to their excellent mechanical properties. They have a young's modulus of 110–220 GPa and tensile strength of 1–10 GPa which is similar to kevlar fiber [112]. CNCs and silver nanowire composites exhibit excellent conductivity (10^4 S/m) and stability at temperatures near 250 °C, making them suitable for wearable electronics [113]. Fungi-based substace called 'Mycelium' is attracting interest due to its significant environmental benefits, as it has low water and energy requirement for production. Mycelium decomposes within four to six weeks in soil, presenting itself as a viable sustainable raw material for the development of eco-friendly textile technologies [114]. Graphene-collagen composites are another potential alternative that provide exceptional flexibility, conductivity, and biodegradability showing a biodegradability of less than 60 h [115].

Plant-Based Biodegradable Polymers for Coatings

Renewable polymers such as polylactic acid (PLA) and polyhydroxyalkanoates (PHA) serve as environmentally sustainable and biodegradable substitutes for polyethylene and polypropylene. PLA-coated conductive fabrics have enhanced mechanical flexibility, a tensile strength between 50 and 70 MPa and elongation at break of 6 to 10%. Furthermore, PLA's biodegradability in industrial composting environments aids in waste reduction, as it decomposes within 6 to 12 months upon disposal. The remarkable tensile strength and flexibility of PLA boost its suitability for applications requiring prolonged durability, whereas the quick biodegradation rate of PHA provides benefits for items designed for short-term use [116].

Conductive Polymers

Advanced smart materials can be synthesized from conductive polymers without the need for metal incorporation, thus enhancing resource efficiency. The main types of these polymers, namely polyaniline (PANI), polypyrrole (PPy), and poly (3, 4- ethylenedioxythiophene) (PEDOT), demonstrate electrical conductivities that span from 10^{-10} to 10^5 S/cm. These polymers combine the electrical properties of metals with the mechanical flexibility and lower environmental impact typical of organic materials, making them suitable for sustainable smart textiles. The synthesis of PEDOT generally requires oxidants like $FeCl_3$, which may pose environmental hazards. Additionally, conductive polymers like PEDOT can be processed in water, reducing the need for toxic solvents and enhancing the overall sustainability of the production process [117] (Fig. 13.3).

13.5.1.3 Sustainable Manufacturing

Sustainable production indicates lower water and energy consumptions, durable inherent property throughout the entire life cycle and recyclability in the final stage of its useful life. The melt spinning process is a simple as well as economical method for polymer fiber production, where melted fiber-forming substances are extruded through a spinneret and solidified with cold-air quench ducts [96]. Since there are no solvents involved, there are no byproducts or recycling concerns, making this an ideal alternative to both wet-spinning and dry-spinning methods. Conductive nanomaterials, like graphene and carbon nanotubes (CNTs) can be integrated into textile

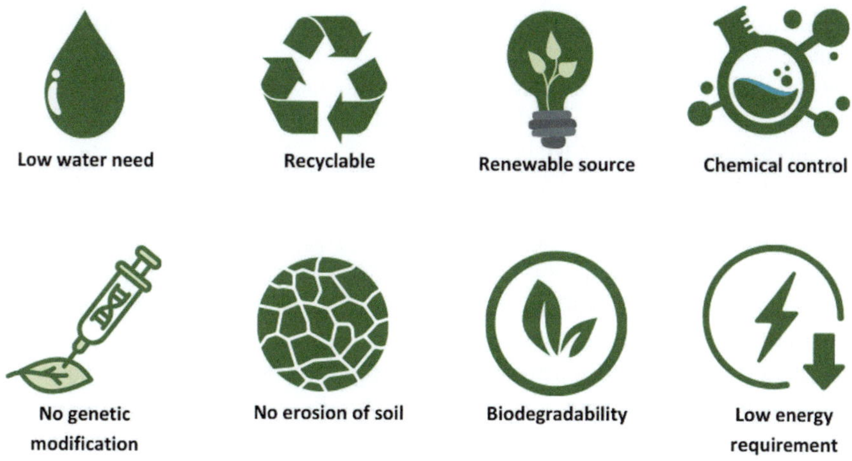

Fig. 13.3 Selection criteria for fiber [96]

structures during the fibre-spinning phase by adding them directly into the polymer solution. This solution is then extruded, forming conductive fibers or filaments. This spinning technique goes well for common fibers made from PET, polyamide (nylon) [118], [3]. Additionally, in the melt-spinning process for e-textiles, PP and PE melts can directly act as initiators, enabling the production of ultrafine fibres [119].

Inkjet printing is currently leading advancements in flexible printed electronics, offering a cost-effective, adaptable, and high-resolution multi-layer printing [96]. Insulating, semi conductive and conductive materials can be utilized in various combinations for inkjet printing, facilitating the fabrication of different heterostructures, bringing sharply designed characteristics to have enhanced performance and diversified functionalities. It represents a sustainable substitute and complementary approach to the multicomponent melt-spinning of multimaterial fibres. Current wearable devices go through reliability and manufacturing related issues that can be easily solved by printing such devices in this process. It offers non-impact and mask-less deposition of controlled amounts of materials in a precise, rapid and reliable way [96].

13.5.1.4 E-waste Collection and Recycling

The treatment of e-waste comes with several challenges, as it consists of materials like gold, copper, silver, and rare earth metals, alongside hazardous compounds, such as cadmium, lead, and halogenated compounds posing great health and ecological risks [120], [121, 122]. These toxic compounds may enter the environment through traditional recycling systems like incineration or landfill leachate [123]. In developing countries like China, India, and Nigeria, e-waste is still processed informally [122].

Burning methods for copper recovery from insulated cables or gold recovery from printed circuit boards can release toxic chemicals like furans, dioxins, and other pollutants [124]. Although certain measures are already taken to mitigate such informal recycling methods through legislation and trade restrictions, practical enforcement remains a significant challenge [124].

A promising method for improving the recyclability of such electrical products could be *Design for Disassembly (DfD)* [125]. This approach focuses on creating easily separable materials enabling recycling, reuse, and maintenance, thus minimizing environmental impact. For example, the ReUse project developed a printed circuit board in which components can be disassembled by simply immersing in hot water [126]. Product service systems (PSSs) has a promising solution by creating closed loop systems, where consumers return product components in exchange for replacement [127]. Beverage industry has successfully applied deposit refunding systems to reduce plastic waste. Manufacturers of wearable devices could implement similar strategies, such as offering trade-in programs to promote sustainable disposal practices [128] (Table 13.5).

13.5.2 Sustainable Practices and Policies in Smart-Textile Manufacturing

In a time when sustainability and technology are key concerns, the fashion industry is at a pivotal point of encountering unique opportunities and continuous challenges [144]. Conventional textile manufacturing process has been infamously resource-draining with excessive water footprint and unavoidable dependence on non-renewable energy sources resulting in environmental damages [145]. Standing at the intersecting point of sustainability and technology, smart textiles represent an optimistic approach with the prospect of thoroughly addressing these challenges.

13.5.2.1 Policy and Regulatory Challenges

Policy and regulatory challenges are the difficulties that an organization faces while navigating the legal frameworks set by governing bodies. Batteries, sensors, and other electrical components are often incorporated with smart textiles which may pose risk to the consumers, if not properly regulated. In the EU, the General Product Safety Directive (2001/95/EC) issues a framework to ensure consumer welfare for all products placed on the market [146]. Smart textiles that contain electronic components must comply with regulations governing electromagnetic fields (EMF). The Electromagnetic Compatibility (EMC) in the EU and the Federal Communications Commission (FCC) guidelines in the U.S. regulate the use of wireless technologies [147]. Many smart textiles collect biometric data, like heart rate, and temperature, particularly in healthcare or fitness. In many jurisdictions, such as the EU's General

Table 13.5 An overview of design implications and themes linking sustainability to wearable technologies [128]

Life cycle categories	Implication in wearable technologies	References
Materials	· Evaluate materials using life cycle assessments with respective eco-costs to understand their environmental impacts · Use recycled and upcycled materials to reduce resource consumption · Design with recyclability and biodegradability such as using bioLogic's natto cells for adaptive functionalities in clothing	[129, 130]
Manufacturing	· Involve end users in assembly using platforms like SparkFun · Design for easy disassembly with modular components to enhance recyclability and employ sustainability-focused toolkits · Leverage on-demand manufacturing methods such as 3D printing, and collaborate with production partners near your market, like 3D Hubs or Studio HILO for localized manufacturing and supply chain efficiency	[131–135]
Distribution	· Adopt open source design for shared design and development responsibility, allowing consumers to complete assembly, and reducing distribution needs. Various examples of open source design include cars, microcontrollers such as BeagleBone, and Adafruit's FLORA platform for wearables · Employ direct digital manufacturing to lessen shipping burdens	[136, 133, 137]
Use	· Focus on designs that require minimal care and are easy to maintain, such as self-maintaining textiles · Instruct users on proper upkeep and repair methods · Integrate alternative energy sources like energy harvesting to make devices self-sufficient · Ensure device compatibility and upgradability to prevent early obsolescence · Address practical needs like functionality, comfort, and aesthetic desires, offering customization options to fit individual needs · Blend wearables into fashion seamlessly, allowing for part transformation and body-conscious designs · Engage users in the design process to confirm the product's practical value through co-design	[138–141]
Disposal	· Promote business models that encourage recycling by users · Guide users in correctly disposing of parts, especially hazardous ones like batteries · Construct wearables to enable component replacement or refurbishment	[1, 142, 143]

Data Protection Regulation (GDPR) or California Consumer Privacy Act (CCPA), there are stringent requirements around data privacy, security, and user consent. So, manufacturers need to ensure that. Regulatory frameworks such as the NIST Cybersecurity Framework in the U.S. provide guidelines for managing cybersecurity risks in connected devices. These standards are particularly relevant for manufacturers of wearables and IoT-enabled textiles. For the case of smart textile production, material safety and disposal are the main focuses. Synthetic material, batteries, and electronics can create environmental challenges. In compliance with environmental regulations, such as the EU's Waste Electrical and Electronic Equipment (WEEE) Directive or Battery Directive, manufacturers must consider the recyclability and safe disposal of electronic components and batteries in smart textiles. The convergence of textile manufacturing and electronics creates a complex patent landscape. Companies may face challenges in determining which parts of a product are patentable and how to protect their intellectual property across different regions. In the case of international trade, companies related to smart textile face challenges in complying with diverse regulations across borders. As some countries like the EU and U.S. may have strict safety standards for electronics, while others might lack these requirements or have any other standards. Besides these, there are some other policy and regulatory challenges like Certification and Standards Compliance, Supply Chain Transparency and Ethical Sourcing, Market Access, and Consumer Acceptance etc.

13.5.2.2 Adoption to 4R

The implementation of 4R (Repair, Recycle, Replacement, Reduction) strategy is crucial in promoting sustainability and environment-friendly approaches in the textile sector, including the development of smart textiles. Reknitting, resewing, reweaving, and self-healable electronic fibers are a few impactful 'fiber level' **repairing** approaches [5]. Active materials are employed through recoating, and respraying for giant 'fabric-level' repairing. E-textile components can be divided into two categories- electronic components and base textiles. Regular e-waste **recycling** streams can be utilized to process the electronic modules, whereas base textiles can be treated in the same manner as regular textiles for **recycling** purposes. **Replacement** emphasizes the minimization of environmental impacts during disposal. For example, adopting naturally abundant and renewable materials for the development of functional fibers. Serving as sustainable and renewable sources of raw materials, biomass fibres, including wool, flax, silk, and hemp hold potential for designing eco-friendly smart textiles [148]. With the advances in the insulating characteristics, bio-based polymers, such as poly(lactic acid), poly(octamethylene maleate (anhydride) citrate), poly hydroxyl butyrate or poly hydroxyl valerate offer potential alternatives to synthetic encapsulation materials [149]. The **reduction** can be analyzed in two ways: firstly, minimizing energy consumption and emissions throughout the production and usage of smart textiles; and secondly, reducing the amount of material used to obtain the desired function, particularly when recycling, repairing and replacing processes may cause high carbon emission or excessive energy costs. By focusing on these four principles, the environmental impact of production and disposal of smart textiles can be restricted [5].

13.5.2.3 Circular Economy

The idea of circular economy highlights the entire product life cycle from production to disposal, more specifically, materials' sustainability, longevity, repairability, reusability and recyclability [150]. The circular economy system promotes sustainable textiles with three primary operations: (a) expanding rental services for short-term clothing, (b) making sustainability more desirable, and (c) growing the utilization of clothing further through brand policies and commitments [150].

Smart textiles hold the capability to provide an eco-friendlier alternative to conventional clothing provided that they give more benefits to the consumer, with increased durability [110]. Research indicates that while one in every ten American adults owns some sort of wearable device, but half of them do not actively use it [151]. Upon investigation, three main factors are found behind such behavior: (1) The product does not match users' distinct self-image and is mostly tailored for specialized users, such as sports enthusiasts or patients, (2) The collected statistics and metrics are not relevant and helpful enough, and (3) The maintenance of such product is considered as burden [151]. Hence, consumer behavior is crucial in determining a product's sustainability. Although a product may be designed based on the maximum standards of sustainability, user's dealings with that product can overrule these attempts. Smart textiles are required to align with real-world applications as well as wearable properties like flexibility and breathability in order to fully unlock their commercial potential [152].

The increasing interest in smart textiles and their applications is driven by their underlying ability to create innovative, versatile, and customizable product experiences. While most of the focus on smart textiles has been on their physical properties and electronic performance, sustainability has always received minimal attention. When considering sustainability in the assessment of smart textiles, it is evident that eco-friendly alternatives can achieve the same "smart" effects, but they often fall short in performance compared to usual fossil fuel-based and non-biodegradable substances. A quantitative assessment of each material's environmental impact throughout its life cycle is essential to make manufacturers confident with eco-friendly alternative materials. This assessment needs to be compatible with existing evaluation methods, allowing a fair comparison of newly developed sustainable materials with traditional options for similar applications. Metrics like biodegradability, life cycle analysis, and recyclability offer valuable insights for assessing sustainability [96]. Furthermore, research continues on the way materials are perceived, the potential of using them interchangeably to create distinct user experiences and the enhancement in performance of sustainable materials [153–155].

Although smart textiles show promising potential in promoting sustainable fashion, there are several hurdles that need to be overcome. With a view to fully utilizing the potential of smart textiles, challenges such as durability, cost, environmental impact of electronic components, as well as the necessity for scalability and accessibility, must be addressed for reaching a broader audience.

13.6 Conclusion

The field of smart textiles is rapidly advancing with great potential for innovation and commercial success. Both in the case of hardware and software, there is still a lot of room for improvement. The integration of conductive materials and sensors into textiles depends on selecting the right materials and integration method. Employed manufacturing techniques ensure durability, wearability, stretchability, and functionality. Advancements in materials, such as self-healing polymers, graphene, nanomaterials, etc., are expected to improve the overall performance of smart textiles. These textiles are unlocking new applications from healthcare and military to consumer products and industrial systems. The e-textiles market is set for significant growth. To seize this opportunity, smart textiles must meet demands for scalability, comfort, durability, safety, innovative design, and cost-effectiveness. However, moving from lab-scale prototypes to market-ready products requires addressing several key challenges. Overcoming challenges related to power generation, material compatibility, production, policy and regulation will be crucial. Ongoing research and collaboration among academics, industry, and government will drive the next phase of innovation in this field.

Declaration The authors declare that they have no known competing financial interests or personal relationships that could have appeared to influence the work reported in this chapter. Generative AI and AI-assisted technologies have been used in the writing process to improve the readability and language of the work only. After using this tool, the authors carefully reviewed and revised the content to ensure accuracy and quality. The authors take full responsibility for the final version of this publication.

References

1. Komolafe, A., Zaghari, B., Torah, R., Weddell, A. S., Khanbareh, H., Tsikriteas, Z. M., Vousden, M., Wagih, M., Jurado, U. T., Shi, J., Yong, S., Arumugam, S., Li, Y., Yang, K., Savelli, G., White, N. M., & Beeby, S. (2021). E-textile technology review-from materials to application. *IEEE Access, 9*, 97152–97179. https://doi.org/10.1109/ACCESS.2021.3094303
2. Afroj, S., Tan, S., Abdelkader, A. M., Novoselov, K. S., & Karim, N. (2020). Highly conductive, scalable, and machine washable graphene-based e-textiles for multifunctional wearable electronic applications. *Advanced Functional Materials, 30*(23), 2000293. https://doi.org/10.1002/adfm.202000293
3. Chen, G., Cuculo, J. A., & Tucker, P. A. (1992). Effects of spinning conditions on morphology and properties of polyethylene terephthalate fibers spun at high speeds. *Journal of Applied Polymer Science, 44*(3), 447–458. https://doi.org/10.1002/app.1992.070440310

4. Islam, M. R., Afroj, S., Yin, J., Novoselov, K. S., Chen, J., & Karim, N. (2024). Advances in printed electronic textiles. *Advanced Science, 11*(6), 2304140. https://doi.org/10.1002/advs.202304140
5. Shi, H. H., Pan, Y., Xu, L., Feng, X., Wang, W., Potluri, P., Hu, L., Hasan, T., & Huang, Y. Y. S. (2023). Sustainable electronic textiles towards scalable commercialization. *Nature Materials, 22*(11), 1294–1303. https://doi.org/10.1038/s41563-023-01615-z
6. Yang, K., McErlain-Naylor, S. A., Isaia, B., Callaway, A., & Beeby, S. (2024). E-Textiles for sports and fitness sensing: Current state, challenges, and future opportunities. *Sensors, 24*(4), 1058. https://doi.org/10.3390/s24041058
7. Yin, J., Wang, S., Di Carlo, A., Chang, A., Wan, X., Xu, J., Xiao, X., & Chen, J. (2023). *Smart textiles for self-powered biomonitoring. Med-X, 1*(1), 3. https://doi.org/10.1007/s44258-023-00001-3
8. Konstantinidis, D., Iliakis, P., Tatakis, F., Thomopoulos, K., Dimitriadis, K., Tousoulis, D., & Tsioufis, K. (2022). Wearable blood pressure measurement devices and new approaches in hypertension management: The digital era. *Journal of Human Hypertension, 36*(11), 945–951. https://doi.org/10.1038/s41371-022-00675-z
9. Fernández-Caramés, T., & Fraga-Lamas, P. (2018). Towards The internet of smart clothing: A review on IoT wearables and garments for creating intelligent connected e-textiles. *Electronics, 7*(12), 405. https://doi.org/10.3390/electronics7120405
10. Baurley, S. (2004). Interactive and experiential design in smart textile products and applications. *Personal and Ubiquitous Computing, 8*(3–4). https://doi.org/10.1007/s00779-004-0288-5
11. Yang, K., Isaia, B., Brown, L. J. E., & Beeby, S. (2019). E-Textiles for healthy ageing. *Sensors, 19*(20), 4463. https://doi.org/10.3390/s19204463
12. Beeby, S. P., Torah, R. N., Wagih, M., Isaia, B., Black, S., Saunders, J., & Yang, K. (2024). Heterogeneous e-textiles: Materials, manufacturing and sustainability. *Advanced Materials Technologies*, 2400844. https://doi.org/10.1002/admt.202400844
13. Islam, M. R., Afroj, S., Beach, C., Islam, M. H., Parraman, C., Abdelkader, A., Casson, A. J., Novoselov, K. S., & Karim, N. (2022). Fully printed and multifunctional graphene-based wearable e-textiles for personalized healthcare applications. *I science, 25*(3). https://doi.org/10.1016/j.isci.2022.103945
14. Idumah, C. I. (2023). Design, fabrication, characterization and properties of metallic and conductive smart polymeric textiles for multifunctional applications. *Nano-Structures and Nano-Objects, 35*, 100982. https://doi.org/10.1016/j.nanoso.2023.100982
15. Ghahremani Honarvar, M., & Latifi, M. (2017). Overview of wearable electronics and smart textiles. *Journal of the Textile Institute, 108*(4), 631–652. https://doi.org/10.1080/00405000.2016.1177870
16. Sharaf, S. M. (2020). Smart conductive textile. In *Advances in Functional and Protective Textiles*. LTD. https://doi.org/10.1016/b978-0-12-820257-9.00007-2
17. Ashayer-Soltani, R., Hunt, C., & Thomas, O. (2016). Fabrication of highly conductive stretchable textile with silver nanoparticles. *Textile Research Journal, 86*(10), 1041–1049. https://doi.org/10.1177/0040517515603813
18. Shahzad, A., Rasheed, A., Khaliq, Z., Qadir, M. B., Khan, M. Q., Hamdani, S. T. A., Ali, Z., Afzal, A., Irfan, M., Shafiq, M., & Kim, I. S. (2019). Processing of metallic fiber hybrid spun yarns for better electrical conductivity. *Materials and Manufacturing Processes, 34*(9), 1008–1015. https://doi.org/10.1080/10426914.2019.1594270
19. Hou, Z., Liu, X., Tian, M., Zhang, X., Qu, L., Fan, T., & Miao, J. (2023). Smart fibers and textiles for emerging clothe-based wearable electronics: Materials, fabrications and applications. *Journal of Materials Chemistry A, 11*(33), 17336–17372. https://doi.org/10.1039/d3ta02617e
20. Cheng, H., Dong, Z., Hu, C., Zhao, Y., Hu, Y., Qu, L., Chen, N., & Dai, L. (2013). Textile electrodes woven by carbon nanotube-graphene hybrid fibers for flexible electrochemical capacitors. *Nanoscale, 5*(8), 3428–3434. https://doi.org/10.1039/c3nr00320e

21. Shim, B. S., Chen, W., Doty, C., Xu, C., & Kotov, N. A. (2008). Smart electronic yarns and wearable fabrics for human biomonitoring made by carbon nanotube coating with polyelectrolytes. *Nano Letters, 8*(12), 4151–4157. https://doi.org/10.1021/nl801495p
22. Sanivada, U. K., Esteves, D., Arruda, L. M., Silva, C. A., Moreira, I. P., & Fangueiro, R. (2022). Joule-heating effect of thin films with carbon-based nanomaterials. *Materials, 15*(12). https://doi.org/10.3390/ma15124323
23. Karim, N., Afroj, S., Leech, D., & Abdelkader, A. M. (2021). Flexible and wearable graphene-based e-textiles. *Oxide Electronics*, 21–49. https://doi.org/10.1002/9781119529538.ch2
24. Alhashmi Alamer, F., Almalki, G. A., & Althagafy, K. (2023). Advancements in conductive cotton thread-based graphene: A new generation of flexible, lightweight, and cost-effective electronic applications. *Journal of Composites Science, 7*(11). https://doi.org/10.3390/jcs7110476
25. Wei, H., Cauchy, X., Navas, I. O., Abderrafai, Y., Chizari, K., Sundararaj, U., Liu, Y., Leng, J., & Therriault, D. (2019). Direct 3D printing of hybrid nanofiber-based nanocomposites for highly conductive and shape memory applications. *ACS Applied Materials and Interfaces, 11*(27), 24523–24532. https://doi.org/10.1021/acsami.9b04245
26. Zhang, Y., Zhang, Y., Zhou, J., Zhang, D., Lin, H., Chen, Y., Li, Y., & Xiong, J. (2023). Stretchable composite conductive fibers for wearables. *Advanced Materials Technologies, 8*(6), 1–25. https://doi.org/10.1002/admt.202201442
27. Lv, X., Tang, Y., Tian, Q., Wang, Y., & Ding, T. (2020). Ultra-stretchable membrane with high electrical and thermal conductivity via electrospinning and in-situ nanosilver deposition. *Composites Science and Technology, 200*(July), 108414. https://doi.org/10.1016/j.compscitech.2020.108414
28. Shin, Y. E., Cho, J. Y., Yeom, J., Ko, H., & Han, J. T. (2021). Electronic textiles based on highly conducting poly(vinyl alcohol)/carbon nanotube/silver nanobelt hybrid fibers. *ACS Applied Materials and Interfaces, 13*(26), 31051–31058. https://doi.org/10.1021/acsami.1c08175
29. Moradi, A., Szewczyk, P. K., & Stachewicz, U. (2023). Bridging a gap in thermal conductivity and heat transfer in hybrid fibers and yarns via polyimide and silicon nitride composites. *Small (Weinheim an der Bergstrasse, Germany), 19*(52), 2305104. https://doi.org/10.1002/SMLL.202305104
30. Shak Sadi, M., & Kumpikaitė, E. (2022). Advances in the robustness of wearable electronic textiles: strategies, stability, washability and perspective. *Nanomaterials, 12*(12). https://doi.org/10.3390/nano12122039
31. Zhu, C., Wu, J., Yan, J., & Liu, X. (2022). Advanced fiber materials for wearable electronics. *Advanced Fiber Materials, 5*(1), 12–35. https://doi.org/10.1007/s42765-022-00212-0
32. Shi, J., Liu, S., Zhang, L., Yang, B., Shu, L., Yang, Y., Ren, M., Wang, Y., Chen, J., Chen, W., Chai, Y., & Tao, X. (2020). Smart textile-integrated microelectronic systems for wearable applications. *Advanced Materials, 32*(5), 1–37. https://doi.org/10.1002/adma.201901958
33. Singha, K., Kumar, J., & Pandit, P. (2019). Recent advancements in wearable & smart textiles: An overview. *Materials Today: Proceedings, 16*, 1518–1523. https://doi.org/10.1016/j.matpr.2019.05.334
34. Lauterbach, C., & Jung, S. (2005). Integrated microelectronics for smart textiles. *Ambient Intelligence*, 31–47. https://doi.org/10.1007/3-540-27139-2_3
35. Ali, I., Dulal, M., Karim, N., & Afroj, S. (2024). 2D Material-based wearable energy harvesting textiles: A review. *Small Structures, 5*(1), 2300282. https://doi.org/10.1002/sstr.202300282
36. Behera, S. A., Panda, S., Hajra, S., Kaja, K. R., Pandey, A. K., Barranco, A., Jeong, S. M., Vivekananthan, V., Kim, H. J., & Achary, P. G. R. (2024). Current trends on advancement in smart textile device engineering. *Advanced Sustainable Systems, 2400344*, 1–23. https://doi.org/10.1002/adsu.202400344
37. Tseghai, G. B., Malengier, B., Fante, K. A., Nigusse, A. B., & Van Langenhove, L. (2020). Integration of conductive materials with textile structures, an overview. *Sensors (Switzerland), 20*(23), 1–28. https://doi.org/10.3390/s20236910
38. Griggs, T., Ahmed, J., Majd, H., Edirisinghe, M., & Chen, B. (2024). A bio-based thermoplastic polyurethane with triple self-healing action for wearable technology and smart textiles. *Materials Advances, 5*(15), 6210–6221. https://doi.org/10.1039/D4MA00289J

39. Biswas, A., Singh, A. P., Rana, D., Aswal, V. K., & Maiti, P. (2018). Biodegradable toughened nanohybrid shape memory polymer for smart biomedical applications. *Nanoscale, 10*(21), 9917–9934. https://doi.org/10.1039/C8NR01438H
40. Sun, S., Chen, C., Zhang, J., & Hu, J. (2023). Biodegradable smart materials with self-healing and shape memory function for wound healing. *RSC Advances, 13*(5), 3155–3163. https://doi.org/10.1039/D2RA07493A
41. Farahani, M., & Shafiee, A. (2021). Wound healing: From passive to smart dressings. *Advanced Healthcare Materials, 10*(16), 2100477. https://doi.org/10.1002/adhm.202100477
42. Shuai, L., Guo, Z. H., Zhang, P., Wan, J., Pu, X., & Wang, Z. L. (2020). Stretchable, self-healing, conductive hydrogel fibers for strain sensing and triboelectric energy-harvesting smart textiles. *Nano Energy, 78*, 105389. https://doi.org/10.1016/j.nanoen.2020.105389
43. Falahati, M., Ahmadvand, P., Safaee, S., Chang, Y.-C., Lyu, Z., Chen, R., Li, L., & Lin, Y. (2020). Smart polymers and nanocomposites for 3D and 4D printing. *Materials Today, 40*, 215–245. https://doi.org/10.1016/j.mattod.2020.06.001
44. Ryan, K. R., Down, M. P., & Banks, C. E. (2021). Future of additive manufacturing: Overview of 4D and 3D printed smart and advanced materials and their applications. *Chemical Engineering Journal, 403*, 126162. https://doi.org/10.1016/j.cej.2020.126162
45. Yu, K., Xin, A., Du, H., Li, Y., & Wang, Q. (2019). Additive manufacturing of self-healing elastomers. *NPG Asia Materials, 11*(1), 7. https://doi.org/10.1038/s41427-019-0109-y
46. Darabi, M. A., Khosrozadeh, A., Mbeleck, R., Liu, Y., Chang, Q., Jiang, J., Cai, J., Wang, Q., Luo, G., & Xing, M. (2017). Skin-inspired multifunctional autonomic-intrinsic conductive self-healing hydrogels with pressure sensitivity, stretchability, and 3D printability. *Advanced Materials, 29*(31), 1700533. https://doi.org/10.1002/adma.201700533
47. Kee, S., Haque, M. A., Corzo, D., Alshareef, H. N., & Baran, D. (2019). Self-healing and stretchable 3D-printed organic thermoelectrics. *Advanced Functional Materials, 29*(51), 1905426. https://doi.org/10.1002/adfm.201905426
48. Li, X., Yu, R., He, Y., Zhang, Y., Yang, X., Zhao, X., & Huang, W. (2019). Self-healing polyurethane elastomers based on a disulfide bond by digital light processing 3D printing. *ACS Macro Letters, 8*(11), 1511–1516. https://doi.org/10.1021/acsmacrolett.9b00766
49. Hansen, C. J., Wu, W., Toohey, K. S., Sottos, N. R., White, S. R., & Lewis, J. A. (2009). Self-healing materials with interpenetrating microvascular networks. *Advanced Materials, 21*(41), 4143–4147. https://doi.org/10.1002/adma.200900588
50. Bernhart, S., Harbour, E., Kranzinger, S., Jensen, U., & Finkenzeller, T. (2023). Wearable chest sensor for stride and respiration detection during running. *Sports Engineering, 26*(1), 19. https://doi.org/10.1007/s12283-023-00410-y
51. Meng, K., Xiao, X., Wei, W., Chen, G., Nashalian, A., Shen, S., Xiao, X., & Chen, J. (2022). Wearable pressure sensors for pulse wave monitoring. *Advanced Materials, 34*(21), 2109357. https://doi.org/10.1002/adma.202109357
52. Karim, N., Afroj, S., Lloyd, K., Oaten, L. C., Andreeva, D. V., Carr, C., Farmery, A. D., Kim, I.-D., & Novoselov, K. S. (2020). Sustainable personal protective clothing for healthcare applications: A review. *ACS Nano, 14*(10), 12313–12340. https://doi.org/10.1021/acsnano.0c05537
53. Schwarz, A., Van Langenhove, L., Guermonprez, P., & Deguillemont, D. (2010). A roadmap on smart textiles. *Textile Progress, 42*(2), 99–180. https://doi.org/10.1080/00405160903465220
54. Meng, K., Zhao, S., Zhou, Y., Wu, Y., Zhang, S., He, Q., Wang, X., Zhou, Z., Fan, W., Tan, X., Yang, J., & Chen, J. (2020). A wireless textile-based sensor system for self-powered personalized health care. *Matter, 2*(4), 896–907. https://doi.org/10.1016/j.matt.2019.12.025
55. Zhang, M., Zhao, M., Jian, M., Wang, C., Yu, A., Yin, Z., Liang, X., Wang, H., Xia, K., Liang, X., Zhai, J., & Zhang, Y. (2019). Printable smart pattern for multifunctional energy-management e-textile. *Matter, 1*(1), 168–179. https://doi.org/10.1016/j.matt.2019.02.003
56. Popescu, M., & Ungureanu, C. (2023). Green nanomaterials for smart textiles dedicated to environmental and biomedical applications. *Materials, 16*(11), 4075. https://doi.org/10.3390/ma16114075

57. Shah, M. A., Pirzada, B. M., Price, G., Shibiru, A. L., & Qurashi, A. (2022). Applications of nanotechnology in smart textile industry: A critical review. *Journal of Advanced Research, 38*, 55–75. https://doi.org/10.1016/j.jare.2022.01.008
58. Abou Elmaaty, T. M., Elsisi, H., Elsayad, G., Elhadad, H., & Plutino, M. R. (2022). Recent Advances in functionalization of cotton fabrics with nanotechnology. *Polymers, 14*(20), 4273. https://doi.org/10.3390/polym14204273
59. Mehravani, B., Ribeiro, A., & Zille, A. (2021). Gold nanoparticles synthesis and antimicrobial effect on fibrous materials. *Nanomaterials, 11*(5), 1067. https://doi.org/10.3390/nano11051067
60. Rohani, R., Dzulkharnien, N. S. F., Harun, N. H., & Ilias, I. A. (2022). Green approaches, potentials, and applications of zinc oxide nanoparticles in surface coatings and films. *Bioinorganic Chemistry and Applications, 2022*(1), 3077747. https://doi.org/10.1155/2022/3077747
61. Wasim, M., Khan, M. R., Mushtaq, M., Naeem, A., Han, M., & Wei, Q. (2020). Surface modification of bacterial cellulose by copper and zinc oxide sputter coating for UV-resistance/antistatic/antibacterial characteristics. *Coatings, 10*(4), 364. https://doi.org/10.3390/coatings10040364
62. Jang, H.-S., Moon, M. S., & Kim, B. H. (2021). Electronic textiles fabricated with graphene oxide-coated commercial textiles. *Coatings, 11*(5), 489. https://doi.org/10.3390/coatings11050489
63. Ortelli, S., Malucelli, G., Blosi, M., Zanoni, I., & Costa, A. L. (2019). NanoTiO2@DNA complex: A novel eco, durable, fire retardant design strategy for cotton textiles. *Journal of Colloid and Interface Science, 546*, 174–183. https://doi.org/10.1016/j.jcis.2019.03.055
64. Mirani, B., Pagan, E., Shojaei, S., Dabiri, S. M. H., Savoji, H., Mehrali, M., Sam, M., Alsaif, J., Bhiladvala, R. B., Dolatshahi-Pirouz, A., Radisic, M., & Akbari, M. (2020). Facile method for fabrication of meter-long multifunctional hydrogel fibers with controllable biophysical and biochemical features. *ACS Applied Materials & Interfaces, 12*(8), 9080–9089. https://doi.org/10.1021/acsami.9b23063
65. Samuel, E., Joshi, B., Kim, M.-W., Kim, Y.-I., Swihart, M. T., & Yoon, S. S. (2019). Hierarchical zeolitic imidazolate framework-derived manganese-doped zinc oxide decorated carbon nanofiber electrodes for high performance flexible supercapacitors. *Chemical Engineering Journal, 371*, 657–665. https://doi.org/10.1016/j.cej.2019.04.065
66. Smart Textiles Production: Smart Textiles Production. (2019). MDPI. https://doi.org/10.3390/books978-3-03897-498-7
67. Ahsan, M., Teay, S. H., Sayem, A. S. M., & Albarbar, A. (2022). Smart clothing framework for health monitoring applications. *Signals, 3*(1), Article 1. https://doi.org/10.3390/signals3010009
68. Zaman, S. U., Tao, X., Cochrane, C., & Koncar, V. (2022). Smart e-textile systems: A review for healthcare applications. *Electronics (Switzerland), 11*(1). https://doi.org/10.3390/electronics11010099
69. Presti, D. Lo, Massaroni, C., Di Tocco, J., Schena, E., Formica, D., Caponero, M. A., Longo, U. G., Carnevale, A., D'Abbraccio, J., Massari, L., & Oddo, C. (2019). Cardiac monitoring with a smart textile based on polymer-encapsulated FBG: Influence of sensor positioning. *Medical Measurements and Applications, MeMeA 2019–Symposium Proceedings*, 1–6. https://doi.org/10.1109/MeMeA.2019.8802157
70. Mordon, S., Thécua, E., Ziane, L., Lecomte, F., Deleporte, P., Baert, G., & Vignion-Dewalle, A. (2020). Light emitting fabrics for photodynamic therapy: Technology, experimental and clinical applications. *Translational Biophotonics, 2*(3), 1–10. https://doi.org/10.1002/tbio.202000005
71. Arquilla, K., Webb, A. K., & Anderson, A. P. (2020). Textile electrocardiogram (Ecg) electrodes for wearable health monitoring. *Sensors (Switzerland), 20*(4), 1–13. https://doi.org/10.3390/s20041013
72. López-Soriano, S., Vilajosana, X., & Melià-Seguí, J. (2024). A battery-less and low-cost RFID sensor for unassisted multi-level detection of sweat loss. *IEEE Sensors Journal, XX*(Xx), 1–10. https://doi.org/10.1109/JSEN.2024.3485592

73. Romano, C., Lo Presti, D., Silvestri, S., Schena, E., & Massaroni, C. (2024). Flexible textile sensors-based smart t-shirt for respiratory monitoring: design, development, and preliminary validation. *Sensors, 24*(6), 2018. https://doi.org/10.3390/s24062018
74. Arcarisi, L., Di Pietro, L., Carbonaro, N., Tognetti, A., Ahluwalia, A., & De Maria, C. (2019). Palpreast-A new wearable device for breast self-examination. *Applied Sciences (Switzerland), 9*(3). https://doi.org/10.3390/app9030381
75. *Xiaomi Mijia Cardiogram T-shirt: Smart T-shirt with ECG measurement.* (n.d.). Retrieved Nov 26, 2024 from https://xiaomiplanets.com/xiaomi-mijia-cardiogram-t-shirt-2/
76. Dong, H., Sun, J., Liu, X., Jiang, X., & Lu, S. (2022). Highly sensitive and stretchable mXene/CNTs/TPU composite strain sensor with bilayer conductive structure for human motion detection. *ACS Applied Materials and Interfaces, 14*(13), 15504–15516. https://doi.org/10.1021/ACSAMI.1C23567/SUPPL_FILE/AM1C23567_SI_001.PDF
77. Ma, Y., Ouyang, J., Raza, T., Li, P., Jian, A., Li, Z., Liu, H., Chen, M., Zhang, X., Qu, L., Tian, M., & Tao, G. (2021). Flexible all-textile dual tactile-tension sensors for monitoring athletic motion during taekwondo. *Nano Energy, 85*(February), 105941. https://doi.org/10.1016/j.nanoen.2021.105941
78. Wang, R., Zhai, Q., An, T., Gong, S., & Cheng, W. (2021). Stretchable gold fiber-based wearable textile electrochemical biosensor for lactate monitoring in sweat. *Talanta, 222*, 121484. https://doi.org/10.1016/j.talanta.2020.121484
79. Angelucci, A., Cavicchioli, M., Cintorrino, I. A., Lauricella, G., Rossi, C., Strati, S., & Aliverti, A. (2021). Smart textiles and sensorized garments for physiological monitoring: A review of available solutions and techniques. *Sensors (Switzerland), 21*(3), 1–23. https://doi.org/10.3390/s21030814
80. Yang, Y. L., Chuang, M. C., Lou, S. L., & Wang, J. (2010). Thick-film textile-based amperometric sensors and biosensors. *The Analyst, 135*(6), 1230–1234. https://doi.org/10.1039/b926339j
81. Revaiah, R. G., Kotresh, T. M., & Kandasubramanian, B. (2020). Technical textiles for military applications. *Journal of the Textile Institute, 111*(2), 273–308. https://doi.org/10.1080/00405000.2019.1627987
82. Liu, X., Miao, J., Fan, Q., Zhang, W., Zuo, X., Tian, M., Zhu, S., Zhang, X., & Qu, L. (2021). Smart textile based on 3D stretchable silver nanowires/MXene conductive networks for personal healthcare and thermal management. *ACS Applied Materials and Interfaces, 13*(47), 56607–56619. https://doi.org/10.1021/ACSAMI.1C18828/SUPPL_FILE/AM1C18828_SI_001.PDF
83. Chen, G., Wang, H., Guo, R., Duan, M., Zhang, Y., & Liu, J. (2020). Superelastic egain composite fibers sustaining 500% tensile strain with superior electrical conductivity for wearable electronics. *ACS Applied Materials and Interfaces, 12*(5), 6112–6118. https://doi.org/10.1021/acsami.9b23083
84. Tseghai, G. B., Malengier, B., Fante, K. A., & Van Langenhove, L. (2021). Dry electroencephalography textrode for brain activity monitoring. *IEEE Sensors Journal, 21*(19), 22077–22085. https://doi.org/10.1109/JSEN.2021.3103411
85. Islam, M. R., Afroj, S., & Karim, N. (2023). Scalable production of 2D material heterostructure textiles for high-performance wearable supercapacitors. *ACS Nano, 17*(18), 18481–18493. https://doi.org/10.1021/acsnano.3c06181
86. Tan, S., Islam, M. R., Li, H., Fernando, A., Afroj, S., & Karim, N. (2022). Highly scalable, sensitive and ultraflexible graphene-based wearable e-textiles sensor for bio-signal detection. *Advanced Sensor Research, 1*(1), 2200010. https://doi.org/10.1002/adsr.202200010
87. Luo, J., Gao, S., Luo, H., Wang, L., Huang, X., Guo, Z., Lai, X., Lin, L., Li, R. K. Y., & Gao, J. (2021). Super hydrophobic and breathable smart MXene-based textile for multifunctional wearable sensing electronics. *Chemical Engineering Journal, 406*, 126898. https://doi.org/10.1016/j.cej.2020.126898
88. Trovato, V., Teblum, E., Kostikov, Y., Pedrana, A., Re, V., Nessim, G. D., & Rosace, G. (2021). Electrically conductive cotton fabric coatings developed by silica sol-gel precursors doped with surfactant-aided dispersion of vertically aligned carbon nanotubes fillers in organic

solvent-free aqueous solution. *Journal of Colloid and Interface Science, 586*(xxxx), 120–134. https://doi.org/10.1016/j.jcis.2020.10.076

89. Wang, N., Sun, H., Yang, X., Lin, W., He, W., Liu, H., Bhat, G., & Yu, B. (2022). Flexible temperature sensor based on RGO/CNTs@PBT melting blown nonwoven fabric. *Sensors and Actuators A: Physical, 339*, 113519. https://doi.org/10.1016/J.SNA.2022.113519
90. Greene, S., Kirchain, R., Miller, T. R., & Olivetti, E. (2015). *Sustainable Apparel Materials.* https://search.issuelab.org/resource/sustainable-apparel-materials.html
91. Hossain, M. T., Shahid, M. A., Ali, M. Y., Saha, S., Jamal, M. S. I., & Habib, A. (2023). Fabrications, classifications, and environmental impact of PCM-incorporated textiles: current state and future outlook. *ACS Omega, 8*(48), 45164–45176. https://doi.org/10.1021/ACSOMEGA.3C05702/ASSET/IMAGES/LARGE/AO3C05702_0007.JPEG
92. Niu, Z., & Yuan, W. (2021). Smart nanocomposite nonwoven wearable fabrics embedding phase change materials for highly efficient energy conversion-storage and use as a stretchable conductor. *ACS Applied Materials and Interfaces, 13*(3), 4508–4518. https://doi.org/10.1021/ACSAMI.0C19674/SUPPL_FILE/AM0C19674_SI_001.PDF
93. Zhai, S., Jiang, W., Wei, L., Karahan, H. E., Yuan, Y., Ng, A. K., & Chen, Y. (2015). All-carbon solid-state yarn supercapacitors from activated carbon and carbon fibers for smart textiles. *Materials Horizons, 2*(6), 598–605. https://doi.org/10.1039/C5MH00108K
94. Hasanbeigi, A., & Price, L. (2015). A technical review of emerging technologies for energy and water efficiency and pollution reduction in the textile industry. *Journal of Cleaner Production, 95*, 30–44. https://doi.org/10.1016/J.JCLEPRO.2015.02.079
95. Eppinger, E., Slomkowski, A., Behrendt, T., Rotzler, S., Marwede, M., Eppinger, E., Slomkowski, A., Behrendt, T., Rotzler, S., & Marwede, M. (2022). Design for recycling of e-textiles: current issues of recycling of products combining electronics and textiles and implications for a circular design approach. In *Recycling Strategy and Challenges Associated with Waste Management towards Sustaining the World. Intech Open.* https://doi.org/10.5772/intechopen.107527
96. Dulal, M., Afroj, S., Ahn, J., Cho, Y., Carr, C., Kim, I. D., & Karim, N. (2022). Toward sustainable wearable electronic textiles. *ACS Nano, 16*(12), 19755–19788. https://doi.org/10.1021/acsnano.2c07723
97. Koehler, A. R. (2008). End-of-life implications of electronic textiles—Assessment of a converging technology. https://core.ac.uk/download/pdf/289935327.pdf
98. Browne, M. A., Crump, P., Niven, S. J., Teuten, E., Tonkin, A., Galloway, T., & Thompson, R. (2011). Accumulation of microplastic on shorelines woldwide: sources and sinks. *Environmental Science & Technology, 45*(21), 9175–9179. https://doi.org/10.1021/es201811s
99. Wäger, P. A., Eugster, M., Hilty, L. M., & Som, C. (2005). Smart labels in municipal solid waste—A case for the precautionary principle? *Environmental Impact Assessment Review, 25*(5), 567–586. https://doi.org/10.1016/j.eiar.2005.04.009
100. EPA. (2024). [Overviews and Factsheets]. Sustainable materials management (SMM) electronics challenge. https://www.epa.gov/smm-electronics/sustainable-materials-management-smm-electronics-challenge
101. Ogunseitan, O. A., Schoenung, J. M., Saphores, J.-D.M., & Shapiro, A. A. (2009). The electronics revolution: From E-wonderland to E-wasteland. *Science, 326*(5953), 670–671. https://doi.org/10.1126/science.1176929
102. Bovea, M. D., Ibáñez-Forés, V., Pérez-Belis, V., & Quemades-Beltrán, P. (2016). Potential reuse of small household waste electrical and electronic equipment: Methodology and case study. *Waste Management, 53*, 204–217. https://doi.org/10.1016/j.wasman.2016.03.038
103. Köhler, A. R., Hilty, L. M., & Bakker, C. (2011). Prospective impacts of electronic textiles on recycling and disposal. *Journal of Industrial Ecology, 15*(4), 496–511. https://doi.org/10.1111/j.1530-9290.2011.00358.x
104. Asimakopoulos, S., Asimakopoulos, G., & Spillers, F. (2017). Motivation and user engagement in fitness tracking: Heuristics for mobile healthcare wearables. *Informatics, 4*(1), 5. https://doi.org/10.3390/informatics4010005

105. Cui, J., & Zhou, S. (2018). Highly conductive and ultra-durable electronic textiles *via* covalent immobilization of carbon nanomaterials on cotton fabric. *Journal of Materials Chemistry C, 6*(45), 12273–12282. https://doi.org/10.1039/C8TC04017F
106. Ometov, A., Shubina, V., Klus, L., Skibińska, J., Saafi, S., Pascacio, P., Flueratoru, L., Gaibor, D. Q., Chukhno, N., Chukhno, O., Ali, A., Channa, A., Svertoka, E., Qaim, W. B., Casanova-Marqués, R., Holcer, S., Torres-Sospedra, J., Casteleyn, S., Ruggeri, G., & Lohan, E. S. (2021). A survey on wearable technology: History, state-of-the-art and current challenges. *Computer Networks, 193*, 108074. https://doi.org/10.1016/j.comnet.2021.108074
107. Weng, W., Chen, P., He, S., Sun, X., & Peng, H. (2016). Smart electronic textiles. *Angewandte Chemie International Edition, 55*(21), 6140–6169. https://doi.org/10.1002/anie.201507333
108. Lund, A., Van Der Velden, N. M., Persson, N.-K., Hamedi, M. M., & Müller, C. (2018). Electrically conducting fibres for e-textiles: An open playground for conjugated polymers and carbon nanomaterials. *Materials Science and Engineering: R: Reports, 126*, 1–29. https://doi.org/10.1016/j.mser.2018.03.001
109. Ahirwar, R., & Tripathi, A. K. (2021). E-waste management: A review of recycling process, environmental and occupational health hazards, and potential solutions. *Environmental Nanotechnology, Monitoring & Management, 15*, 100409. https://doi.org/10.1016/j.enmm.2020.100409
110. Ossevoort, S. H. W. (2013). Improving the sustainability of smart textiles. In: *Multidisciplinary Know-How for Smart-Textiles Developers* (pp. 399–419). Elsevier. https://doi.org/10.1533/9780857093530.3.399
111. Provin, A. P., Regina de Aguiar Dutra, A., Machado, M. M., & Vieira Cubas, A. L. (2021). New materials for clothing: Rethinking possibilities through a sustainability approach—A review. *Journal of Cleaner Production, 282*(xxxx), 124444. https://doi.org/10.1016/j.jclepro.2020.124444
112. Abitbol, T., Rivkin, A., Cao, Y., Nevo, Y., Abraham, E., Ben-Shalom, T., Lapidot, S., & Shoseyov, O. (2016). Nanocellulose, a tiny fiber with huge applications. *Current Opinion in Biotechnology, 39*(1), 76–88. https://doi.org/10.1016/j.copbio.2016.01.002
113. Liu, K., Liu, W., Li, W., Duan, Y., Zhou, K., Zhang, S., Ni, S., Xu, T., Du, H., & Si, C. (2022). Strong and highly conductive cellulose nanofibril/silver nanowires nanopaper for high performance electromagnetic interference shielding. *Advanced Composites and Hybrid Materials, 5*(2), 1078–1089. https://doi.org/10.1007/S42114-022-00425-2/METRICS
114. Lazaro Vasquez, E. S., & Vega, K. (2019). Myco-accessories: Sustainable wearables with biodegradable materials. *In: Proceedings– International Symposium on Wearable Computers, ISWC*, 306–311. https://doi.org/10.1145/3341163.3346938
115. Teng, M., Luo, X., Qin, R., Feng, J., Zhang, P., Wang, P., Zhang, X., & Wang, X. (2024). Biocompatible and biodegradable 3D graphene/collagen fiber hybrids for high-performance conductive networks and sensors. *ACS Applied Materials and Interfaces, 16*(26), 34213–34228. https://doi.org/10.1021/ACSAMI.4C07412/SUPPL_FILE/AM4C07412_SI_001.PDF
116. Naser, A. Z., Deiab, I., & Darras, B. M. (2021). Poly(lactic acid) (PLA) and polyhydroxyalkanoates (PHAs), green alternatives to petroleum-based plastics: A review. *RSC Advances, 11*(28), 17151–17196. https://doi.org/10.1039/D1RA02390J
117. Grancarić, A. M., Jerković, I., Koncar, V., Cochrane, C., Kelly, F. M., Soulat, D., & Legrand, X. (2018). Conductive polymers for smart textile applications. *Journal of Industrial Textiles 48*(3). https://doi.org/10.1177/1528083717699368
118. Bankar, V. G., Spruiell, J. E., & White, J. L. (1977). Melt spinning of nylon 6: Structure development and mechanical properties of as-spun filaments. *Journal of Applied Polymer Science, 21*(9), 2341–2358. https://doi.org/10.1002/app.1977.070210905
119. Zhang, L.-H., Duan, X.-P., Yan, X., Yu, M., Ning, X., Zhao, Y., & Long, Y.-Z. (2016). B. *RSC Advances, 6*(58), 53400–53414. https://doi.org/10.1039/C6RA09558E
120. Izatt, R. M., Izatt, S. R., Bruening, R. L., Izatt, N. E., & Moyer, B. A. (2014). Challenges to achievement of metal sustainability in our high-tech society. *Chemical Society Reviews, 43*(8), 2451–2475. https://doi.org/10.1039/C3CS60440C

121. Köhler, A., & Erdmann, L. (2004). Expected environmental impacts of pervasive computing. *Human and Ecological Risk Assessment: An International Journal, 10*(5), 831–852. https://doi.org/10.1080/10807030490513856
122. Williams, E. (2011). Environmental effects of information and communications technologies. *Nature, 479*(7373), 354–358. https://doi.org/10.1038/nature10682
123. Böni, H., Schluep, M., & Widmer, R. (2015). Recycling of ICT equipment in industrialized and developing countries. In Hilty, L. M., & Aebischer, B. (Eds.), *ICT Innovations for Sustainability* (pp. 223–241). Springer International Publishing. https://doi.org/10.1007/978-3-319-09228-7_13
124. Kahhat, R., & Williams, E. (2012). Materials flow analysis of e-waste: Domestic flows and exports of used computers from the United States. *Resources, Conservation and Recycling, 67*, 67–74. https://doi.org/10.1016/j.resconrec.2012.07.008
125. Boothroyd, G., & Alting, L. (1992). Design for assembly and disassembly. *CIRP Annals, 41*(2), 625–636. https://doi.org/10.1016/S0007-8506(07)63249-1
126. *Phys Org*. (2012). Recyclable printed circuit boards. https://phys.org/news/2012-06-recyclable-circuit-boards.html
127. Roy, R. (2000). Sustainable product-service systems. *Futures, 32*(3), 289–299. https://doi.org/10.1016/S0016-3287(99)00098-1
128. Gurova, O., Merritt, T. R., Papachristos, E., & Vaajakari, J. (2020). Sustainable solutions for wearable technologies: mapping the product development life cycle. *Sustainability, 12*(20). https://doi.org/10.3390/su12208444
129. Prahl, A. (2018). Exploring circular design opportunities for wearable technology. *Designing for the Circular Economy*, 248–259. https://doi.org/10.4324/9781315113067-23/Exploring-Circular-Design-Opportunities-Wearable-Technology-Anne-PraHL
130. Yao, L., Ou, J., Cheng, C. Y., Steiner, H., Wang, W., Wang, G., & Ishii, H. (2015). Biologic: Natto cells as nanoactuators for shape changing interfaces. *Conference on Human Factors in Computing Systems - Proceedings*, 1–10. https://doi.org/10.1145/2702123.2702611/SUPPL_FILE/P1.MP4
131. An E-textile interface kit for persons with limited mobility. (n.d.). Retrieved Nov 23, 2024 from https://flex-ability.org/
132. Buechley, L., & Eisenberg, M. (2008). The lilypad arduino: Toward wearable engineering for everyone. *IEEE Pervasive Computing, 7*(2), 12–15. https://doi.org/10.1109/MPRV.2008.38
133. Chen, D., Heyer, S., Ibbotson, S., Salonitis, K., Steingrímsson, J. G., & Thiede, S. (2015). Direct digital manufacturing: Definition, evolution, and sustainability implications. *Journal of Cleaner Production, 107*, 615–625. https://doi.org/10.1016/J.JCLEPRO.2015.05.009
134. *Studio Hilo*. (n.d.). Retrieved 23 November 2024, from https://www.studio-hilo.com/
135. Timmins, M. (2009). Environmental and waste issues concerning the production of smart clothes and wearable technology. *Smart Clothes and Wearable Technology*, 319–331. https://doi.org/10.1533/9781845695668.3.319
136. Bonvoisin, J. (2016). Implications of open source design for sustainability. *Smart Innovation, Systems and Technologies, 52*, 49–59. https://doi.org/10.1007/978-3-319-32098-4_5
137. Raasch, C., Herstatt, C., & Balka, K. (2009). On the open design of tangible goods. *R&D Management, 39*(4), 382–393. https://doi.org/10.1111/J.1467-9310.2009.00567.X
138. Cherenack, K., & Van Pieterson, L. (2012). Smart textiles: Challenges and opportunities. *Journal of Applied Physics, 112*(9). https://doi.org/10.1063/1.4742728/372800
139. Juhlin, O., Zhang, Y., Wang, J., & Andersson, A. (2016). Fashionable services for wearables-inventing and investigating a new design path for smart watches. *ACM International Conference Proceeding Series, pp 23–27*. https://doi.org/10.1145/2971485.2971505
140. Kohler, A. R. (2013). Anticipatory eco-design strategies for smart textiles. https://doi.org/10.4233/uuid:850be7ae-1f9e-4b3f-b49f-242488bab216
141. Thielen, M., Sigrist, L., Magno, M., Hierold, C., & Benini, L. (2017). Human body heat for powering wearable devices: From thermal energy to application. *Energy Conversion and Management, 131*, 44–54. https://doi.org/10.1016/J.ENCONMAN.2016.11.005

142. Kuusk, K., Niinimäki, K., Wensveen, S., & Tomico, O. (2014). Smart textile products and services in sustainability context. In Proceedings of the Ambience'14&10i3m, 19 July 2014, Tampere, Finland
143. Pan, Y. (2014). Fashion thinking and sustainable HCI. *Conference on Human Factors in Computing Systems– Proceedings*, 355–358. https://doi.org/10.1145/2559206.2559967
144. Shen, B. (2014). Sustainable fashion supply chain: Lessons from H&M. *Sustainability, 6*(9), 6236–6249. https://doi.org/10.3390/su6096236
145. McNeill, L., & Moore, R. (2015). Sustainable fashion consumption and the fast fashion conundrum: Fashionable consumers and attitudes to sustainability in clothing choice. *International Journal of Consumer Studies, 39*(3), 212–222. https://doi.org/10.1111/ijcs.12169
146. *Directive 2001/95—General product safety—EU monitor*. (2001). https://www.eumonitor.eu/9353000/1/j4nvk6yhcbpeywk_j9vvik7m1c3gyxp/vitgbgi960yh
147. *Directive 2014/30/EU of the European Parliament and of the Council of 26 February 2014 on the harmonisation of the laws of the Member States relating to electromagnetic compatibility (recast) Text with EEA relevance*. (2014). Europian Union. https://eur-lex.europa.eu/legal-content/EN/TXT/?uri=CELEX%3A32014L0030
148. Sanjay, M. R., Madhu, P., Jawaid, M., Senthamaraikannan, P., Senthil, S., & Pradeep, S. (2018). Characterization and properties of natural fiber polymer composites: A comprehensive review. *Journal of Cleaner Production, 172*, 566–581. https://doi.org/10.1016/j.jclepro.2017.10.101
149. Boutry, C. M., Beker, L., Kaizawa, Y., Vassos, C., Tran, H., Hinckley, A. C., Pfattner, R., Niu, S., Li, J., Claverie, J., Wang, Z., Chang, J., Fox, P. M., & Bao, Z. (2019). Biodegradable and flexible arterial-pulse sensor for the wireless monitoring of blood flow. *Nature Biomedical Engineering, 3*(1), 47–57. https://doi.org/10.1038/s41551-018-0336-5
150. Veske, P., & Ilén, E. (2021). Review of the end-of-life solutions in electronics-based smart textiles. *The Journal of The Textile Institute, 112*(9), 1500–1513. https://doi.org/10.1080/00405000.2020.1825176
151. Lazar, A., Koehler, C., Tanenbaum, T. J., & Nguyen, D. H. (2015). Why we use and abandon smart devices. *In: Proceedings of the 2015 ACM International Joint Conference on Pervasive and Ubiquitous Computing*, 635–646. https://doi.org/10.1145/2750858.2804288
152. Kan, C.-W., & Lam, Y.-L. (2021). Future trend in wearable electronics in the textile industry. *Applied Sciences, 11*(9), 3914. https://doi.org/10.3390/app11093914
153. Karana, E. (2012). Characterization of 'natural' and 'high-quality' materials to improve perception of bio-plastics. *Journal of Cleaner Production, 37*, 316–325. https://doi.org/10.1016/j.jclepro.2012.07.034
154. Karana, E., & Nijkamp, N. (2014). Fiberness, reflectiveness and roughness in the characterization of natural and high quality materials. *Journal of Cleaner Production, 68*, 252–260. https://doi.org/10.1016/j.jclepro.2014.01.001
155. Van Der Velden, N. M., Kuusk, K., & Köhler, A. R. (2015). Life cycle assessment and eco-design of smart textiles: The importance of material selection demonstrated through e-textile product redesign. *Materials & Design, 84*, 313–324. https://doi.org/10.1016/j.matdes.2015.06.129
156. Chen, G., Li, Y., Bick, M., & Chen, J. (2020). Smart textiles for electricity generation. *Chemical Reviews, 120*(8), 3668–3720. https://doi.org/10.1021/acs.chemrev.9b00821
157. Islam, M. R., Afroj, S., Novoselov, K. S., & Karim, N. (2022). Smart electronic textile-based wearable supercapacitors. *Advanced Science, 9*(31), 2203856. https://doi.org/10.1002/advs.202203856
158. *Overview | Getting Started with FLORA | Adafruit Learning System*. (n.d.). Retrieved Nov 23 2024, from https://learn.adafruit.com/getting-started-with-flora/flora-projects
159. Revaiah, R. G., Kotresh, T. M., & Kandasubramanian, B. (2020). Technical textiles for military applications. *The Journal of The Textile Institute, 111*(2), 273–308. https://doi.org/10.1080/00405000.2019.1627987
160. Wang, J., He, J., Ma, L., Zhang, Y., Shen, L., Xiong, S., Li, K., & Qu, M. (2020). Multifunctional conductive cellulose fabric with flexibility, superamphiphobicity and flame-retardancy for all-weather wearable smart electronic textiles and high-temperature warning device. *Chemical Engineering Journal, 390*, 124508. https://doi.org/10.1016/j.cej.2020.124508

161. Wang, L., Fu, X., He, J., Shi, X., Chen, T., Chen, P., Wang, B., & Peng, H. (2020). *Advanced Materials, 32*(5), 1901971. https://doi.org/10.1002/adma.201901971

MIX
Papier aus verantwortungsvollen Quellen
Paper from responsible sources
FSC® C105338
FSC
www.fsc.org

If you have any concerns about our products,
you can contact us on
ProductSafety@springernature.com

In case Publisher is established outside the EU,
the EU authorized representative is:
Springer Nature Customer Service Center GmbH
Europaplatz 3, 69115 Heidelberg, Germany

Printed by Libri Plureos GmbH
in Hamburg, Germany